AF686655

RNAi Technologies

RNAi Technologies

Edited by:

Prerna Pandey

Delve
PUBLISHING

www.delvepublishing.com

RNAi Technologies

Prerna Pandey

Delve Publishing
2010 Winston Park Drive,
2nd Floor
Oakville, ON L6H 5R7
Canada
www.delvepublishing.com
Tel: 001-289-291-7705
001-905-616-2116
Fax: 001-289-291-7601
Email: orders@arclereducation.com

© 2018 Delve Publishing

ISBN: 978-1-77361-147-1 (Hardcover)

This book contains information obtained from highly regarded resources. Reprinted material sources are indicated. Copyright for individual articles remains with the authors as indicated and published under Creative Commons License. A Wide variety of references are listed. Reasonable efforts have been made to publish reliable data and views articulated in the chapters are those of the individual contributors, and not necessarily those of the editors or publishers. Editors or publishers are not responsible for the accuracy of the information in the published chapters or consequences of their use. The publisher assumes no responsibility for any damage or grievance to the persons or property arising out of the use of any materials, instructions, methods or thoughts in the book. The editors and the publisher have at- tempted to trace the copyright holders of all material reproduced in this publication and apologize to copyright holders if permission has not been obtained. If any copyright holder has not been acknowledged, please write to us so we may rectify.

Notice: Registered trademark of products or corporate names are used only for explanation and identification without intent of infringement.

Delve Publishing publishes wide variety of books and eBooks. For more information about Delve Publishing and its products, visit our website at www.delvepublishing.com.

DECLARATION

Some content or chapters in this book are open access copyright free published research work, which is published under Creative Commons License and are indicated with the citation. We are thankful to the publishers and authors of the content and chapters as without them this book wouldn't have been possible.

ABOUT THE EDITOR

Dr Prerna Pandey, a biotechnologist with several years of wet lab research experience, worked at International Center for Genetic Engineering and Biotechnology, New Delhi. Her field of research in PhD included isolation and molecular characterization of geminiviruses, genome sequencing, gene annotation, and gene silencing using the RNA interference technology. She has also worked at Transasia Biomedicals and Advance Enzyme Technologies as a scientist. Prerna has published papers in peer-reviewed journals, and has submitted a number of annotated Geminiviral genome sequences in the GenBank, including two novel ones. She has also completed her editing and proof-reading courses from Society for the promotion of Editors and Proof-readers, UK and now works as a freelance scientific editor and writer. When Prerna is not busy with her assignments, she loves to explore historical places.

TABLE OF CONTENTS

LIST OF CONTRIBUTORS

Prerna Pandey

Dr Prerna Pandey, a biotechnologist with several years of wet lab research experience, worked at International Center for Genetic Engineering and Biotechnology, New Delhi. Her field of research in PhD included isolation and molecular characterization of geminiviruses, genome sequencing, gene annotation, and gene silencing using the RNA interference technology. She has also worked at Transasia Biomedicals and Advance Enzyme Technologies as a scientist. Prerna has published papers in peer-reviewed journals, and has submitted a number of annotated Geminiviral genome sequences in the GenBank, including two novel ones. She has also completed her editing and proof-reading courses from Society for the promotion of Editors and Proof-readers, UK and now works as a freelance scientific editor and writer. When Prerna is not busy with her assignments, she loves to explore historical places

Marc S. Weinberg

Department of Molecular and Experimental Medicine, The Scripps Research Institute, La Jolla, CA 92037, USA

Wits/SAMRC Antiviral Gene Therapy Research Unit, School of Pathology, University of the Witwatersrand, WITS 2050, South Africa

HIV Pathogenesis Research Unit, Department of Molecular Medicine and Haematology, School of Pathology, University of the Witwatersrand, WITS 2050, South Africa

Kevin V. Morris

Department of Molecular and Experimental Medicine, The Scripps Research Institute, La Jolla, CA 92037, USA

Center for Gene Therapy, City of Hope – BeckmanResearch Institute; Duarte, CA 91010, USA

School of Biotechnology and Biomedical Sciences, University of New South Wales, Kensington, NSW, 2033 Australia

Joshua M. Thornbrough

Department of Biology, University of Louisville, Louisville, KY, 40292, United States of America

Adarsh Gopinath

Department of Biology, University of Louisville, Louisville, KY, 40292, United States of America

Tom Hundley

Department of Biology, University of Louisville, Louisville, KY, 40292, United States of America

Micah J. Worley

Department of Biology, University of Louisville, Louisville, KY, 40292, United States of America

Department of Microbiology and Immunology, University of Louisville, Louisville, KY, 40202, United States of America

Trung Anh Trieu

Departmento de Gene´tica y Microbiologı´a, Facultad de Biologı´a, Universidad de Murcia, Spain

Marı´a Isabel Navarro-Mendoza

Departmento de Gene´tica y Microbiologı´a, Facultad de Biologı´a, Universidad de Murcia, Spain

Carlos Pe´rez-Arques

Departmento de Gene´tica y Microbiologı´a, Facultad de Biologı´a, Universidad de Murcia, Spain

Marta Sanchis

Unidad de Microbiologı´a, Universitat Rovira i Virgili, IISPV, Tarragona, Spain

Javier Capilla

Unidad de Microbiologı´a, Universitat Rovira i Virgili, IISPV, Tarragona, Spain

Patricia Navarro-Rodriguez

Unidad de Microbiologı´a, Universitat Rovira i Virgili, IISPV, Tarragona, Spain

Loida Lopez-Fernandez

Unidad de Microbiologı´a, Universitat Rovira i Virgili, IISPV, Tarragona, Spain

Santiago Torres-Martı´nez

Departmento de Gene´tica y Microbiologı´a, Facultad de Biologı´a, Universidad de Murcia, Spain

Victoriano Garre

Departmento de Gene´tica y Microbiologı´a, Facultad de Biologı´a, Universidad de Murcia, Spain

Rosa Marı´a Ruiz-Va´zquez

Departmento de Gene´tica y Microbiologı´a, Facultad de Biologı´a, Universidad de Murcia, Spain

Francisco E. Nicola´s

Departmento de Gene´tica y Microbiologı´a, Facultad de Biologı´a, Universidad de Murcia, Spain

Ayushi Kamthan

National Institute of Plant Genome Research, New Delhi, India

Abira Chaudhuri

National Institute of Plant Genome Research, New Delhi, India

Mohan Kamthan

Indian Institute of Toxicology Research, Lucknow, India

Asis Datta

National Institute of Plant Genome Research, New Delhi, India

Aline Koch

Institute for Phytopathology, Centre for BioSystems, Land Use and Nutrition, Justus Liebig University, Giessen, Germany

Dagmar Biedenkopf

Institute for Phytopathology, Centre for BioSystems, Land Use and Nutrition, Justus Liebig University, Giessen, Germany

Alexandra Furch

Institute of General Botany and Plant Physiology, Friedrich-Schiller-University, Jena, Germany

Lennart Weber

Institute for Microbiology and Molecular Biology, Centre for BioSystems, Land Use and Nutrition, Justus Liebig University, Giessen, Germany

Oliver Rossbach

Institute of Biochemistry, Centre for BioSystems, Land Use and Nutrition, Justus Liebig University, Giessen, Germany

Eltayb Abdellatef

Institute for Phytopathology, Centre for BioSystems, Land Use and Nutrition, Justus Liebig University, Giessen, Germany

Lukas Linicus

Institute for Phytopathology, Centre for BioSystems, Land Use and Nutrition, Justus Liebig University, Giessen, Germany

Jan Johannsmeier

Institute for Phytopathology, Centre for BioSystems, Land Use and Nutrition, Justus Liebig University, Giessen, Germany

Lukas Jelonek

Institute for Bioinformatics and Systems Biology, Centre for BioSystems, Land Use and Nutrition, Justus Liebig University, Giessen, Germany

Alexander Goesmann

Institute for Bioinformatics and Systems Biology, Centre for BioSystems, Land Use and Nutrition, Justus Liebig University, Giessen, Germany

Vinitha Cardoza

BASF Plant Science LP, Research Triangle Park, Durham, North Carolina, United States of America

John McMillan

BASF Plant Science LP, Research Triangle Park, Durham, North Carolina, United States of America

Tobias Mentzel

BASF SE, Limburgerhof, Germany

KarlHeinz Kogel

Institute for Phytopathology, Centre for BioSystems, Land Use and Nutrition, Justus Liebig University, Giessen, Germany

Unnikrishnan Unniyampurath

Program on Emerging Infectious Diseases, Duke-NUS Medical School, 8 College Road, Singapore 169857, Singapore

Rajendra Pilankatta

Department of Biochemistry and Molecular Biology, School of Biological Sciences, Central University of Kerala, Nileshwar 671328, India

Manoj N. Krishnan

Program on Emerging Infectious Diseases, Duke-NUS Medical School, 8 College Road, Singapore 169857, Singapore

Alessandra Guidi

Department of Infection and Immunity, London School of Hygiene and Tropical Medicine, London, United Kingdom

Nuha R. Mansour

Department of Infection and Immunity, London School of Hygiene and Tropical Medicine, London, United Kingdom

Ross A. Paveley

Department of Infection and Immunity, London School of Hygiene and Tropical Medicine, London, United Kingdom

Ian M. Carruthers

Division of Biological Chemistry and Drug Discovery, College of Life Sciences, University of Dundee, Dundee, United Kingdom

Jérémy Besnard

Division of Biological Chemistry and Drug Discovery, College of Life Sciences, University of Dundee, Dundee, United Kingdom

Andrew L. Hopkins

Division of Biological Chemistry and Drug Discovery, College of Life Sciences, University of Dundee, Dundee, United Kingdom

Ian H. Gilbert

Division of Biological Chemistry and Drug Discovery, College of Life Sciences, University of Dundee, Dundee, United Kingdom

Quentin D. Bickle

Department of Infection and Immunity, London School of Hygiene and Tropical Medicine, London, United Kingdom

PREFACE

RNA interference (RNAi), one of the most exciting discoveries of the last decade, is also known as Post-Transcriptional Gene Silencing (PTGS). It is a conserved host defense response via double-stranded RNA against both endogenous and foreign nucleic acids, and occurs at the levels of transcription, post-transcription and translation. RNA induced gene silencing has been used, with considerable success as a reverse genetics tool for the study of gene function. This technology has revolutionized the gene expression analysis studies because of the specificity and speed it offers. The small RNA molecules central to RNAi are microRNA (miRNA) and small interfering RNA (siRNA). These molecules bind to other RNAs specifically, such as messenger RNA (mRNA), to regulate their activities.

In this book, we will discuss how gene silencing methodology has been successfully applied as therapeutic tool against metabolic disorders, cancer, and infectious diseases. This tool has also been successfully used to improve crops resistance against plant pathogens. In contrast to conventional knock-out mutagenesis techniques, the "gene-silencing" vectors, that allow the study of functions of genes whose complete knock-out would otherwise be lethal to the plant.

Editor

CHAPTER 1

INTRODUCTION TO RNAi

Prerna Pandey

Dr Prerna Pandey, a biotechnologist with several years of wet lab research experience, worked at International Center for Genetic Engineering and Biotechnology, New Delhi. Her field of research in PhD included isolation and molecular characterization of geminiviruses, genome sequencing, gene annotation, and gene silencing using the RNA interference technology. She has also worked at Transasia Biomedicals and Advance Enzyme Technologies as a scientist. Prerna has published papers in peer-reviewed journals, and has submitted a number of annotated Geminiviral genome sequences in the GenBank, including two novel ones. She has also completed her editing and proof-reading courses from Society for the promotion of Editors and Proof-readers, UK and now works as a freelance scientific editor and writer. When Prerna is not busy with her assignments, she loves to explore historical places

Mattick and Makunin, 2006 defined "non-coding RNA" as RNA that is not translated into a protein. These molecules are increasingly being described as having several important functions in biological systems.

The silencing of genes to achieve several ends for the betterment of life has eluded man since long. This quest came to a direction with the discovery of RNAi. RNA-mediated interference (RNAi) achieves silencing of genes rapidly with ease. RNAi is a process that occurs in cells under normal conditions that is associated with mRNA degradation to achieve gene silencing.

Definition: "RNA silencing involves control of genes at the level of their transcripts that entails degradation of the RNA" (Agrawal et al., 2003).

HISTORY

The story started with the use of antisense RNA by Ecker and Davis, 1986 where antisense RNA was expressed in plant cells. The experiments involved the construction of plant promoters to chloramphenicol acetyltransferase gene in sense and antisense orientation. Upon electroporation into plant protoplasts, they reported that transcription of antisense RNA was found to effectively block the expression of target genes.

This was followed by an experiment of serendipity that led to the discovery of "mRNA suppression". Napoli et al., 1990 attempted to overexpress chalcone synthase (CHS) in petunia flowers with a view to make the flowers more purple. This enzyme is involved in the production of anthocyanin pigments. Instead of darker flowers, however, the introduction of chimeric petunia CHS gene led to a block in anthocyanin biosynthesis. They reported that, "forty-two percent of plants with the introduced CHS gene produced totally white flowers and/or patterned flowers with white or pale nonclonal sectors on a wild-type pigmented background; none of hundreds of transgenic control plants exhibited such phenotypes." (Napoli et al., 1990) The term co-suppression was coined by Jorgensen: one of the research members to describe the loss of mRNAs of both the endo- and the transgene.

This was followed by a group who coined the term "quelling" upon experiments on *Neurospora crassa.* Romano and Macino, 1992 reported the expression of certain endogenous genes: *albino-3 (al-3)* and *albino-1 (al-1)* when sequences were supplied externally. They found the phenomenon to be phenotypically reversible and termed it as "quelling", as the numbers of exogenous sequences were decreased; the inhibition of the expression of the genes was also uplifted.

Kooter *et al.*, 1999 reviewed that experiments involving cosuppression resulted in the degradation of endogene and transgene RNAs following nuclear transcription. Keeping the above observation in mind, the term "posttranscriptional gene silencing (PTGS)" was introduced to describe the above.

Cogoni et al., 1996 worked on quelling when they wanted to enhance orange pigment made by the gene *al1* of the fungus *Neurospora crassa.* A N. *crassa* strain containing a wild-type orange *al1*$^+$ gene was transformed with a plasmid containing a 1,500-bp fragment of the coding sequence of the *al1* gene. They reported quelling or few transformants exhibited albino phenotypes. They reported that the level of unspliced *al1* mRNA in the *al1*-quelled strain was similar to that of the wild-type strain, however the native *al1* mRNA was reduced. This was suggestive of quelling rather than transcription rate.

As the introduction of exogenous RNA interfered with the expression of specific genes, Fire et al., 1998 investigated the requirements for structure and delivery of the interfering RNA in the nematode *Caenorhabditis elegans*. They reported that when compared to either strand alone the use of double-stranded RNA achieved a higher level of interference. Single strands exhibited modest effects whereas double stranded mixtures produced specific and pointed interference. Additionally, the effects of the interference were reported in the organisms injected as well as their progeny.

The experiments involved injection of dsRNA corresponding to a 742-nucleotide segment of *unc22* (codes for a non-essential protein of myofilament) into either the gonad or body cavity region of an adult nematode and the decrease in *unc22* activity was expected to produce a severe twitching phenotype. The injected animal showed weak twitching, while the progeny displayed strong twitching. Similar loss-of-function individuals could also be generated with dsRNAs corresponding to four other nematode genes.

This phenomenon of silencing a functional gene by exogenous application of dsRNA was named RNA interference (RNAi). Timmons and Fire, 1998 reported that placing *C. elegans* along with double stranded RNA in the form of soaking or *Escherichia coli* expressing dsRNAs led to RNAi.

PLANT VIRUSES AND SILENCING

Lindbo and Dougherty, 1992 reported that virus-derived transgenes gave protection against the challenge viruses. The experiment involved the expression of tobacco etch virus potyvirus (TEV) coat protein (CP) gene sequence in tobacco plants. It was observed that of 7 transgenics; one could quell the disease upon exposure to the virus. When untranslatable sense transcripts were observed 30% of the plants could resist the disease.

They suggested the phenomenon was "mediated by an RNA molecule" and that the resistance was specific.

Al-Kaff et al., 1998 worked on *Brassica napus* (oilseed rape) and DNA virus cauliflower mosaic virus. They observed lesions at the site of virus entry were 5 to 7 days following inoculation of *Brassica napus* was inoculated with cauliflower mosaic virus. The symptoms were most prominent at 30 to 40 days following inoculation however they declined thereafter (recovery of plants was seen).

A PTGS-like mechanism was reported as the 19S and 35S RNAs encoded by the virus were degraded while cauliflower mosaic virus DNA was still replicating in the nucleus. The induction of PTGS was visualized if the cauliflower mosaic virus infection and subsequent recovery were followed up in a transgenic *B. napus* with a p35S-GUS (β-glucuronidase) transgene. At the site of inoculation, GUS silencing associated with local lesions was first observed 7 days post inoculation. GUS silencing eventually spread systemically, and the GUS activity of the entire plant was suppressed by 50 days post inoculation. Additionally, the virus itself was also the target of the induced PTGS, since 19S and 35S RNAs were found degraded.

MECHANISMS

Lu et al., 2008 attributed RNAi in several vital processes such as cell growth, differentiation of tissue, formation of heterochromatin, and cell proliferation. This implicates the anomalies of RNAi can be associated to cardiovascular disease, neurological disorders several cancers.

MacFarlane and Murphy, 2010 reported that 5% of the human genome is associated with encoding and producing the >1,000 miRNAs that regulate at least 30% of the genes. Instead of coding for proteins, these non-coding act through complementary base pairing with target RNAs. Together with their associated protein called the Argonaute proteins, short RNAs form the RNA-induced silencing complexes (RISCs).

Hamilton, A. J. and Baulcombe, 1999 studied gene silencing by transgene and viruses to report the presence of antisense RNA complementary to a target mRNA. These antisense RNA molecules were close to twenty-five nucleotides long and were putatively synthesized from an RNA molecule.

Due to their potential as gene regulators, two classes small interfering RNAs (siRNAs) and microRNAs (miRNAs) are being studied extensively. The two classes of RNAs possess certain notable distinctive features. small interfering RNAs:siRNAs: They are a 21-23 nucleotide RNA duplex with 2 nucleotides 3'overhang. They are completely complementary to their target mRNA: usually one in number. They find application as a therapeutic agent. microRNAs: miRNAs: The first such molecules were identified in *C. elegans* genome while examining the role of developmental genes (Lee et al., 1993).

They are 19-25 nucleotide RNA duplex with 2 nucleotides 3'overhang. They are partially complementary to target mRNA: multiple in number: targeting the 3' untranslated region of mRNA. They have been reported to be of utility as agents of therapy and biomarkers (Reviewed by Lam et al., 2015).

Another distinctive feature between siRNAs and miRNAs is their origin: miRNAs are derived from the genome, while siRNAs may be endogenous or exogenous sources like viral infection or others (Carthew and Sontheimer, 2009).

BIOGENESIS OF RNAI

siRNA

The process starts with a double stranded RNA: dsRNA (as already mentioned above: either cellular or exogenous sources). An enzyme: Dicer acts on the dsRNA to shorten its length. Dicer is a form of ribonuclease (RNase) III. This generated shorter form of the dsRNA is between 21–23 nucleotides and 3' overhangs and is termed as siRNA. This molecule associates with RNA-induced silencing complex (RISC) to facilitate its activation.

This RISC possesses an endonuclease component called argonaute 2 (AGO2) which is responsible for the cleaving of sense strand of the siRNA. This sense strand is also referred to as passenger strand; the complementary antisense strand; also called guide strand stays with the RISC. This guide strand facilitates the path of RISC to its RNA target for its degradation by AGO2. While the siRNA causes silencing of genes, the guide strand causes binding to mRNA with 100% complementarily (Lam et al., 2015).

miRNA

The transcription of miRNA is performed by RNA polymerase II in the nucleus to produce a primary miRNA (pri-miRNA) that is at least 1000 nucleotides long with the features: 5′ capped, 3′ polyadenylated RNA as well as a stem-loop structure with two strands. The pri-miRNA is then cleaved by a microprocessor complex that comprises Drosha 9 an RNase III family enzyme) and DiGeorge syndrome critical region gene 8 (DCGR8: it is known as Pasha in invertebrates). The DCGR8 is involved in the recognition of the pri-mRNA to facilitate the cleavage that is performed by Drosha. This leads to the formation of a 65-70 nucleotide precursor mRNA called pre-mRNA that is transported by Exportin 5 and Ran-GTP to be transported from the nucleus to the cytoplasm.

Here, the preRNA is processed by the Dicer to yield a miRNA duplex that associates with RISC. In the case of siRNA, AGO2 was involved in the cleavage of the passenger strand, in the case of miRNAs the miRNA duplex is unwound following which the passenger strand is discarded. (Lam et al., 2015 and Ross Wilson and Jennifer, 2013).

Convergent Pathway for Processing siRNA and miRNA in the Cytoplasm

Endogenous miRNAs and typically exogenous siRNAs show a converged pathway for processing in the cytoplasm. Both the classes of RNAs are processed to the duplex dsRNA by Dicer enzyme for subsequent loading onto an argonaute protein. Dicers are endoribonucleases that possess a helicase domain and an internally dimerized pair of RNase III domains, whose composition can vary between organisms.

Argonaute proteins can bind short single strand RNAs: sometimes they cleave a bound complementary strand. They possess PAZ (PIWI, Argonaute, and Zwille) domains involved in the generation of RISC. The product of processing at the Dicer is a RNA duplex of 21- to 25-nt strands, with a 2-nucleotide overhang at each 3' terminus and a phosphate group at each recessed 5'.

Once the endonucleic cleavage has occurred, the duplex is loaded onto Argonaute. The function of Dicer may involve an accessory dsRNA-binding protein (dsRBP) such as TAR RNA-binding protein (TRBP). It is reported that three proteins (Dicer, Argonaute, and a dsRBP) comprise a minimal RISC-loading complex (RLC) to generate diced dsRNA and loading it onto Argonaute.

Following the presentation of the dsRNA to Argonaute, the 3' of the guide strand binds to PAZ domain and 5' phosphate to MID domain, respectively, to form the RISC. One strand of the duplex is a guide strand that directs the silencing while the other strand is called passenger stand that is discarded (Ross Wilson and Jennifer, 2013).

THE VARIOUS COMPONENTS OF THE PROCESSING PATHWAY

Let us know examine the various components involved in the processing of the small RNAs:

Dicer

Belongs to class 3 RNase III that functions in the generation of a ds duplex for both siRNA and miRNA. This ds duplex as discussed above is presented to Argonaute protein. A distinguishing characteristic of Dicer has been identified as a superfamily 2 helicase domains found at its N terminus (Ross Wilson and Jennifer, 2013).

Dicer 2 of Drosophila processes long, paired siRNA precursors, while Dicer-1 is associated with the shorter and imperfect pre-miRNA. (Cenik et al., 2011).

Human systems possess a single Dicer that processes the precursors of both siRNA and miRNA (Ross Wilson and Jennifer, 2013).

dsRBPs

These are a group of two to three proteins that aid in the dicer activity as well as the formation of RISC. In humans, TRBP has been characterized extensively. While *A. thaliana* possesses five dsRBPs serially named DRB1–5, Drosophila has 2 sets: Dicer-1 pairs with the PB isoform of dsRBP Loquacious (Loqs-PB) to process pre-miRNAs and Dicer-2 binds either R2D2 or Loqs-PD to process siRNAs of exogenous or endogenous origin. The human complex is comprised of the proteins Dicer and two dsRBPs, TRBP and PACT.

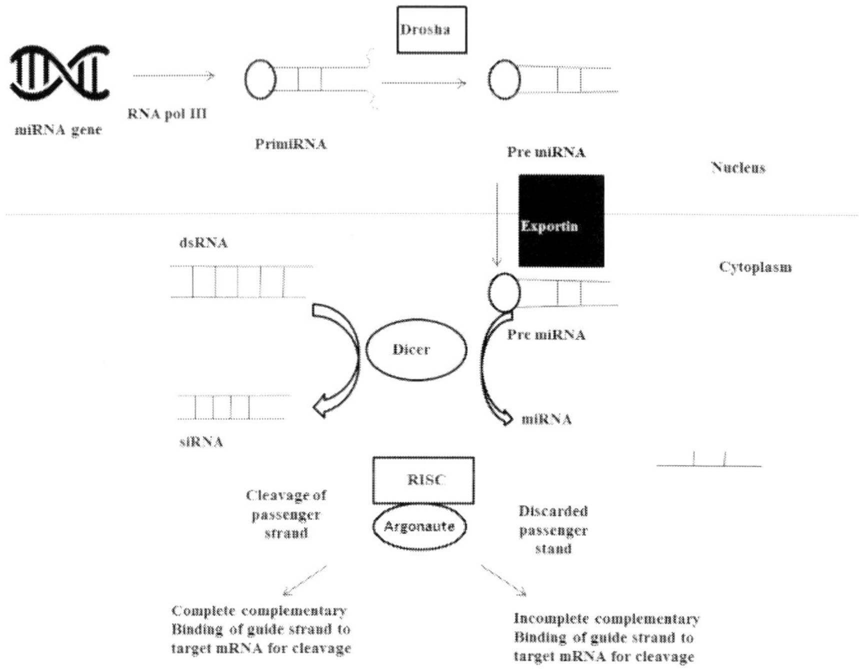

Figure 1.1: Schematic representation of the processing of siRNAs and miRNAs.

Argonaute

This effector protein is bound to a strand of silencing RNA and forms the RISC. This RISC performs the task cellular surveillance and silences ssRNA sequences complementary to its bound guide strand. The argonaute proteins of bacteria and archea are yet to be defined. (Ross Wilson and Jennifer, 2013)

The proteins of eukaryotes are placed in two clades based on sequence similarity: the Ago clade and the Piwi (P-element induced wimpy testis) clade. While the former interact with siRNAs and miRNAs to perform gene silencing, the latter interact with piRNAs to regulate mobile genetic elements in germlines. (Carmell, 2002).

MRNA TARGET RECOGNITION BY SIRNA AND MIRNA

siRNA

siRNA is fully complementary to its target mRNA sequence. Subsewuent to the pairing of bases, the AGO cuts mRNA by functioning as a phosphodiesterase between ten and eleven base pairs from the 5'end of the guide (Elbashir et al., 2001). The fragments of the mRNA generated are further processed by other nucleases. (Valencia-Sanchez et al., 2006). Due to complementarily of siRNA and its target sequence, specific gene silencing is effected by siRNA. (Agrawal et al., 2003).

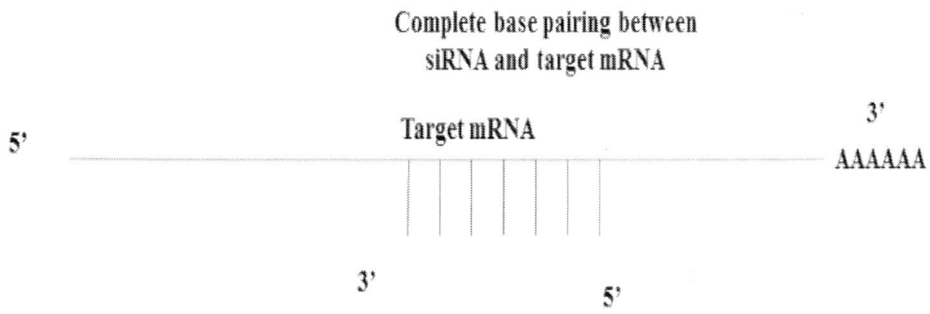

Figure 1.2: Schematic figure of siRNA and its target mRNA

miRNA

As previously discussed, there is incomplete base pairing between miRNAs and their target mRNA with pairing occurring between nucleotides that are numbered two to seven from the 5'end of the miRNA and 3′ untranslated region (UTR) of the target mRNA (Bartell, 2009 and Lim et al., 2005).

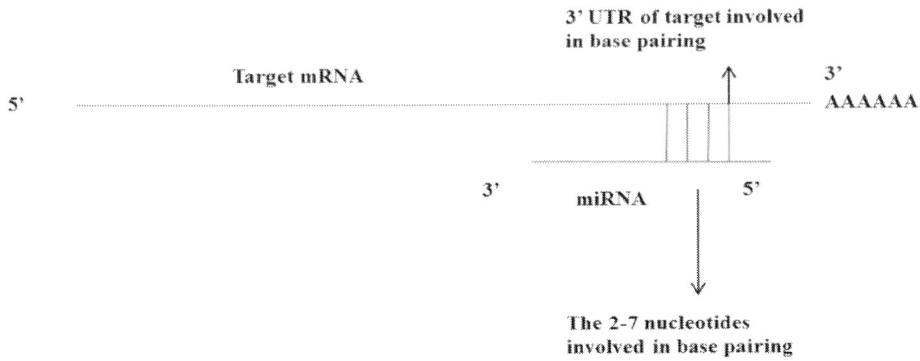

Figure 1.3: Schematic figure of miRNA and its target mRNA

Due to the imperfect complemetarity, a single miRNA can recognize several target mRNAs thus, possessing multiple targets. The study by Lim et al., 2005 revealed that an miRNA enriched in the brain called miRNA-124 represses the expression of 174 annotated genes.

As miRNA has imperfect pairing with its target; the process does not involve AGO. Hence, the process involves mechanisms such as degradation by deadenylation or repression of translation or exonuclease cleavage (Huntzinger and Izaurralde, 2011)

However, Kim et al., 2009 have reported that similar to siRNA mediated silencing, a high proportion of complementary bases between the miRNA and target mRNA can lead to activation of AGO component of RISC and cleavage of the target mRNA.

piRNAs

Piwi-interacting RNAs: piRNAs: These piRNAs (Piwi-interacting RNAs) are reported as a complex group of small RNAs enriched in the germline tissues of several organisms. Lin and Spradling, 1997 reported a novel gene, piwi, whose mutations abolish germline stem cell division in *Drosophila* ovarian germline. Piwi (P-element induced wimpy testis) interacting RNAs have been attributed to silencing of mobile elements. As the insertion of mobile elements could introduce DNA changes, piRNA are involved in their regulation by cleavage of the target. (Weik and Miska, 2014) piRNAs are a class of small non-coding RNAs originally isolated from mammalian germline cells. Their distribution is diverse: it is reported that over 80 percent map to unique genome locations : of which 5 percent mapped to loci with multiple transposons and 5 percent mapped to protein-coding genes in Drosophila. (Zuo et al., 2016)

Features of piRNAs

The piRNAs possess certain features such as: A mature molecule is of length 24–32 in comparison with siRNAs and miRNAs that are shorterThey also exhibit a preference for 5' uridine residue probably due to the binding of their first 10 nucleotides with Piwi or Aub (Aubergine: a distinct group of Argonaute family protein found in Drosophila) (Zuo et al., 2016). Apart from the 5' U they also possess a a 3′-most sugar that is 2′-O-methylated. (Weik and Miska, 2014)

Weik and Miska, 2014 have reviewed that both D. melanogaster and vertebrates show piRNAs that are between 26 and 30 nucleotides (nt) in length. However, in C. elegans the length is 21 nucleotides but display the other similar features. Mice possess piwi proteins called MIWI (PIWIL1), MILI (PWIL2) and MIWI2 (PWIL4): of which pre-pachytene piRNAs bind to MILI and MIWI2 while pachytene piRNAs bind with MILI and MIWI.

Biogenesis of piRNAs

Benjamin Czech and Gregory J. Hannon, 2016 have reviewed the pathways involved in biogenesis of piRNAs. Small RNA profiling coupled with genetics and biochemical studies have revealed pathways for the synthesis. One involves the production of silencers that identify mobile elements of the genome under control of the genetic material. Another pathway involves the amplification of specific piRNAs in order to target movile elements; referred to as ping-pong cycle.

- The overall process possesses the following steps:
- Generation of piRNA cluster transcripts
- Production of piRNA molecules
- Export to the cytoplasmic processing sites
- piRNA intermediates are produced from the clusters.

- Loading on into PIWI proteins
- Trimming and methylation to yield mature piRNA-PIWI complexes.

Computational studies have revealed clustering of multiple piRNAs at distinct genomic loci in *Drosophila* and mammalian genomes; the location is in heterochromatin for the former and in euchromatin in the latter.

Site of Production

The production of piRNA takes place in the Yb bodies in follicle cells. piRNA production occurs in nuage or multi-protein complexes on near perinuclear locations. It was also researched that cluster transcripts are enriched in gonadal cells.

Transcription and Processing

Several research teams have unearthed that uni-strand clusters are RNA Polymerase II transcripts with 5' caps, polyadenylation and terminal sites and display splicing mechanisms (Li et al., 2013, Goriaux et al., 2014 and Mohn et al., 2014). Once, the cluster is exported to the cytoplasmic site, the clusters are processed into piRNA intermediates. Such transcripts in mice have been traced to MOV10L1 that is equivalent to *Drosophila* Armitage (Vourekas et al., 2015).

Benjamin Czech and Gregory J. Hannon, 2016 have reviewed that the current model of biogenesis reveals that the loading of piRNA intermediates into PIWI clade of proteins determines the size of the molecules. This step is followed by trimming that is performed by a Mg2+-dependent 30 to 50 exonuclease, trimmer whose activity is yet to be characterized. As an alternative, Zuc can facilitate cleavage to form a 3' end to be followed by trimming.

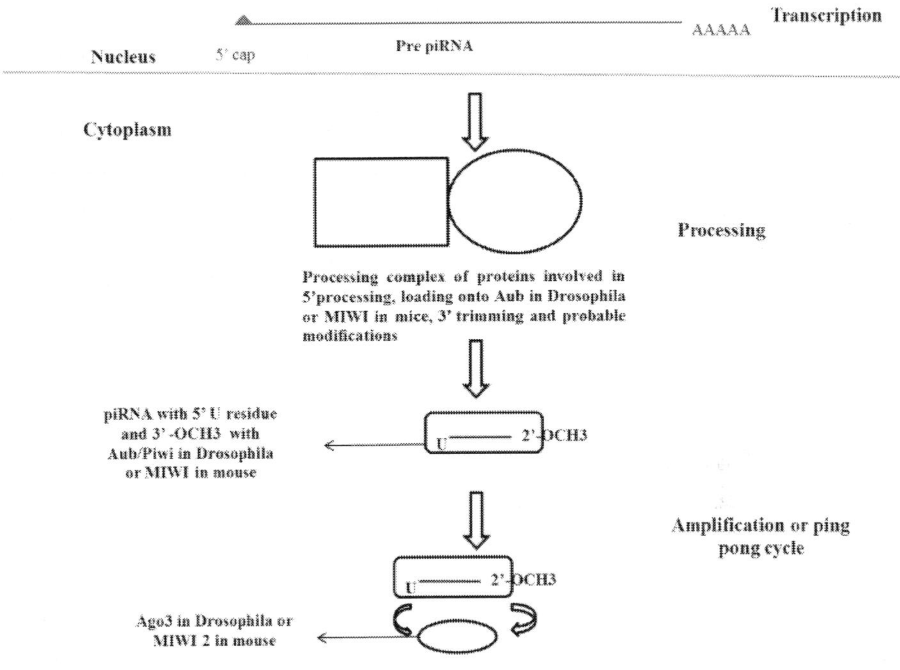

Figure 1.4: Schematic representation of piRNA transcription, processing and its amplification. The respective proteins reported in *Drosophila* and mouse has been mentioned.

Ping Pong Cycle

This phenomenon was first described in *Drosophila* and is considered as piRNA-dependent PTGS mechanism; similar cycles have also been reported from mice, fish and silkworm. This cycle displays a distinctive character of combining both the piRNA generation and its silencing of its targets.

In germ cells of Drosophila, Aub along with cluster derived transcript piRNA is involved in recognition and cleaving of active transposon transcripts. 5' end of piRNAs are thus generated as an outcome that are in sense to mobile genetic elements. This step is followed by loading onto Ago 3 (Argonaute3) which is followed by the steps of processing by the uncharacterized nuclease or the 3' end formation discussed above. These Ago-piRNAs complexes further cleave and generate more piRNA transcripts.

This amplification cycle of new piRNA incorporated into Ago3 which proceeds to generate Aub-bound piRNAs from piRNA cluster transcripts has been reported also in zebra fish and mice. (Weik and Miska, 2014).

Zuo et al., 2016 have reviewed the potential functions of piRNAs in suppressing the activities of transposable elements at genomic and epigenetic levels, their role in spermatogenesis and maintenance of fertility. Additionally, research has revealed The Piwi/piRNA complex is involved in the regulation of some protein-coding genes: for instance, those that control spine shape, playing roles in spine morphogenesis.

Transcriptional Gene Silencing (TGS)

Transcriptional gene silencing (TGS) is an epigenetic mechanism that involves inactivation of promoters to silence a gene. DNA methylation is a hallmark of TGS and is essential for normal plant and animal development (Mette et al., 2000). In plants, several repeated sequences are involved that are substrates of methylation which in turn suppresses their expression (Wassenegger, 2005).

In *Arabidopsis*, rasiRNAs have been identified that are a type of siRNAs. These RNAs are involved in the facilitating of DNA methylation to achieve silencing of transcription of repeated sequences (Chan et al., 2005). The steps involved are:

- A protein with RNA polymerase activity transcribes a wrong RNA from the methylated DNA.
- dsRNA and further RNA with errors are produced from the above RNA molecule.
- This cycle maintains itself (Vaucheret, 2005).

This is followed by processing of dsRNA into the rasiRNAs.

Several other molecules are involved such as DNA methyltransferase, PolIV as well as SNF2-chromatin remodelling protein (Cao et al., 2003; Zilberman et al., 2003; Kanno et al., 2004, 2005). Based on the location of methylated cytosine, the methylation status is maintained by methyltransferases.

SUMMARY

The several mechanisms underlying the history, biogenesis and mode of action of the effectors of RNAi: siRNAs, miRNAs and piRNAs were discussed in the preceding pages. The mechanisms of RNAi have several applications like reverse genetics where a gene is silenced to study the phenotype. A powerful application is therapeutics that has a chapter dedicated to it.

REFERENCES

1. Agrawal N, Dasaradhi PVN, Mohmmed A, Malhotra P, Bhatnagar RK, Mukherjee SK. (2003) RNA Interference: Biology, Mechanism, and Applications. Microbiology and Molecular Biology Reviews. 67(4):657-85. doi:10.1128/MMBR.67.4.657-685.2003.
2. Al-Kaff NS, Covey SN, Kreike MM, Page AM, Pinder R, Dale PJ. (1998 Mar) Transcriptional and posttranscriptional plant gene silencing in response to a pathogen. Science. 279(5359):2113-5.
3. Bartel, DP. (2009) MicroRNAs: target recognition and regulatory functions. Cell. 136: 215–33.
4. Benjamin Czech and Gregory J. Hannon. (2016) One Loop to Rule Them All: The Ping-Pong Cycle and piRNA-Guided Silencing. Trends in Biochemical Sciences. 41(4):324-37.
5. Carmell, M. A. (2002) The Argonaute family: tentacles that reach into RNAi, developmental control, stem cell maintenance, and tumorigenesis. Genes and Devevelopment. 16:2733-42. doi:10.1101/gad.1026102.
6. Carthew RW, Sontheimer EJ. (2009) Origins and mechanisms of miRNAs and siRNAs. Cell 136(4):642– 55
7. Cenik ES, Fukunaga R, Lu G, Dutcher R, Wang Y, et al. (2011) Phosphate and R2D2 restrict the substrate specificity of Dicer-2, an ATP-driven ribonuclease. Molecular Cell 42:172–84.
8. Cogoni C, Irelan JT, Schumacher M, Schmidhauser TJ, Selker EU, Macino G. (1996) Transgene silencing of the al-1 gene in vegetative cells of Neurospora is mediated by a cytoplasmic effector and does not depend on DNA-DNA interactions or DNA methylation. EMBO Journal. 15(12):3153-63.

9. Ecker JR, Davis RW; Davis (1986). Inhibition of gene expression in plant cells by expression of antisense RNA. Proceedings of the National Academy of Sciences, USA. 83 (15): 5372–76.

10. Elbashir, SM, Lendeckel, W and Tuschl, T. (2001) RNA interference is mediated by 21- and 22-nucleotide RNAs. Genes and Development. 15: 188– 200.

11. Eva-maria Weick and Eric. A. (2014) Miska. piRNAs: from biogenesis to function. Development. 141: 3458-71.

12. Fire A, Xu S, Montgomery MK, Kostas SA, Driver SE, Mello CC. (1998) Potent and specific genetic interference by double-stranded RNA in Caenorhabditis elegans. Nature. 391(6669):806-11

13. Goriaux, C. et al. (2014) Transcriptional properties and splicing of the flamenco piRNA cluster. EMBO Reports. 15:411–418.

14. Hamilton, A. J. and Baulcombe, D. C. (1999) A species of small antisense RNA in posttranscriptional gene silencing in plants. Science. 286:950-952.

15. Huntzinger, E and Izaurralde, E. (2011) Gene silencing by microRNAs: contributions of translational repression and mRNA decay. Nature Review Genetics. 12:99–110.

16. Jenny K W Lam , Michael Y T Chow, Yu Zhang, Susan W S Leung. (2015) siRNA Versus miRNA as Therapeutics for Gene Silencing. Molecular Therapy Nucleic acids. 4, Article e252.

17. JS Mattick, IV Makunin. (2006) Non-coding RNA. Human Molecular Genetics. 15(1):R17–R29.

18. Kim, VN, Han, J and Siomi, MC. (2009) Biogenesis of small RNAs in animals. Nature Reviews Molecular Cell Biology. 10: 126–139.

19. Kooter JM, Matzke MA, Meyer P. (1999) Listening to the silent genes: transgene silencing, gene regulation and pathogen control. Trends in Plant Science. 4(9):340-347.

20. Li XZ, et al. (2013) An ancient transcription factor initiates the burst of piRNA production during early meiosis in mouse testes. Molecular Cell. 50:67–81.

21. Lim LP, Lau NC, Garrett-Engele P, Grimson A, Schelter JM, Castle J, et al. (2005) Microarray analysis shows that some microRNAs

downregulate large numbers of target mRNAs. Nature. 433:769–773.

22. Lin H, Spradling AC. (1997) A novel group of pumilio mutations affects the asymmetric division of germline stem cells in the Drosophila ovary. Development. 124:2463-2476.
23. Lindbo JA, Dougherty WG. (Aug 1998) Untranslatable transcripts of the tobacco etch virus coat protein gene sequence can interfere with tobacco etch virus replication in transgenic plants and protoplasts. Virology. 189(2):725-33.
24. Lu M, Zhang Q, Deng M, Miao J, Guo Y, et al. (2008) An analysis of human microRNA and disease associations. PLoS ONE. 3(10):e3420
25. MacFarlane L-A, Murphy PR. (2010) MicroRNA: biogenesis, function and role in cancer. Curr. Genomics. 11(7):537–561.
26. Mohn F, et al. (2014) The rhino-deadlock-cutoff complex licenses noncanonical transcription of dual-strand piRNA clusters in Drosophila. Cell. 157:1364–1379.
27. Napoli C, Lemieux C, Jorgensen R. (1990) Introduction of a Chimeric Chalcone Synthase Gene into Petunia Results in Reversible CoSuppression of Homologous Genes in trans. Plant Cell. 2(4):279-289.
28. Lee RC, Feinbaum RL, Ambros V. (1993) The C. elegans heterochronic gene lin-4 encodes small RNAs with antisense complementarity to lin-14. Cell. 75:843–854.
29. Romano N, Macino G. (1992) Quelling: transient inactivation of gene expression in Neurospora crassa by transformation with homologous sequences. Molecular Microbiology. 6(22):3343–3353.
30. Ross C W, Jennifer AD. (2013) Molecular Mechanisms of RNA Interference. Annual Review of Biophysics. 42:217–239.
31. Timmons L, Fire A. (1998) Specific interference by ingested dsRNA. Nature. 395(6705):854.
32. Valencia-Sanchez MA, Liu J, Hannon GJ, Parker, R. (2006) Control of translation and mRNA degradation by miRNAs and siRNAs. Genes and Development. 20:515–524.
33. Vourekas, A. et al. (2015) The RNA helicase MOV10L1 binds piRNA

precursors to initiate piRNA processing. Genes and Development. 29, 617–629.

34. Zuo L, Wang Z, Tan Y, Chen X, Luo X. (2016) piRNAs and Their Functions in the Brain. International journal of human genetics. 16(1-2):53-60.

CHAPTER 2

RNAi: DESIGNING PLATFORMS, DELIVERY SYSTEMS AND CHALLENGES

Prerna Pandey

Dr Prerna Pandey, a biotechnologist with several years of wet lab research experience, worked at International Center for Genetic Engineering and Biotechnology, New Delhi. Her field of research in PhD included isolation and molecular characterization of geminiviruses, genome sequencing, gene annotation, and gene silencing using the RNA interference technology. She has also worked at Transasia Biomedicals and Advance Enzyme Technologies as a scientist. Prerna has published papers in peer-reviewed journals, and has submitted a number of annotated Geminiviral genome sequences in the GenBank, including two novel ones. She has also completed her editing and proof-reading courses from Society for the promotion of Editors and Proof-readers, UK and now works as a freelance scientific editor and writer. When Prerna is not busy with her assignments, she loves to explore historical places.

The siRNA molecules, which are only 21 nucleotides long, suppresses the m RNA expression of the intended gene which are perfectly complementary to the siRNA guide strand. Although the silencing should be ideally specific, siRNAs can downregulate non-target genes too which have partial complementarity to the seven-nucleotide seed region of siRNA (see chapter 2), also known as off-target effect. To understanding the target gene functions accurately and for successful therapeutic applications, it is critical to select a siRNA with minimum or no off-target effect. In this chapter, a few siRNA design software tools and various modes of delivery of these effector molecules have been discussed.

DESIGNING OF TARGET SPECIFIC SIRNAS

One of the most critical step to have a significant impact on the efficiency and efficacy of the RNAi of a particular gene target is the selection of siRNA (Elbashir et al., 2002). Although several algorithms have been published, they are inefficient, obscure, or costly.

There are many online algorithms available for proper siRNA design (Reynolds et al., 2004)

- The selected regions must be translational or regulatory protein-free coding sequence regions (located nearly 50 – 100 nucleotide downstream of the start codon).
- The 5' UTR and 3' UTR should be avoided.
- The G+C content of the target sequence should be between 35% - 60%.
- Stretches of four or more nucleotide repeats must be avoided.
- Importantly, to avoid off target effects, the sequences which share certain degree of homology with other related or unrelated genes must be avoided.
- Thermodynamic stability
- Other structural features

On an average, 65-75% of the siRNA duplexes designed by conventional methods cause 50- 65% gene silencing, with variable gene knockdown efficacy (Bernstein et al., 2001a), which, might not be therapeutically significant in many cases. Thus, there is a constant need for additional techniques that design effector molecules to ensure more robust gene-silencing ability.

In a study by Hsieh et al., (2004) the siRNAs designed based on the functional parameters using an algorithm showed better results compared to those of the conventionally designed siRNAs. Unexpectedly, the benefit was found on pooling the rationally designed siRNAs, and on average, the efficacy required to silence a particular gene target was greater than that of conventionally designed siRNAs (Holen et al., 2002; Martinez et al., 2002; Huang et al., 2004).

Developers like Ambion, Dhermacon, GenScript, Clontech, EMBOSS, Promega, Qiagen, siDirect, siRNA Whizard, OptiTNAi, siSearch, etc., have designed softwares that use their own algorithms, have mixed rules, or follow previously published studies.

DELIVERY AND STABILITY OF SIRNA

The delivery of siRNA molecules to the desired cell type, tissue or organ is a major bottleneck in the development of siRNA therapies. Due to their negative charge and size, siRNAs are not able to cross the cellular membrane readily.

Liposome based Strategies: chemically synthesized or *in vitro* transcribed siRNAs can be delivered in the cell usually via cationic liposome based strategies. Lipid based delivery in vivo has the disadvantage of

- rapid liver clearance and
- lack of target tissue specificity.

The siRNAs duplex and the liposomes are complexed *in vitro* and through the endosomal pathway, they enter the cells. siRNAs are then released into the cytoplasm where they associate with RISC.

Behelke, 2006, has reported the chemical modifications of siRNA backbone for their optimized *in vivo* stability. A few of these strategies are:

- an increased stability of siRNAS in serum has been observed with 2'F, 2'O-Me and 2'H substitutions in the RNA backbone
- longer RNA species (25-27mer) may have increased potency (Kim etl., 2005; Rose et al., 2005; Siolas et al., 2005), thus reducing the required concentrations of siRNAs for optimum therapeutic effect.
- conjugation of cholesterol to the siRNA sense strand (Soutschek et al., 2004),
- antibody-protamine fusions that bind siRNAs via cell-surface receptors (Song et al., 2005)
- cyclodextrin nanoparticles (Hu-Lieskovan et al., 2005)
- aptamer-siRNA conjugates (McNamara et al., 2006)

DELIVERY AND EXPRESSION OF SHORT HAIRPIN RNAS

Use of Promoters

DNA-based expression cassettes that express short hairpins (shRNA) or separate sense and antisense 21mers from Pol III promoters can be constructed (Brummelkamp et al., 2002; Paddison et al., 2002; Lee et al., 2002).

- These **promoters** have well defined transcriptional start sites and terminators, which makes them a preferred choice for DNA-based gene silencing.
- A **dual promoter system** which express two individual RNA strands separately, hybridize and produce functional siRNAs can also be used (sui et al., 2002; Miyagishi and Taira, 2002).
- The produced duplex RNAs are substrates for exportin-5 nuclear export by the pathway. The hairpins yield functional siRNA duplexes after being processed by Dicer, while the siRNAs enter RISC directly.
- However, the PolIII based promoters may not be therapeutically favorable, although most of them give long term and robust silencing in cell culture systems, because they might saturate the endogenous RNAi pathway.

Use of Inducible Systems

Theoritically, inducible systems are tightly controlled by the inducer and the transcription is dose-dependent, prompt, and reversibly regulated.

- PolII/miRNA-based systems, and a number of advanced inducible or polycistronic systems have been developed and published (Chung et al., 2006; Xia et al., 2006; Sun et al., 2006). Although the inducible mammalian transcription systems are somewhat leaky even when uninduced, improved technologies may help to achieve tighter control of gene expression.
- Cre-loxP system can be used to generate non-reversible genetic switches, but the ectopic expression of the Cre recombinase protein can be quite complicated.

Vector Mediated Delivery

One of the major problems associated with DNA-based shRNAs is that negatively charged nucleic acids do not readily pass the cellular membrane in absence of carriers.

- Direct delivery through **plasmids** has been done in mice using tail vein injections, this strategy is not suitable for clinical use.
- Vector-based shRNA systems usually deliver using viral vectors. An efficient delivery for transient shRNA can be achieved through **adenovirus- and adeno-associated virus-** (AAV) derived vectors.
- **Retroviruses** encoded proteins mediate stable silencing of the target genes. Murine leukemia virus (MLV) or lentivirus based (e.g. HIV, FIV or EIAV) are a few retroviral vectors that have generated clinical interest.

FACTORS AFFECTING SIRNA EFFICIENCY

The siRNA efficacy depends on many factors, such as

- the thermodynamic stability at the 59-antisense end of the duplex,
- the silencing ability is greatly reduced with ability to form internal hairpins in the RISC complex,
- base preference for an A, not a C, at the 19th or 20th base position of the sense strand (required for the unwinding/ activation step in the RISC), GC content,
- preferences for A at position 3, U at position 10 and any base except G at position 13 (Khvorova et al., 2003).

Attempts to prolong the siRNAs gene-silencing activity have been attempted by introducing various chemical modifications to increase the stability of the molecules.

- replacing one of the non-bridging oxygen atoms in the phosphodiester bond by sulfur: phosphorothioate modifications (Amarzguioui et al., 2003).
- single- or multiple-siRNA duplexes which target both single as well as multiple genes yield good gene silencing outcome

(>more than 80%), compared with the pooling of conventionally designed siRNAs.

- Relatively long double-stranded RNAs (25-30 bp in length) have significantly higher gene-silencing capability than the conventional 21-bp long siRNAs

RNAI DELIVERY: ANIMALS

Unmodified siRNAs in early clinical trials can be injected into the eye and other organs, however, they are degraded by enzymes and are unable to cross cell membranes.

Due to the similar physicochemical properties, carriers for DNA can also be applied to siRNA, which can be broadly divided into two categories, i.e., viral and non-viral. Non-viral vectors have become increasingly popular alternatives because of the potential toxicities of viral vectors.

Non-viral siRNA Vectors involve:

- **complexing** siRNA with a positively charged vector like cationic lipids, cationic cell penetrating peptides, and cationic polymers;
- **conjugating** siRNA with small molecules, antibodies, polymers, and RNAs;
- **encapsulating** siRNA in nanoparticulate formulations.

Peptides that are Cationic and Cell Penetrating (CPP)

CPP have been used for the intracellular delivery of macromolecules (Meade and Dowdy, 2007) and may enter cells directly by crossing the cell membrane.

- Through electrostatic interactions, the CPP and siRNA form positively charged non-covalent complexes (non-covalent CPP–siRNA (Unnamalai, 2004). This approach is that it does not require a chemical modification of siRNA, thus preserving siRNA activity and reducing the purification procedures.
- CPP and siRNA can also form covalent cross-links through disulfide bonds, which cleaves inside the cells and releases siRNA into the cytoplasm (Muratovska and Eccles, 2004).

- conjugation of siRNA and CPP affects cellular localization (Chiu et al., 2004), possibly in endosomes and lysosomes that are localized perinuclearly.

Polymeric Carriers

Linear or Branched Cationic Polymers: The positively charged polymers, through electrostatic interactions, form complexes with the negatively charged phosphates of DNA (Agarwal, 2005). This process protects plasmids from nuclease digestion. Similarly, siRNA–polymer polyplexes have been used. The size, surface charge, and structural properties of polyplexes are dependent on the ratio of the number of phosphate groups on siRNA to the positive charges of cationic polymers.

Cationic Polymers protect degradation of oligonucleotides because their high charge density stimulates escape from endosomes (Tseng et al., 2009). PEI attracts and sustains substantial proton influx, induces osmotic swelling and rupture of endosomes, thus leading to the release of DNA in the cytoplasm, thereby preventing DNA transport to lysosomes for subsequent degradation (Boussif et al., 1995; Abdallah et al., 1996).

Another copolymer-carrier for controlled drug release, PLGA is most extensively investigated. It is less toxic compared to cationic- lipids and polymers. To encapsulate DNA, antisense oligodeoxynucleotides, and siRNA, nanocapsules have been used for improving the stability and ensuring sustained release PLGA (Khan et al., 2004).

Dendrimers are synthetic macromolecules with well-defined and highly branched peripheral chain ends. These are synthesized in step-by-step fashion, algorithmically, and have been used as carriers of large biomolecules and small molecule drugs. Dendrimers with surface groups that are positively charged are used to deliver antisense oligonucleotides, genes, and siRNA.

Cyclodextrin-containing cationic polymers: Cyclodextrins (CD) are cyclic oligosaccharides which occur naturally and their derivatives enhance the solubility, stabilization, and absorption of small molecule drugs and proteins or peptides, thus are used in pharmaceutical formulations. The CD-containing cationic polymers, for example, those including linear and branched PEI and PAMAM are readily PEGylated and have reduced cytotoxicity (Li and Loh, 2008).

Conjugation of siRNAs with Bioconjugates

The stability, cellular internalization, or cell-specific active targeting delivery is improved by conjugation of siRNA with a variety of small molecules including cholesterol, bile acids, and lipids, peptides, polymers, antibodies, and aptamers (Jeong, 2009).

- Cholesterol can be conjugated *via* a pyrrolidone linkage to the 3′-terminus of siRNA sense strand (Soutschek et al., 2004).
- siRNA conjugated to amphipathic poly(vinyl ether) PBAVE, also a endosomolytic agent, shows enhanced endosomal escape after endocytosis (Rozema, 2007).
- Conjugation or complexation of siRNA to antibody increases the specificity of siRNA delivery and minimize the off-target effects (Song et al., 2005).
- The siRNA - RNA aptamer linkage also help in achieving cell-type specificity (Dassie et al., 2009).

Lipid Based Carriers

The siRNAs can be embedded in lipid nanoparticles and carried via the blood to the diseased cells.

The liposomes, solid lipid nanoparticles, and micelles form a part of lipid-based siRNA delivery systems. Liposomes are biocompatible and biodegradable globular vesicles with a phospholipid bilayer, an aqueous core, and are comprised of natural constituents like sterols and lipids. a wide variety of hydrophobic and hydrophilic drugs can be incorporated to the amphipathic nature of liposomes (Oh and Park, 2009).

The key components are cationic lipids for lipid-based siRNA delivery systems even when neutral liposomes can successfully deliver siRNA *in vitro* and *in vivo* (Halder et al., 2006).

Cationic Lipids consist of a cationic head group, a lipophilic tail group, and a connecting linker (Tsenfg et al., 2009). On mixing the cationic lipid with the anionic siRNA, multilamellar structures are spontaneously formed. The formation of such lipoplexes protects siRNA from degradation, facilitates cellular uptake through endocytosis, enhances the release of siRNA from endosomal/lysosomal entrapment, and thereby promotes siRNA accumulation in the cytosol (Akhtar and Benter, 2007).

RNAI DELIVERY: PLANTS

RNAi has been used in plant transformation techniques as a tool to study gene expression, to bring new traits such as improved nutritional quality, better resistance to disease, and stress resistance in various crops. RNAi transient delivery systems include:

Virus-Induced Gene Silencing (**VIGS** vectors): Viral based vectors have been developed for VIGS in many plant species such as tomato (Ramegowda et al. 2014); barley (Bruun-Rasmussen et al., 2007); and the experimental plants like Arabidopsis (Jupin, 2013) and tobacco (Senthil-Kumar and Mysore, 2014). The engineered plant viruses have been used for production of dsRNA in both transient and stable transformation studies targeting endogenous plant transcripts and controlling pests (Senthil-Kumar and Mysore 2014). However, the viral vectors exhibit different levels of stability in different plant species; moreover, the using some viruses for VIGS becomes limiting because of biosafety regulations (Shang et al. 2007).

Agrobacterium -Mediated Transient Assays

Particle Bombardments (biolistics)

This process has been used for direct delivery of dsRNA constructs in stable transformations, and also in functional transient assays (Shang et al. 2007). Particle bombardment on mature plants is limited to the targeted tissue and random in showing a change in phenotype. Thus, biolistics are preferred for specific tissues such as flower petals or developing seeds, and microprojectile bombardment is used on monocot plants where Agrobacterium -mediated transformation is mostly limited (Schweizer et al. 2000).

Agrobacterium Tumerfaciens Infiltration (Agroinfiltration)

It is a widely used for RNAi transient assays because it is fast, and convenient for delivering silencing constructs to vegetative tissues and obtaining a high level of localized changes in the phenotype. Agroinfiltration transient assays have been optimized for many plant species, including monocots such as rice (Andrieu et al. 2012).

SIRNA DELIVERY: CHALLENGES AND BARRIERS

Systemic and Regional Delivery

The barriers to siRNA delivery are multiple and depend on the targeted organs and the administration routes. Generally, local delivery of siRNA faces fewer barriers compared to systemic delivery. For instance, intranasal inhalation of siRNA against respiratory syncytial virus, either embeded in polycationic liposomes, or naked, were nearly equally effective in reducing the infection (Bitko, 2005).

Systemic Elimination of siRNA

The siRNA molecules, drug delivery standpoint, have certain unfavorable physicochemical properties such as negative charges, large molecular weight and size, and instability. Naked siRNA have a short plasma half-life of <10 min because it is degraded by endonucleases in the serum and removed by glomerular filtration (van de Water, 2006).

Entrapment in Reticluo-endothelial System

The siRNA-loaded carriers include nanospheres, nanocapsules, liposomes, micelles, microemulsions, conjugates, and other nanoparticulates. These systems offer efficient means to deliver small-molecular-weight compounds such as proteins, peptides, or siRNA, and protect siRNA from undesirable interactions, and degradation (Jang et al., 2003).

The uptake and biodistribution by RES is affected by several factors, including

Surface Charge: a negative surface charge increases the clearance of particulates from systemic circulation relative to neutral or positively charged particles (Gregoriadis, 1974).

Surface Characteristics: Surface modifications using flexible polyethylene glycol (e.g., pegylation), which is hydrophilic, and other surfactant copolymers result in particles that have prolonged systemic circulation (Illum, 1984). These modifications protect the vectors against opsonization,and reduce the complement activation. Pegylation may neutralize the positive surface charge that is required for siRNA uptake into cells (Santel, et al., 2006).

The Vascular Endothelial Barriers

The siRNA or carriers undergo extravasation within an organ. siRNA transport across vessel walls can occur *via*

- Diffusion: driven by the concentration gradient
- Convection through capillary pores, and
- Transcytosis.

Due to vessel leakiness in tumors, the drug transport across tumor microvascular wall occurs by extravasation *via* diffusion and/or convection through the discontinuous endothelial junctions. Leakiness in tumor vessels promotes siRNA/carrier extravasation, elevates pressure of interstitial fluid and reduces transvascular fluid transport.

Transport in Tissues

Transport of small molecules occurs mainly by diffusion, and that of large molecules is mainly by convection in the interstitial space.

Cellular Uptake

Naked siRNA has high molecular weight, large size, and negative charges of the phosphate backbone; thus, they cannot readily cross the anionic cell membrane through passive diffusion, and the endocytosis becomes major mode of internalization. Ligands and antibody coatings with carriers can promote the carrier-specific binding to cell membrane. (Goula et al., 2000).

Intracellular Transport

Following cellular internalization, the siRNA–carrier complex (in endocytic vesicles) is transported along microtubules to lysosomes that are co-localized with the microtubule organizing center. The endocytotic vesicles sequentially fuse with early endosomes which mature into late endosomes before fusing with lysosomes, where they are recycled to the cell surface, degraded inside lysosomes, or released to other intracellular compartments including the cytosol (Mukherjee and Ghosh, 1997). The low transfection efficiency is caused mainly due to endosomal entrapment and lysosomal degradation of siRNA (Tseng et al., 2009; Endoh and Ohtsuki, 2009).

CASE STUDIES

Lipid-based carriers have been successfully used to deliver siRNA to target sites in the endothelium, RES organs (e.g., liver), and solid tumors. The following findings indicate liposomes as a viable carriers for siRNAs.

- Although the naked siRNA when injected, localized in the kidney, the siRNAs that were loaded in liposomes of cationic lipids, sustained presence of siRNA and decreased levels of the mRNA and protein of two targets, *Tie2*, and *CD31* genes was observed (Santel et al., 2006).
- When the liposomal anti-TNF-alpha siRNA (cationic) was injected intraperitoneally in mice, it inhibited TNF-alpha gene expression, which is induced by lipopolysaccharide (Sorensen et al., 2003).
- A single intravenous injection of pegylated solid nucleic acid lipid particle-encapsulated ApoB siRNA, resulted in reduced ApoB mRNA and protein in the in cynomolgus monkey liver in dose-dependent manner followed by reductions in serum cholesterol and low-density lipoprotein levels (Zimmermann et al., 2006).
- selective delivery to solid tumors, followed by silencing of the targeted epidermal growth factor receptor was observed when the siRNA complexed with peptide protamine (positively charged) and coated with surface pegylated liposomes (cationic) were injected in mice, gene (Li and Loh, 2008).

CONCLUSION

RNAi represents a powerful tool for studies of gene function in plants and animals. This technique enables to silence virtually any gene and utilise the cellular machinery for efficiently targeting complementary transcripts. In vivo studies to prove the hypotheses have demonstrated that viral as well as non-viral delivery methods can provide potent and selective target gene suppression without any clear toxic effects. Although the results from studies on a large number of animal model are promising, the challenges like, safety, effectiveness, and ease of production are important considerations for selecting the appropriate siRNA carriersn and finally fully harnessing RNAi a therapeutic option.

REFERENCES

1. Abdallah B, Hassan A, Benoist C, Goula D, Behr JP, Demeneix BA. (1996) A powerful nonviral vector for in vivo gene transfer into the adult mammalian brain: polyethylenimine. Hum Gene Ther. 7(16):1947–54.
2. Agarwal A, Unfer R, Mallapragada SK. (2005) Novel cationic pentablock copolymers as non-viral vectors for gene therapy. J Control Release. 103(1):245–58.
3. Akhtar S, Benter I. (2007) Toxicogenomics of non-viral drug delivery systems for RNAi: potential impact on siRNA-mediated gene silencing activity and specificity. Adv Drug Deliv Rev. 59(2–3):164–82.
4. Amarzguioui M, Prydz H. (2004) An algorithm for selection of functional siRNA sequences. Biochem Biophys Res Commun. 316:1050–58.
5. Andrieu, A., Breitler, J. C., Siré, C., Meynard, D., Gantet, P., & Guiderdoni, E. (2012). An *in planta*, *Agrobacterium*-mediated transient gene expression method for inducing gene silencing in rice (*Oryza sativa* L.) leaves. Rice, 5: 23'.
6. Behlke MA. (2006) Progress towards in vivo use of siRNAs. Mol Ther. 13:644–670.
7. Bernstein E, Caudy AA, Hammond SM, Hannon GJ (2001a). Role for a bidentate ribonuclease in the initiation step of RNA interference. Nature 409:363-366. Bernstein E, Denli AM, Hannon GJ. (2001b). The rest is silence. RNA 7:1509-1521.
8. Bitko V, Musiyenko A, Shulyayeva O, Barik S. (2005) Inhibition of respiratory viruses by nasally administered siRNA. Nat Med. 11(1):50–5. doi: 10.1038/nm1164.

9. Boussif O, Lezoualc'h F, Zanta MA, Mergny MD, Scherman D, Demeneix B, et al. (1995) A versatile vector for gene and oligonucleotide transfer into cells in culture and in vivo: polyethylenimine. Proc Natl Acad Sci USA. 92(16):7297–301.
10. Brummelkamp TR, Bernards R, Agami R. (2002) A system for stable expression of short interfering RNAs in mammalian cells. Science. 296:550–553.
11. Bruun-Rasmussen M, Madsen CT, Jessing S, Albrechtsen M. (2007) Stability of Barley stripe mosaic virus-induced gene silencing in barley. Mol Plant Microbe Interact. 20(11):1323-31.
12. Chiu YL, Ali A, Chu CY, Cao H, Rana TM. (2004) Visualizing a correlation between siRNA localization, cellular uptake, and RNAi in living cells. Chem Biol. 11(8):1165–75.
13. Chung KH, Hart CC, Al-Bassam S, Avery A, Taylor J, Patel PD, Vojtek AB, Turner DL. (2006) Polycistronic RNA polymerase II expression vectors for RNA interference based on BIC/miR-155. Nucleic Acids Res. 34:e53.
14. Dassie JP, Liu XY, Thomas GS, Whitaker RM, Thiel KW, Stockdale KR, et al. (2009) Systemic administration of optimized aptamer–siRNA chimeras promotes regression of PSMA-expressing tumors. Nat Biotechnol. 27(9):839–49.
15. Elbashir SM, Harborth J, Weber K, Tuschl T (2002). Analysis of gene function in somatic mammalian cells using small interfering RNAs. Methods 26:199-213.
16. Endoh T, Ohtsuki T. (2009) Cellular siRNA delivery using cell-penetrating peptides modified for endosomal escape. Adv Drug Deliv Rev. 61(9):704–9.
17. Goula D, Becker N, Lemkine GF, Normandie P, Rodrigues J, Mantero S, et al. (2000) Rapid crossing of the pulmonary endothelial barrier by polyethylenimine/DNA complexes. Gene Ther. 7(6):499–504.
18. Gregoriadis G, Neerunjun DE. (1974) Control of the rate of hepatic uptake and catabolism of liposome-entrapped proteins injected into rats. Possible therapeutic applications. Eur J Biochem. 47(1):179–85.
19. Halder J, Kamat AA, Landen CN, Jr, Han LY, Lutgendorf SK, Lin YG, et al. (2006) Focal adhesion kinase targeting using in vivo short

interfering RNA delivery in neutral liposomes for ovarian carcinoma therapy. Clin Cancer Res. 12(16):4916–24.

20. Hsieh AC, Bo R, Manola J, Vazquez F, Bare O, Khvorova A, et al. (2004). A library of siRNA duplexes targeting the phosphoinositide 3-kinase pathway: determinants of gene silencing for use in cell-based screens. Nucleic Acids Res 32:893-901.
21. Huang F, Khvorova A, Marshall W, Sorkin A (2004). Analysis of clathrinmediated endocytosis of epidermal growth factor receptor by RNA interference. J Biol Chem 279:16657-16661.
22. Hu-Lieskovan S, Heidel JD, Bartlett DW, Davis ME, Triche TJ. (2005) Sequence-specific knockdown of EWS-FLI1 by targeted, nonviral delivery of small interfering RNA inhibits tumor growth in a murine model of metastatic Ewing's sarcoma. Cancer Res. 65:8984–8992.
23. Illum L, Davis SS. (1984) The organ uptake of intravenously administered colloidal particles can be altered using a non-ionic surfactant (Poloxamer 338) FEBS Lett. 167(1):79–82.
24. Jang SH, Wientjes MG, Lu D, Au JL. (2003) Drug delivery and transport to solid tumors. Pharm Res. 20(9):1337–50.
25. Jeong JH, Mok H, Oh YK, Park TG. (2009) siRNA conjugate delivery systems. Bioconjug Chem. 20(1):5–14.
26. Jupin I. (2013) A protocol for VIGS in Arabidopsis thaliana using a one-step TYMV-derived vector. Methods Mol Biol. 975:197-210.
27. Khan A, Benboubetra M, Sayyed PZ, Ng KW, Fox S, Beck G, et al. (2004) Sustained polymeric delivery of gene silencing antisense ODNs, siRNA, DNAzymes and ribozymes: in vitro and in vivo studies. J Drug Target. 12(6):393–404.
28. Khvorova A, Reynolds A, Jayasena SD. (2003) Functional siRNAs and miRNAs Exhibit Strand Bias. Cell. 115:209–216.
29. Kim DH, Behlke MA, Rose SD, Chang MS, Choi S, Rossi JJ. (2005) Synthetic dsRNA Dicer substrates enhance RNAi potency and efficacy. Nat Biotechnol. 23:222–226.
30. Lee NS, Dohjima T, Bauer G, Li H, Li MJ, Ehsani A, Salvaterra P, Rossi J. (2002) Expression of small interfering RNAs targeted against HIV-1 rev transcripts in human cells. Nat Biotechnol. 20:500–505.

31. Li J, Loh XJ. (2008) Cyclodextrin-based supramolecular architectures: syntheses, structures, and applications for drug and gene delivery. Adv Drug Deliv Rev. 60(9):1000–17.

32. McNamara JO, 2nd, Andrechek ER, Wang Y, Viles KD, Rempel RE, Gilboa E, Sullenger BA, Giangrande PH. (2006) Cell type-specific delivery of siRNAs with aptamer-siRNA chimeras. Nat Biotechnol.

33. Meade BR, Dowdy SF. (2007) Exogenous siRNA delivery using peptide transduction domains/cell penetrating peptides. Adv Drug Deliv Rev. 59(2–3):134–40.

34. Miyagishi M, Taira K. (2002) U6 promoter driven siRNAs with four uridine 3′ overhangs efficiently suppress targeted gene expression in mammalian cells. Nat Biotechnol. 20:497–500.

35. Mukherjee S, Ghosh RN, Maxfield FR. (1997) Endocytosis. Physiol Rev. 77(3):759–803.

36. Muratovska A, Eccles MR. (2004) Conjugate for efficient delivery of short interfering RNA (siRNA) into mammalian cells. FEBS Lett. 558(1–3):63–8.

37. Oh YK, Park TG. (2009) siRNA delivery systems for cancer treatment. Adv Drug Deliv Rev. 61(10):850–62.

38. P.N. Pushparaj1,3*, J.J. Aarthi1, J. Manikandan1, and S.D. Kumar (2008). siRNA, miRNA, and shRNA: in vivo Applications. J Dent Res 87(11):992-1003.

39. Paddison PJ, Caudy AA, Bernstein E, Hannon GJ, Conklin DS. (2002) Short hairpin RNAs (shRNAs) induce sequence-specific silencing in mammalian cells. Genes Dev. 16:948–958.

40. Ramegowda, V., Mysore, K. S., & Senthil-Kumar, M. (2014). Virus-induced gene silencing is a versatile tool for unraveling the functional relevance of multiple abiotic-stress-responsive genes in crop plants. Frontiers in Plant Science, 5, 323.

41. Reynolds A et al. (2004) 22:326. Rational siRNA design for RNA interference. Nat Biotechnol. 2004 Mar;22(3):326-30. Epub 2004 Feb 1.

42. Rose SD, Kim DH, Amarzguioui M, Heidel JD, Collingwood MA, Davis ME, Rossi JJ, Behlke MA. (2005) Functional polarity is introduced by Dicer processing of short substrate RNAs. Nucleic Acids Res. 33:4140–4156.

43. Rozema DB, Lewis DL, Wakefield DH, Wong SC, Klein JJ, Roesch PL, et al. (2007) Dynamic PolyConjugates for targeted in vivo delivery of siRNA to hepatocytes. Proc Natl Acad Sci USA. 104(32):12982–7.

44. Santel A, Aleku M, Keil O, Endruschat J, Esche V, Fisch G, et al. (2006) A novel siRNA-lipoplex technology for RNA interference in the mouse vascular endothelium. Gene Ther. 13(16):1222–34.

45. Senthil-Kumar M, Mysore KS. (2014) Tobacco rattle virus-based virus-induced gene silencing in Nicotiana benthamiana. Nat Protoc. 9(7):1549-62.

46. Siolas D, Lerner C, Burchard J, Ge W, Linsley PS, Paddison PJ, Hannon GJ, Cleary MA. (2005) Synthetic shRNAs as potent RNAi triggers. Nat Biotechnol. 23:227–31.

47. Song E, Zhu P, Lee SK, Chowdhury D, Kussman S, Dykxhoorn DM, Feng Y, Palliser D, Weiner DB, Shankar P, Marasco WA, Lieberman J. (2005) Antibody mediated in vivo delivery of small interfering RNAs via cell-surface receptors. Nat Biotechnol. 23:709–17.

48. Sorensen DR, Leirdal M, Sioud M. (2003) Gene silencing by systemic delivery of synthetic siRNAs in adult mice. J Mol Biol. 327(4):761–6.

49. Soutschek J, Akinc A, Bramlage B, Charisse K, Constien R, Donoghue M, et al. (2004) Therapeutic silencing of an endogenous gene by systemic administration of modified siRNAs. Nature. 432:173–178.

50. Soutschek J, Akinc A, Bramlage B, Charisse K, Constien R, Donoghue M, et al. (2004) Therapeutic silencing of an endogenous gene by systemic administration of modified siRNAs. Nature. 432(7014):173–8.

51. Sui G, Soohoo C, Affar el B, Gay F, Shi Y, Forrester WC. (2002) A DNA vector-based RNAi technology to suppress gene expression in mammalian cells. Proc Natl Acad Sci U S A. 99:5515–20.

52. Sun D, Melegari M, Sridhar S, Rogler CE, Zhu L.(2006) Multi-miRNA hairpin method that improves gene knockdown efficiency and provides linked multi-gene knockdown. Biotechniques. 41:59–63.

53. Tseng YC, Mozumdar S, Huang L. (2009) Lipid-based systemic delivery of siRNA. Adv Drug Deliv Rev. 61(9):721–31.

54. Unnamalai N, Kang BG, Lee WS. (2004) Cationic oligopeptide-mediated delivery of dsRNA for post-transcriptional gene silencing in plant cells. FEBS Lett. 566(1–3):307–10.

55. van de Water FM, Boerman OC, Wouterse AC, Peters JG, Russel FG, Masereeuw R. (2006) Intravenously administered short interfering RNA accumulates in the kidney and selectively suppresses gene function in renal proximal tubules. Drug Metab Dispos. 34(8):1393–7.

56. Xia XG, Zhou H, Xu Z. (2006) Multiple shRNAs expressed by an inducible pol II promoter can knock down the expression of multiple target genes. Biotechniques. 41:64–68.

57. Zimmermann TS, Lee AC, Akinc A, Bramlage B, Bumcrot D, Fedoruk MN, et al.(2006) RNAi-mediated gene silencing in non-human primates. Nature. 441(7089):111–4.

CHAPTER 3

RNAi AND HUMAN DISEASES

Prerna Pandey

Dr Prerna Pandey, a biotechnologist with several years of wet lab research experience, worked at International Center for Genetic Engineering and Biotechnology, New Delhi. Her field of research in PhD included isolation and molecular characterization of geminiviruses, genome sequencing, gene annotation, and gene silencing using the RNA interference technology. She has also worked at Transasia Biomedicals and Advance Enzyme Technologies as a scientist. Prerna has published papers in peer-reviewed journals, and has submitted a number of annotated Geminiviral genome sequences in the GenBank, including two novel ones. She has also completed her editing and proof-reading courses from Society for the promotion of Editors and Proof-readers, UK and now works as a freelance scientific editor and writer. When Prerna is not busy with her assignments, she loves to explore historical places.

As previously discussed in the first chapter, RNAi was discovered by Fire et al, 1998 in *Caenorhabditis elegans*, there has been research directed towards the use of the interference to serve as an arsenal against human diseases.

Zamecnik et al, 1978 and Stephenson and Zamecnik, 1978 reported the use of a 13-nucleotide designed complementary to 3 nucleotides of the 3'- and 5'-reiterated terminal sequences of Rous sarcoma virus 35S RNA. They reported the inhibition of virus production in chick fibroblasts. This spawned the use of antisense oligonucleotides in the field of therapy.

FIRST REPORT OF USE OF SIRNA FOR DISEASE

Song et al, 2003 employed the potential of RNAi for silencing of genes in mice. Cell death mediated by Fas is implicated in a broad spectrum of liver diseases, where inhibiting hepatocyte death is life-saving. The team tested small interfering RNA (siRNA) duplexes against gene Fas (also known as Tnfrsf6), encoding the Fas receptor, to protect mice from liver failure and fibrosis in two models of autoimmune hepatitis. They reported that the injection of Fas siRNA decreased the levels of Fas mRNA and expression of Fas protein in mouse hepatocytes, additionally, these results were constant for 10 days. They also injected agonistic Fas-specific antibody to induce fulminant hepatitis, they reported that 82% of mice treated with siRNA effectively silenced Fas survived for 10 days of observation, while all control mice died within 3 days.

CASE STUDY OF USE OF SIRNA: FIRST CLINICAL TRIAL

The first clinical trial involving RNAi was carried out by Acuity Pharmaceuticals (now Opko) in 2004 for the treatment of age-related macular degeneration (AMD). siRNA was injected by intravitreal route in patients with AMD to target the vascular endothelial growth factor (VEGF) messenger RNA.

The siRNA drug compound was called Bevasiranib: was the first siRNA in the industry to make it to Phase III clinical trials. However, the drug was recalled in 2009 by the firm as it was considered not meeting its endpoint of treating the condition. (Lares et al, 2010)

DESIGN

Chemical Modifications

One prominent difference between mammalian RNAi and RNAi in other eukaryotes is the lack of an amplification system for long-term persistence of RNAi in mammalian cells. For example, in *Drosophila*, ~35 molecules of dsRNA can silence ~1000 copies of the targeted mRNA per cell and continues for many generations. A research team used chemically modified nucleotides to enhance the longevity of siRNA molecules. They showed that 2′-OH was not required for RNAi, suggesting a priority of structural over chemical properties of siRNA–

mRNA duplexes and RISC did not require the 2′-OH for ribonuclease activity. 2′-modified siRNAs also increased the persistence of RNAi in human cells.

They observed that odifications like the 2′-fluoro and P–S linkages both increased the half-life of siRNAs upon exposure to cytoplasmic extracts. Studies in vivo with 2′-FU, 2′-FC siRNAs showed that increasing the half-life of siRNAs prolonged the effects of RNAi. (Chiu and Rana, 2003)

Shukla et al, 2010 have examined the reasons for chemical modifications of siRNAs to serve as therapeutic agents. It is reported that the use of such modifications increase resistance of siRNAs to nucleases, prevent immune activation, decrease off-target effects, and improve pharmacokinetic and pharmacodynamic properties. Additionally, modifications in the bases or sugar backbone improve RNAi silencing and increase the melting temperature of siRNA duplexes.

The use of 2′-*O*-methyl (2′OMe)-modified uridine or guanosine nucleoside into a strand of the siRNA duplex has developed a non-inflammatory siRNA.

Examples of modifications are summarized below: (Lares et al, 2010)

Modifications in bases: Methyl group attached to 5' or 3' position of uridine, isomer C-glycoside of uridine, pseudouridine and Sulfur atom attached to uridine at 2' position.

Modifications in Sugar: Methyl group at 2'-OH position of ribose, Flouro at 2'-OH position of ribose, Methyl or ethyl group at 2'-OH, Oxygen at 4' replaced with Sulfur, Locked Nucleic Acid that has a bridge between 2' and 4'.

Modifications in backbone: Peptide Nucleic acids: where a peptide backbone of N-2 amino ethyl glycine are attached to bases by methylene carbonyl bonds, phosphorothiorate where nonbridging Oxygen atoms are replaced with sulfur.

2′OMe modification generally reduces the toxicity of synthetic dsRNAs when introduced into mammalian cells. It was reported that modification with 2′OMe residues at the 5′-end of the guide strand and full passenger strand modification prevented degradation and enhanced its life in serum. Another common modification is the 2′F modification:

with research suggesting substitution of a fluorine at the 2′-position is generally well-tolerated *in vivo*.

The potential of Locked Nucleic acids (LNAs) have also been examined: they are bicyclic nucleic acids where the 2′-position of the ribose is linked to the 4′-position via a methylene bridge. This modification locks the sugar backbone in the 3′-endo conformation (RNA form) that leads to increased duplex stability and improved resistance to nuclease (Retting and Behlke, 2013).

It was shown that siRNAs that are minimally modified with LNA residues can maintain their potency and enhanced stability in serum beyond 72 hours when compared to the same sequence in unmodified form is completely degraded in just 5 hours in mouse serum (Mook et al, 2007).

Another study investigated DNA bases into the seed region (bases 2–8 from the 5′-end of the guide strand: refer Chapter 1). The potency of the siRNA was reported in addition to lessening of off target effects (Ui-Tei et al, 2008).

Novel chemical modifications have been reported by Odadzic et al, 2008 who used aminoethyl-, guanidinoethyl-, cyanoethyl- and allyl-modifications at the 2′ oxygen enhanced serum stability.

Other studies have reported the use of UNAs (Unlocked Nucleic acids) that are acyclic, lacking a C2–C3 bond in the ribose ring. The combination of UNAs and LNAs were found to enhance the relative stability of siRNA termini.

Selective placement of LNA residues may help reduce off-target effects: OTEs more efficiently than 2′OMe modifications. Several experiments employing the use of LNA and 2′OMe modifications are stabilizing, *i.e.*, they increase the T_m of the siRNA duplex. Introduction of UNA residue in the seed region (at position 7 from the 5′-end of the guide strand) was particularly effective in decreasing OTEs (Retting and Behlke, 2013).

Overcoming Innate Immnunity

Most of these receptors preferentially recognize long RNAs better than short RNAs, and siRNA designs which employ shorter sequences will generally have a lower risk of triggering an immune response than longer sequences. 2′OMe RNA modification in 20% of the residues can help detection escape detection by the immune system. 2′F and LNA modifications have also been reported. (Retting and Behlke, 2013)

Overcoming Adaptive Immunity

Nucleic acids they do not efficiently elicit antibodies: however, when conjugated with protein may develop antibody. There have been no reports of antibodies against siRNAs conjugated to proteins, however, there is no clarity as to whether such antibodies have been investigated. (Anders Wittrup and Judy Lieberman, 2015).

However, Rusconi, 2015 has reported anaphylactic reactions developed in response to PEG component of a PEG-RNA aptamer against factor IX by Regado Biosciences, which is used as an anti coagulant during cardiac surgery. This called for the investigation to be terminated.

DELIVERY OF THE RNAI TOOLS

Liposomes

Proprietary lipid compositions can be employed for the delivery of siRNA molecules. The use of vehicles of dioleoylphosphotidylethanolamine (DOPE), 1,2-dioleoyl-3-(trimethylammonium) propane (DOTAP) and cholesterol have reported to have low toxicity. (Retting and Behlke, 2013)

Principle and Construction: Delivery systems have also employed lipid nanoparticles (LNPs): The basic principle of LNP formation is self-assembly that is an integral property of smart nanosystems. As cationic lipids are inadequate delivery vehicles for siRNA, and require so-called helper lipids should be added to form the LNP. Fusogenic phospholipids that enhance the transfection activity of LNPs by destabilizing the lipid bilayer structure in cell membranes and PEG of optimal proportions decrease the immune response by increasing colloidal stability and shielding LNPs from macrophages. The use of cholesterol in LNPs stabilizes nanoparticles by filling the gaps between lipids and enhances the activity of cationic lipids.

The size and charge of the lipids must be regulated to minimize immune response and renal clearance. The size of LNPs is controlled mainly by concentration of the components during mixing and the fraction of PEG lipids contained in the mixture.

Research indicates neutral carriers are more efficient during circulation in the blood and penetration in tumor tissues, while cationic carriers are preferable during cellular uptake and escape from endosome.

The use of melittin, a pore-forming peptide: has been reported to function as an effector molecule (Zatsepin et al, 2016).

SNALP: Retting and Behlke, 2013 have reviewed various lipids that employed for the purpose of delivery. The SNALPs: cationic lipid nanoparticle system from Tekmira Pharmaceuticals (Burnaby, British Columbia, Canada; formerly Protiva Biotherapeutics) have been used for siRNAs *in vivo* to hepatic targets. The SNALP was employed in the first study to non-human primates where anti-*Apob* siRNA was administered to cynomolgus monkeys with a view to target apoB levels. They reported an 80% decrease in mRNA and significant reduction of cholesterol levels.

SNALP was used to target anti-*PLK1* and anti-*KIF11* (KSP1) siRNAs for tumour therapy in mice. Apoptosis was induced when the siRNAs were administered to xenogeneic human Hep3B tumor cells implanted in the livers of nude mice or syngeneic Neuro2A tumor cells implanted in the livers of immune-competent A/J mice.

anti-Ebola siRNAs generated against Zaire Ebola virus was administered to macaque monkeys to yield ed hope for this high mortality disease.

Novel modifications: Lipid-based amphoteric nanoparticle delivery system based on dialkylated arginine (DiLA^2 compounds) have been reported by Marina Biotech.

The use of conjugated moieties to regulate the biodistribution has been reported. Liposomes made of *O,O′*-ditradecanoyl-*N*-(α-trimethylammonioacetyl)diethanolamine chloride (DC-6-14) with cholesterol and DOPE were conjugated to Vitamin A to deliver siRNAs against collagen chaperone heat shock protein 47 gene (*Serpinh1*, gp46) in mice . Fibrosis was prevented and circumventing potential cirrhosis in such mice was reported (Retting and Behlke, 2013).

Results of Clinical Trials Using LNPs

- A lipid-formulated (AtuPLEX) siRNA: Atu027 was being used by Silence Therapeutics in 2009 to target *PKN3* in solid tumors. In 2011 results of a satisfactory safety profile and limited toxicity was discussed.
- University of Duisburg–Essen tested administration of siRNAs for fusion oncogene *BCR-ABL* in an anionic lipid (Lipovenös) formulation in a single-patient phase I safety study. The siRNA therapy administered to a chronic myeloid leukemia patient resulted in inhibited BCR-ABL and cell death of the myeloid leukemia cells (Retting and Behlke, 2013).

Case Studies

- siRNAs with LNPs were used to target apoB production for treatment of hypercholesterolemia in Clinical trials. However, due to immunogenicity, the project was terminated.

Alnylam Pharmaceuticals used ALN-PCS02 trial for targeting proprotein convertase subtilisin/kexin type 9 (*PCSK9*) transcript. This study reported no adverse side effects with a similar approach as above. This suggests that the lipid components of the delivery vessels must be selected properly to prevent such immunogenicity. (Zatsepin et al, 2016)

- Aleku et al, 2008 designed Atu027 that is a 23-nt, 2′OMe-modified blunt design siRNA and targets protein kinase N3 (*PKN3*) and employs lipid nanoparticle AtuFECT01.
- Khoury et al, 2006 used cationic lipid RPR209120 and DOPE to deliver 21-nt siRNA targeting *Tnf* .There was reduction in tumor necrosis factor (TNF) levels in an artgiritis model studied.

Polyplex Systems

Several nonlipid cationic polymers along with uncharged, hydrophilic groups to minimize complexities of charge have been used to systemically deliver siRNAs. For instance, PEI, a cationic polymer can condense nucleic acids and aids in escape from endosomes has been investigated. Polyamidoamine (PAMAM) is a dendrimer that forms nanocomplexes with siRNA similar to PEI. This PAMAM was used for an anti HIV treatment module.

Other polymers have been successfully employed in the construction of siRNA-nanoparticles include chitosan, polypeptides, and cyclodextrin. Atelocollagen, a water-soluble, pepsin digestion product of collagen, is another biopolymer that has been successfully used as a siRNA delivery vehicle. PepFect6 is a novel cell-penetrating peptide (CPP) that has shown very promising results.

Result of Clinical Trial of Using Polyplex System

Calando Pharmaceuticals in 2008 used CALAA-01: a siRNA targeting *RRM2* delivered in an amantadine-cyclodextrin vehicle into patients who had solid tumors that were refractory to therapy. A patient demonstrated cleavage of the RRM mRNA, providing the first proof of RNAi mechanism of action in humans (Retting and Behlke, 2013).

Naked administration:

siRNAs have been covalently attached to ligands to facilitate their direct uptake in tissues. This technique necessitates the use of such modifiers to protect the siRNAs in the in-vivo environments.

The use of aptamers: folded single-stranded nucleic acid molecules with conformations that bind a target molecule with high affinity has been investigated. Zhou et al, 2011 studied the efficacy of a developed siRNA against HIV viral load. developed an anti-gp120 aptamer was fused with a 27-nt DsiRNA specific for the HIV *tat/rev* gene. Following uptake by the cells and processing by Dicer, an siRNA was formed that reduced the levels of *HIV* mRNA whose effect lasted for weeks.

Tammali et al, 2011 used chemically-stabilized (siSTABLE) anti-*ALDR1* siRNA in mice and demonstrated a role for aldose reductase (*ALDR1*) in colon cancer metastasis.

Clinical Trials

I5NP is a siRNA drug that is a p53 inhibitor is being studied in Phase I/ Phase II in clinical trials sponsored by Quark Pharmaceuticals (Retting and Behlke, 2013).

Gold Nanoparticles

Lu et al, 2010 used siRNAs targeting *RELA* (nuclear factor-κB p65 subunit) with gold nanospheres in nude mice bearing HeLa cell implants. They reported reduction in the tmour load and complete inhibition when these particles were used with radiation.

TARGETED DELIVERY

siRNAs can also be targeted for uptake in target tissues or cells by using high affinity antibody or antibody fragments, aptamers (nucleic acids selected for high-affinity binding), or receptor ligands. The advantages of this method are effectivity at low concentrations and lowered OTEs.

Subcutane ous administration of siRNAs conjugated to trivalent N-acetylgalactosamine (GalNAc), which mediates hepatocyte uptake causes wholeliver, durable gene knockdown without limiting toxicity. These GalNAc–siRNA conjugates are currently being evaluated in a Phase III study for the treatment of a rare genetic neurodegenerative disease (Anders Wittrup and Judy Lieberman, 2015).

miRNAs AS BIOMARKERS

Anti-miRNA oligonucleotides (AMOs), which are enginnered to be complementary to oncogenic miRNAs, can specifically inhibit miRNA activity in tumours. However, overexpression of miRNAs that act as tumour suppressors might also be beneficial for anticancer therapy. MicroRNAs not only serve as potential therapy approaches for cancer, but also for many other diseases like virus infections or cardiovascular diseases, in which they are also involved as gene regulators (Angaji et al, 2010).

A comprehensive study of miRNA revealed that the results of a recent large-scale miRNA study 50% of miRNA genes are located in cancer-associated genomic regions. For example: The genes encoding mir-15 and mir-16 are present on chromosome 13q14, a region that is deleted in the majority of B-cell chronic lymphocytic leukaemias. The miRNAs, miR-143 and miR-145, are downregulated in colonic adenocarcinoma samples (Angaji et al, 2010).

RNAI AND DISEASES

Cancer

Mirna Therapeutics tested LNPs to deliver miRNA miR-34: a tumor cell suppressor in Phase I studies. Two patients (one patient with primary hepatocellular carcinoma [HCC] and another patient with melanoma, both showing metastasis of Stage IV) displayed 30% tumor shrinkage after treatment with MRX34.

Arbutus Biopharma Corporation has finished Phase I studies and is currently conducting Phase II trials in patients with gastrointestinal neuroendocrine tumors, adrenocortical carcinoma, and HCC with a view to inhibit polo-like kinase 1 (PLK1): that is over expressed in cancers. The phase I studies were reported to have displayed disease response and a 19.3% reduction in tumor size.

Dicerna is conducting Phase I studies with DCR-MYC, a Dicer substrate-based RNAi targeting the MYC oncogene, in combination with proprietary EnCore Dicerna's liposomal delivery technology in 72 patients with solid and hematological cancers, and focusing on HCC.

MD Anderson Cancer Center is involved a Phase I study involving administerting EphA2 siRNA in 1,2-dioleoyl-*sn*-glycero-3-phosphatidylcholine neutral liposomes with 40 patients with advanced solid tumors. The Epha2 is tyrosine kinase associated with cancers. The study has reported antineoplastic activity along with reduced expressionof EphA2.

Control of Cholesterol Levels

PCSK9 is a protease causing hypercholesterolemia and is associated with degradation of low density lipoprotein (LDL) clearing receptors in hepatocytes. Patients were administered with siRNA against PCSK9 formulated in LNPs: they displayed lower LDL-C levels by 60% in the absence of statins (a class of drug used to control cholesterol). The effect lasted for 3 months and opens up an avenue for siRNA based approaches to enter the field of regulation of cholesterol.

Rare Disorders

Primary hyperoxaluria type I (PH1) is caused by a mutation in alanine glyoxylate aminotransferase (AGT) protein and results with buildup of oxalate and kidney damage. Dicerna worked with RNAi inhinition of Glycolate Oxidase that converts glycolate into the precursor of oxalate: glyoxylate. Studies in a mouse model have reported nearly normal levels of oxalate while an 85% reduction in the Glycolate Oxidase has been reported in primates.

TTR: Transthyretin (TTR)

TTR is implicated in a rare genetic disease: amyloidosis of TTR (ATTR) in which mutated protein causes amyloid deposition in nerve endings and in the heart. Alnylam Pharmaceuticals has reported that the LNP formulations ALN-TTR01 and ALN-TTR02 suppressed the production of both mutant and nonmutant forms of TTR. It has now reached Phase III clinical trials (reviewed by Zatsepin et al, 2016).

HIV

HIV/AIDS is a chronic and debilitating disease that cannot be cured with current antiretroviral drugs. Combinatorial antiretroviral therapy can control replication of HIV-1 but, the virus undergoes mutagenesis to escape drugs. This warrants the need of several new therapies.

The first clinical demonstration of the use of RNAi against HIV-1 was a phase 0 safety and feasibility clinical trial conducted by the City of Hope Hospital (Duarte, CA, USA) and Benitec, Inc. (Sydney, Australia). DiGiusto et al, 2010 transduced CD 34 cells with a tat/rev shRNA using a lentiviral vecto. The lentiviral-based therapy was well tolerated and no drug-related adverse events were observed in the four patients.

An anti-HIV vctor with CCR5 shRNA and HIV gp41 dervied peptide was tested by Calimmune, Inc. (Los Angeles, CA, USA). The project in its Phase I/II involves transduction of the vector into Haematopoetic cells and CD4 T cells and transplanted to the HIV patient (Wolstein et al, 2014).

A G5 PAMAM (generation 5 polyamidoamine) dendrimer was used to deliver three DsiRNA engineered against HIV-1 *tat/rev ORF, CD4 receptor and Transportin 3 involved in entry of the preintegration complex into the nucleus. The team reported potent suppression in a humanized mouse model. The treated animals* displayed viral loads reduced by approximately three times when compared to various negative controls. (Zhou et al, 2011)

Bobbin et al, 2015 have reviewed the various strategies to check HIV-1 infection. For instance, antibodies have been used for targeted therapy: *gag* siRNAs complexed with a protamine-gp160 antibody fusion protein and siRNAs against the CCR5 co-receptor. siRNAs or aptamer siRNAs coupled to microbicides have been designed to limit the virus transmission. Prophylaxis against HIV-1 can include topical application of CD4 aptamer-siRNA chimeras.

Sexually Transmitted Infections (STIs)

HIV-1, HSV-2 and HPV are three major viruses that are sexually transmitted. Katakowski et al, 2010 have reviewed siRNAs as microbicides for these morbid diseases. siRNAs targeting the essential HSV-2 genes were complexed to lipid transfection reagent and conferred up to 80% protection from infection with limited side effects.

siRNA AND CLINICAL STUDIES

(Anders Wittrup and Judy Lieberman, 2015)

- In initial clinical studies 1,2dilinoleyloxy3dimethylaminopropane (DLinDMA), a first-generation LNP, potently knocked down liver gene expression in rodents and non-human primates, and caused some toxicity.
- The second generation LNPs have shown significant gene knockdown in the liver at reduced doses compared with the first-generation products.
- GalNAc conjugates: in early-phase clinical trials, although the GalNAc-conjugated siRNAs seem to provide robust knockdown assecond-generation LNPs and provide an acceptable delivery solution for the liver. The advantages are:
    - subcutaneous administration,
    - tolerated better and steroid pretreatment not required.
    - Production of conjugated siRNAs are much simpler than

LNP-formulated siRNAs.

- In a Phase I study in healthy volunteers, injections of up to 10 mg per kg weekly of GalNAc conjugated siRNA targeting TTR (ALN-TTRsc; known as revusiran) showed sustained serum TTR protein knockdown of > 90%; the main adverse reaction was only mild erythemas at the injection site, that too, in a minority of subjects.
- Recent proprietary siRNA modifications, the GalNAc-conjugated ESC-modified siRNAs, developed by Alnylam Pharmaceuticals, enhanced stabilization chemistry (ESC), improved gene knockdown potency by as much as 50fold. The clinical trials are in process for
    - haemophilia (ALNAT3, targeting AT3): Instead of supplying the missing clotting factors, the treatment seeks to enhance thrombin generation and clotting by knocking down AT3. Preliminary data from the Phase I trial in haemophilia patients showed > 60% reduction in circulating AT3, increased thrombin generation and improved blood clotting using a dose as low as 45 μg per kg.
    - complement mediated diseases: (ALNCC5, targeting complement component and porphyrias (ALNAS1, targeting 5aminolevulinic acid synthase 1 (ALAS1), the first enzyme in haem biosynthesis).
- Novel strategy for HBV: The Dynamic PolyConjugates (DPCs) strategy improves the endosomal release by using non-lipid reversibly masked endosome-disrupting peptides or polymers. The DPC is based on a masked amphipathic peptide covalently conjugated to GalNAc-targeting ligands, which bind to the hepatocytes and PEG. siRNAs are conjugated to cholesterol, which facilitates cellular uptake, and injected separately. As the endosome acidifies, the peptide is unmasked to enhance endosomal release.

For HBV, a cocktail of two cholesterol-conjugated HBV-targeted siRNAs is coinjected with NAG–MLP (ARC520). In a Phase I dose-escalation trial in healthy volunteers and a Phase IIa dose-escalation trial in patients with chronic HBV infection, the results suggest that all NAG–MLP doses tested were well tolerated in 36 healthy volunteers receiving the active compound (0.01- 4 mg per kg), with no changes in laboratory parameters or severe drug-related adverse reactions. However, a few cases of potential hypersensitivity reactions were noted, and the risk of immune reactions against the PEGylated endosomolytic agent - especially with repeat dosing - is a potential concern (Anders Wittrup and Judy Lieberman, 2015).

Table 3.1: A summary of RNAi therapeutics in gene knockdown or efficacy in humans.

<table>
<tr><td>Drug and company that developed it</td><td>Target</td><td>Disease</td><td>Clinical trial Phase</td></tr>
<tr><td>Revusiran and ALN-TTR02 by Alnylam Pharmaceuticals</td><td>Transthyretin (TTR)</td><td>TTR-mediated amyloidosis</td><td>Phase III</td></tr>
<tr><td>ARC-520 by Arrow-head Research</td><td>HBV</td><td rowspan="3">HCV</td><td rowspan="3">Phase II</td></tr>
<tr><td>Miravirsen by San-taris Pharma/Roche</td><td rowspan="2">miR-122</td></tr>
<tr><td>Regulus Therapeutics</td></tr>
</table>

Reference: (Anders Wittrup and Judy Lieberman, 2015).

SUMMARY

Through the mechanism of gene silencing and knockdown of target genes, RNAi has seen a surge in its usage for treatment of several genetic and nongenetic diseases. Though reports have been promising, safe delivery and further research would help minimize the side effects of the therapy. Nevertheless, the use of RNAi seems to be poised to overcome its current challenges with the tremendous research on its different aspects in order to serve as a novel treatment tool and promise for patients suffering with anomalies.

REFERENCES

1. Aleku M, Schulz P, Keil O, Santel A, Schaeper U, Dieckhoff B, et al. (2008 Dec) Atu027, a liposomal small interfering RNA formulation

targeting protein kinase N3, inhibits cancer progression. Cancer Res. 68(23):9788-98

2. Anders Wittrup and Judy Lieberman. (Sep 2015) Knocking down disease: a progress report on siRNA therapeutics. Nature Reviews: Genetics. 16:543
3. Angaji S. A., Hedayati S. S., Poor R. H., Madani S., Poor S. S. and Panahi S. (2010) Application of RNA interference in treating human diseases. J Genet. 89:527–37
4. Bobbin ML, Burnett JC, Rossi JJ. (2015) RNA interference approaches for treatment of HIV-1 infection. Genome Medicine. 7(1):50.
5. Chiu YL, Rana TM. (2003) siRNA function in RNAi: A chemical modification analysis. RNA. 9(9):1034-1048. doi:10.1261/rna.5103703.
6. DiGiusto DL, Krishnan A, Li L, Li H, Li S, Rao A, et al. (2010 Jun) RNA-based gene therapy for HIV with lentiviral vector-modified CD34(+) cells in patients undergoing transplantation for AIDSrelated lymphoma. Sci Transl Med. 2(36):36ra43.
7. Katakowski JA, Palliser D. (2010) siRNA-based topical microbicides targeting sexually transmitted diseases. Curr Opin Mol Ther. 12(2):192-202.
8. 8. Khoury M, Louis-Plence P, Escriou V, Noel D, Largeau C, Cantos C, et al. (2006 Jun) Efficient new cationic liposome formulation for systemic delivery of small interfering RNA silencing tumor necrosis factor alpha in experimental arthritis. Arthritis Rheum. 54(6):1867-77.
9. Lares MR, Rossi JJ, Ouellet DL. (2010) RNAi and small interfering RNAs in human disease therapeutic applications. Trends Biotechnol. 28(11):570-579.
10. Lu W, Zhang G, Zhang R, Flores LG 2nd, Huang Q, Gelovani JG, Li C. (2010 Apr) Tumor site-specific silencing of NF-kappaB p65 by targeted hollow gold nanosphere-mediated photothermal transfection. Cancer Res. 70(8):3177-88.
11. Mook OR, Baas F, de Wissel MB, Fluiter K. (2007) Evaluation of locked nucleic acid-modified small interfering RNA in vitro and in vivo. Mol Cancer Ther. 6(3):833-43.

12. Neff CP, Zhou J, Remling L, Kuruvilla J, Zhang J, Li H, et al. (2011 Jan) An aptamer-siRNA chimera suppresses HIV-1 viral loads and protects from helper CD4(+) T cell decline in humanized mice. Sci Transl Med. 3(66):66ra6.
13. Odadzic D, Bramsen JB, Smicius R, Bus C, Kjems J, Engels JW. (2008 Jan) Synthesis of 2'-O-modified adenosine building blocks and application for RNA interference. Bioorg Med Chem. 16(1):518-29.
14. Rettig GR, Behlke MA. (2012) Progress Toward In Vivo Use of siRNAs-II. Mol Ther. 20(3):483-512. doi:10.1038/ mt.2011.263.
15. Rusconi, C. (2015) Phase 3 evaluation of revolixys kit: study summary and lessons learned. Celebrating the 25th Anniversary of Selex, ASGCT meeting May 10–12, 2015, New Orleans
16. Shukla S, Sumaria CS, Pradeepkumar PI. (2010) Exploring Chemical Modifications for siRNA Therapeutics: A Structural and Functional Outlook. Chem Med Chem. 5(3):328–49.
17. Song E, et al. (2003) RNA interference targeting Fas protects mice from fulminant hepatitis. Nat. Med. 9, 347–51.
18. Stephenson ML, Zamecnik PC. (1978) Inhibition of Rous sarcoma viral RNA translation by a specific oligodeoxyribonucleotide. Proc Natl Acad Sci USA. 75:285–88
19. Tammali R, Reddy AB, Saxena A, Rychahou PG, Evers BM, Qiu S, et al. (2011 Aug) Inhibition of aldose reductase prevents colon cancer metastasis. Carcinogenesis. 32(8):1259-67.
20. Ui-Tei K, Naito Y, Zenno S, Nishi K, Yamato K, Takahashi F, et al. (2008) Functional dissection of siRNA sequence by systematic DNA substitution: modified siRNA with a DNA seed arm is a powerful tool for mammalian gene silencing with significantly reduced off-target effect. Nucleic Acids Res. 36(7):2136-51.
21. Wolstein O, Boyd M, Millington M, Impey H, Boyer J, Howe A, et al. (2014) Preclinical safety and efficacy of an anti-HIV-1 lentiviral vector containing a short hairpin RNA to CCR5 and the C46 fusion inhibitor. Mol Ther Methods Clin Dev. 1:11.

22. Zamecnik, P. C. & Stephenson, M. L. (1978) Inhibition of Rous sarcoma virus replication and cell transformation by a specific oligodeoxynucleotide. Proc Natl Acad Sci USA. 75:280–84.
23. Zatsepin TS, Kotelevtsev YV, Koteliansky V. (2016) Lipid nanoparticles for targeted siRNA delivery – going from bench to bedside. Int J Nanomedicine. 11:3077-3086. doi:10.2147/IJN. S106625.
24. Zhou J, Neff CP, Liu X, Zhang J, Li H, Smith DD, et al. (2011 Dec) Systemic administration of combinatorial dsiRNAs via nanoparticles efficiently suppresses HIV-1 infection in humanized mice. Mol Ther. 19(12):2228-38.

CHAPTER 4

TRANSCRIPTIONAL GENE SILENCING IN HUMANS

Marc S. Weinberg[1,2,3] and Kevin V. Morris[1,4,5]

[1]Department of Molecular and Experimental Medicine, The Scripps Research Institute, La Jolla, CA 92037, USA

[2]Wits/SAMRC Antiviral Gene Therapy Research Unit, School of Pathology, University of the Witwatersrand, WITS 2050, South Africa

[3]HIV Pathogenesis Research Unit, Department of Molecular Medicine and Haematology, School of Pathology, University of the Witwatersrand, WITS 2050, South Africa

[4]Center for Gene Therapy, City of Hope – BeckmanResearch Institute; Duarte, CA 91010, USA

[5]School of Biotechnology and Biomedical Sciences, University of New South Wales, Kensington, NSW, 2033 Australia

ABSTRACT

It has been over a decade since the first observation that small non-coding RNAs can functionally modulate epigenetic states in human cells to achieve functional transcriptional gene silencing (TGS). TGS is mechanistically distinct from the RNA interference (RNAi) gene-silencing pathway. TGS can result in long-term stable epigenetic modifications to gene expression that can be passed on to daughter cells during cell division, whereas RNAi does not. Early studies of TGS have been largely overlooked, overshadowed by subsequent discoveries of small RNA-directed post-TGS and RNAi. A reappraisal of early work has been brought about by recent findings in human cells where endogenous long non-coding RNAs function to regulate the epigenome. There are distinct and common overlaps between the proteins involved in small and long non-coding RNA transcriptional regulatory mechanisms, suggesting that the early studies using small non-coding RNAs to modulate transcription were making use of a previously unrecognized endogenous mechanism of RNA-directed gene regulation. Here we review how non-coding RNA plays a role in regulation of transcription and epigenetic gene silencing in human cells by revisiting these earlier studies and the mechanistic insights gained to date. We also provide a list of mammalian genes that have been shown to be transcriptionally regulated by non-coding RNAs. Lastly, we explore how TGS may serve as the basis for development of future therapeutic agents.

INTRODUCTION

The History of RNA-Directed Transcriptional Gene Silencing (TGS)

Almost three decades ago, Marjorie Matzke *et al.* observed that over-expression of a transgene led to DNA hypermethylation and transcriptional silencing in doubly transformed tobacco plants (1) (Figure 1). Mechanistically, this type of silencing in plants was found to be the result of small non-coding RNAs directing epigenetic changes, specifically DNA methylation, to those loci containing sequences homologous to the small RNA. The phenomenon was termed small RNA-directed transcriptional gene silencing (TGS). TGS was later shown in *Arabidopsis* to require the action of RNA-dependent DNA methylation (2,3) and members of the Argonaute protein family (4). A few years later RNA interference (RNAi), mediated by double-stranded RNAs, was discovered as a powerful post-TGS (PTGS) system against messenger RNAs (mRNAs) in plants (5), and a few months later in *Caenorhabditis elegans* (6).

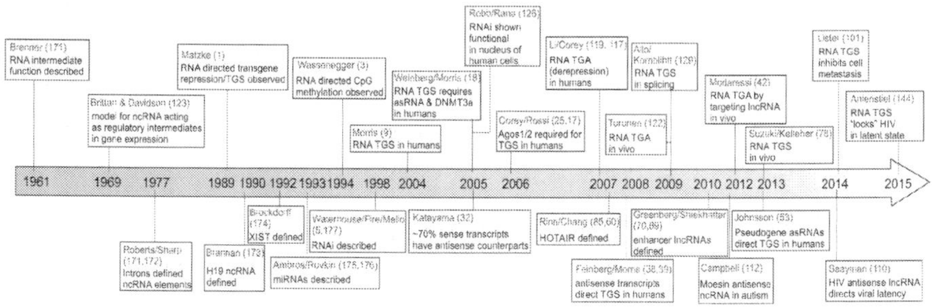

Figure 1: Regulatory non-coding RNA timeline. A timeline of some important observations in RNA biology are shown leading up to our collective understanding of non-coding RNA-directed transcriptional gene silencing (TGS). (5,71,170–177).

Transcriptional Gene Silencing in Humans

The study of small non-coding RNA-directed TGS has been carried out in various model organisms such as plants (*Arabidopsis thaliana*), yeast (*Saccharomyces pombe*), flies (*Drosophila melanogaster*) and worms (*C. elegans*) (reviewed extensively in (7,8)). A decade ago, the first report of RNA-directed TGS in human cells was observed when exogenous siRNAs were used to silence a transgenic elongation factor 1 α promoter driving a Green Fluorescent Protein (GFP) reporter gene (9) (Figure 1). Importantly, the observed silencing was clearly at the transcriptional level, as indicated by nuclear run-on analysis. Moreover, silencing was also epigenetic: inhibition was abrogated by 5′ Aza-cytadine (5′ AzaC) and Trichostatin A (TSA), compounds involved in inhibiting DNA methylation and histone deacetylalion, respectively (9). This early observation was soon followed by other studies (10,11), all of which confirmed that small non-coding RNAs could functionally control gene transcription and epigenetic states in human cells. But the underlying mechanism of action remained unknown.

Mechanisms of Small Non-Coding RNA-directed TGS

TGS is mechanistically distinct from the abundantly studied PTGS pathway of RNAi. One notable difference is that TGS results in long-term stable epigenetic modifications to gene expression that can be passed on to daughter cells during cellular division (reviewed in (12)). Early observations postulated that siRNA-directed TGS functioned through an epigenetic nuclear mechanisms distinct from RNAi-mediated PTGS in the cytoplasm (13). For instance 5′ AzaC and TSA were functional in reverting the siRNA targeted TGS, indicating epigenetic modes of gene regulation were at play in siRNA-directed TGS, and not via a PTGS-based mechanism (9). Indeed, recent studies have observed that two different siRNAs, one targeted to the promoter and one targeted to exon 1 of the coding transcript, can functionally repress the targeted gene in a TGS or PTGS based manner (14). A lot has been gleaned over the last decade regarding the mechanism of action for RNA-directed TGS in human cells. Studies carried out to determine the underlying mechanism of siRNA-directed TGS revealed that RNA-mediated TGS is operative through RNA-directed methylation of histone 3 lysine's 9 and 27 (H3K9 and H3K27, respectively) and DNA methylation at the targeted promoter (9,11,15–21) (Figure 2). These promoter-directed siRNAs interact with a low level expressed (~1–2%) promoter associated RNA, which is essentially the 5′ UTR of the protein coding gene (16,22) (Figure 2). It is worth noting that most genes and gene promoters appear to be transcribed to some extent (23,24) and experimental observations suggest that non-coding RNAs interact with target loci via Watson–Crick-based RNA:RNA hybridization (16,22) and not by double-stranded DNA invasion. Temporal studies have determined that exogenously introduced siRNAs targeted to a promoter region interact first with Argonautes 1 and 2 (AGO1 and AGO2) (17,25,26). SiRNA and AGO interactions is found within the first 24 h, at the siRNA targeted promoter and is followed shortly thereafter with the recruitment of the H3K9me2 and H3K27me3 silent state epigenetic marks (17), and later by the recruitment of DNA methyltransferase and DNA methylation at 72–96 h for some genes (14). It should be noted, however, that the role of DNA methylation in TGS in human cells is not as clearly understood as in plants; DNA methylation at the targeted promoter is not always observed in human TGS applications (Table 1). These effects may be explained by the duration of RNA targeting to the promoter, the presence of robust siRNA targeting (e.g. delivery to the nucleus), the presence and abundance of promoter-occupied RNAs and/or the dynamic interplay of proteins interacting with the promoter. Despite differences in the various experimental observations, a key consistent feature has been the observations that promoter-directed small RNAs can modulate gene transcription and that some level of epigenetic based silencing is ongoing in the observed silenced genes.

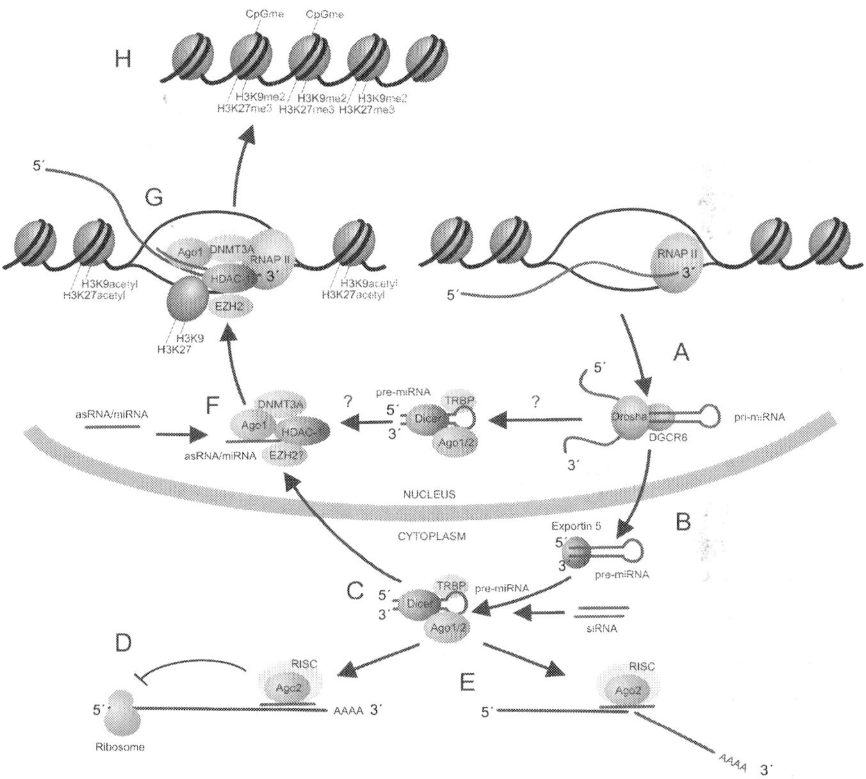

Figure 2: Small non-coding RNA pathways in human cells. Small non-coding RNAs can be generated as priRNAs where they are (A) processed by Drosha and DGCR8 into miRNAs which are (B) exported from the nucleus and (C) loaded into RISC where they can affect mRNA expression by (D) binding and blocking mRNA translation or (E) cutting the target mRNA. Some miRNAs may also be retained in the nucleus (F) where they can interact with epigenetic remodeling proteins and (G) recruit the complexes to target loci in the genome resulting in (H) localized chromatin compaction and epigenetic silencing.

Table 1: Mammalian genes transcriptionally regulated by non-coding RNAs

Gene(s)	Gene symbol	Effector RNA	Cell line	Therapeutic relevance	References
Eukaryotic translation elongation factor 1 α	EEF1A1	siRNA	HEK293T		(9,16,18)
HIV-1/SIV	5′ LTR	siRNA, sasRNA	Jurkat, Tzmb, T-cells, in vivo (mouse)	Regulation of HIV-1	(18–21,26,79, 99,100, 104, 140–144)
nitric-oxide synthase	NOS	siRNA	HAEC	Cardiac disease	(145)
E-cadherin	CDH1	siRNA	HCT116; MCF7	Cancer, tumor suppressor	(10)
BCL-2 (oncogene)	BCL-2	sasRNA	HeLa, 293	Cancer, oncogene	(146)
Fibronectin	FN1	siRNA	HeLa		(129,147)
Huntingtin gene	HTT	siRNA	Glioblastoma	Monogenetic diseases	(148)
Non-sense codon-containing immunoglobulin minigenes	(Ig)-mu and Ig-gamma	siRNA	HeLa	Immunologic diseases	(149)
INK4B/Cyclin-dependent kinase inhibitor 2B/p15 + ARF + INK4A/ Cyclin-dependen kinase inhibitor 2A isoform 3/p16	CDKN2B+ CDKN2A	siRNA	HEK293T	Cancer, tumor suppressor	(150)
INK4A/Cyclin-dependen kinase inhibitor 2A isoform 3/p16	CDKN2A	siRNA	HEK293T	Cancer, tumor suppressor	(151)
Plasminogen activator, urokinase	PLAU	siRNA	PC3 and in vivo	Cancer	(152)
Chemokine receptor 5	CCR5	siRNA	HEK293T	HIV-1 co-receptor	(16,17)
Breast cancer-associated gene 1	BRCA1	siRNA	T47D	Cancer, oncogene	(153)
Progesterone receptor	PGR	siRNA	T47D	Cancer	(11,25,119,153,154)
Huntingtin	HD	siRNA	T47D	Monogenetic diseases	(25)
Androgen receptor	AR	siRNA	T47D	Cancer, spinal bulbar muscular atrophy	(25,155)
v-myc avian myelocytomatosis viral oncogene homolog	c-MYC	siRNA/ sasRNA	PC3, HCT113, 293, Hela, MCF7	Cancer, oncogene	(22,97,156,157)
Papillomavirus-16 oncogenes	HPV-16	siRNA	HeLa	HPV	(158)
v-akt murine thymoma viral oncogene homolog 1	AKT-1	siRNA	293HEK	Cancer, oncogene	(156)

Kirsten rat sarcoma viral oncogene	KRAS	siRNA	293HEK	Cancer, oncogene	(156)
Dual specificity phosphatase 6	DUSP6	siRNA	CFPAC	Cancer, tumor suppressor	(156)
Myostatin	MSTN	siRNA	C2C12 mouse myoblasts	muscle hypertrophy	(159)
Runt-related transcription factor 3	RUNX3	siRNA	stomach carcinoma cell line SGC7901	osteoarthritis	(160)
Small nuclear 7sk (RNAi functional in the nucleus of human cells)	7SK	siRNA	Hela, 293HEK cells	7SK/P-TEFb control of HIV-1	(126)
met proto-oncogene (hepatocyte growth factor receptor)	c-Met	siRNA/asRNA	SKHep1C3 cells	Cancer, oncogene	(161)
Periostin	POSTN	siRNA/ sasRNA	PC3	Cancer and metastasis	(101)
Heparanase (endo-h-D-glucuronidase)	HPA	siRNA	PC3, EJ and SGC-7901 cells	Cancer, angiogenesis	(162)
phosphoglycerate kinase 1 promoter driving GFP	pgk-1	siRNA	293, HeLa		(98)
Interleukin 2	IL2	shRNA	Jurkat	Immunologic	(163)
Ubiquitin C	UBC	siRNA; shRNA	HEK293GT		(14)
Transforming growth factor-β receptor II	TGFβII	shRNA	rat SBC10	Cancer, angiogenesis	(102)
Vascular endothelial growth factor	VEGF-A	shRNA	mouse C166 and in vivo	Cancer, angiogenesis	(107,122)
Ras association domain family 1	RASSF1A	shRNA	HeLa	Cancer, oncogene	(15,17)
Tubulin folding cofactor E-like	TBCEL/ LRRC35	miRNA	HCT116	Kenny-Caffey syndrome (KCS)	(150)
Ras p21 protein activator 2	RASA2	miRNA	HCT116	Cancer, tumor suppressor	(150)
Rhophilin, Rho GTPase binding protein 2	RHPN2	miRNA	HCT116		(150)
Wolf-Hirschhorn syndrome candidate 1	WHSC1	miRNA	HCT116	Wolf-Hirschhorn syndrome	(150)
Homeobox D4	HOXD4	miRNA	MCF7; MDA-MB-231		(164)
HIV-1	LTR	miRNA	Jurkat, T-cells	HIV-1 infection	(28)
HIV-1	TAR	miRNA	Tzmb, Jurkat, T-cells	HIV-1 infection	(29,165)
OCT4 and Nanog (pluripotent factor)	OCT4 and Nanog	lncRNA (antisense, pseudogene)	MCF7	Cancer, pluripotency	(54)
PTENpg1 asRNA alpha	PTEN	lncRNA (Trans-antisense, pseudogene)	293, Hela, Jurkat	Cancer, tumor suppressor	(53)

P21 tumor suppressor	P21	lncRNA (antisense)	MCF7	Cancer, tumor suppressor	(39)
P15 tumor suppressor	P15	lncRNA (antisense)	HL-60, KG-1, Kasumi-1, DG-75, Raji and Ramos	Cancer, tumor suppressor	(38)
alpha-globin gene	HBA2	lncRNA	Embryonic stem cells	Alpha thalassemia	(40)
DM1 insulator	DM1/SIX5	lncRNA	Primary fibroblasts		(166)
Herpes	LAT	lncRNA	in vivo (mice)	HPV	(167)
P53, lncRNA-p21	hnRNP-K	lncRNA	MEF, in vivo (mice)	Cancer, oncogene	(57)
lncRNA HOTAIR	HOX	lncRNA	MDA-MB-231, SK-BR-3, MCF-10A, MCF-7, HCC1954, T47D and MDA-MB-453 cell lines. Human tissue samples	Cancer	(86)
HIV-1	HIV-1	lncRNA (antisense)	Jurkat, T-cells, Tzmb, 293HEK	HIV-1 infection	(49–52)
v-myc avian myelocytomatosis viral oncogene neuroblastoma derived homolog	MYCN MYCNOS	lncRNA (Cis-antisense)	Lan6	Cancer, oncogene, neuroblastoma	(43)
neuroblastoma associated transcript 1	NBAT-1	lncRNA	Neuroblastoma primary tumors	Cancer, neuroblastoma	(48)
brain-derived neurotrophic factor	BDNF	lncRNA (antisense)	Human brain, mouse in vivo	Huntington disease, Alzheimer disease and Parkinson disease	(41,42)
short-chain dehydrogenase/ reductase family member 4	DHRS4	lncRNA (antisense)	HepG2 and HL7702 cell		(44)
potassium voltage-gated channel, KQT-like subfamily, member 1	Kcnq1ot1	lncRNA (antisense)	Human placenta-derived JEG-3 cells	Romano-Ward syndrome, Jervell and Lange-Nielsen syndrome and familial atrial fibrillation	(168)
moesin at 5p14.1 in Autism	Moesin	lncRNA (antisense)	Human brain (postmortem cerebral cortex)	Autism spectrum	(113)
C-terminal binding protein 1	CTBP1	lncRNA (antisense)	LNCaP, VCaP, DU145	Prostate cancer	(169)

The Endogenous Pathway of TGS in Human Cells; Rise of Long Non-Coding RNAs

While small RNAs were observed early on to regulate gene transcription in human cells by the targeting of epigenetic silencing complexes to those loci containing complementarity to the small RNAs (Figure 1), the endogenous mechanism(s) driving this form of gene regulation in the context of human cells remained largely unknown. MicroRNAs (miRNAs) have been shown to be endogenous drivers of TGS with some genes in human cells (27–30)(Table 1). In 2005, through the efforts of the FANTOM and ENCODE consortia, it started to become apparent that a large fraction of the human genome was generating long non-coding RNAs (lncRNAs) and that many of these transcripts were antisense to protein-coding counterparts (31,32). Several of these sense/antisense or bidirectionally-transcribed genes are evolutionarily conserved, suggesting some functional cues for retention of these elements (33,34). Indeed, studies with imprinted genes and X-inactivation found that *cis* acting long non-coding RNAs (lncRNAs) were actively involved in epigenetic regulation of these, dosage-dependent, regulated loci (35).

In 2008, a number of important studies confirmed the role of lncRNAs as endogenous drivers of TGS in human cells, in particular those, which were antisense to their protein-coding counterparts (reviewed in (36,37)). Antisense lncRNAs were shown to regulate the p15 (38) and p21 (39) tumor suppressor genes (Table 1). The over-expression of these antisense lncRNAs resulted in TGS of their protein-coding counterpart while their repression resulted in the de-repression/transcriptional activation (38,39). Support for the role of antisense lncRNAs as active endogenous regulators of gene expression was evident in an earlier understated study, which indicated that antisense transcripts might also be involved in CpG methylation in thalassemia (40). Antisense lncRNAs are now known to affect TGS for genes such as BDNF (41,42), MYCN (43), DHRS4 (44), KCNQ1 (45–47), NBAT (48) and HIV-1 (49–52) (Table 1). Interestingly, non-coding transcripts derived from pseudogenes of Phosphatase and tensin homolog (PTEN) (53) and OCT4 (54), which contain significant homology to their protein-coding counterparts, have also been observed to be involved in directing TGS and subsequent PTEN and OCT4 suppression (Table 1). Also, the PTEN antisense pseudogene-directed TGS of PTEN is one of the first *bona fide*examples of a trans-functional lncRNA (53). LncRNAs direct TGS and CpG methylation (55). *Cis*regulation and antisense lncRNAs have also been observed to epigenetically regulate ribosomal genes (56) as well as p21 (57) and MYCN (43), genes involved in cell regulation and cancer. The plethora of lncRNA functions are extensive and the range includes: protein modifiers (58,59), scaffolds for tethering proteins (43,60,61), miRNAs (62–64), splicing modifiers (65), cellular body transformation (66–69), enhancer function and gene activation (70–73), and epigenetic modifiers (53,74,75), (reviewed in (76)). Collectively, the observations to date suggest that we are only now just beginning to realize the complexity and pervasiveness of lncRNA functional regulation in epigenetic and transcriptional states.

Mechanisms of Small and Long Non-Coding RNA-directed TGS

To date there are ~55 reports of small RNA-directed TGS and ~10 of antisense lncRNA-directed TGS (Table 1). Mechanistically, much of what we know about how small non-coding RNAs, such as siRNAs, miRNAs and small antisense RNAs (sasRNAs)-directed TGS, have been determined from cell culture studies. The promoter-targeted small RNAs interact with various proteins to guide TGS, beginning in the first 24 h, with direct interactions with AGO1 and AGO2 (17,18,25) followed shortly thereafter by interactions at the targeted promoter with DNMT3a (14,18,77,78), HDAC1 (14,20) and resulting ultimately in histone 3 lysine 9 di-methylation and histone 3 lysine 27 tri-methylation (H3K9me2 and H3K27me3, respectively)(14,16–18,20,79,80) (Figure 3). SiRNA-directed TGS has also been observed to occur in the absence of DNA methylation, suggesting that alternative routes may be present for RNA-mediated transcriptional and epigenetic silencing (10). Small RNA-directed TGS appears to require a template or target transcript at the corresponding targeted promoter (16,22), similar to the method by which plants utilize RNA Polymerase V transcribed and processed siRNAs to regulate DNA methylation and TGS (reviewed in (81)). Notably, in plants there is a requirement for RNA-dependent RNA polymerase (RdRP) activity to amplify RNA polymerase V transcript-directed TGS (81), whereas humans lack such a polymerase, which opens up a methodology for specific RNA-directed epigenetic modes of regulation. Curiously, this is exactly what lncRNAs appear to be doing in human cells via *cis* and trans-specific targeting of epigenetic complexes to particular loci (Figure33 and Table 1), similar to what is also observed in *Saccharomyces cerevisiae*, which also lacks RdRP activity (82,83).

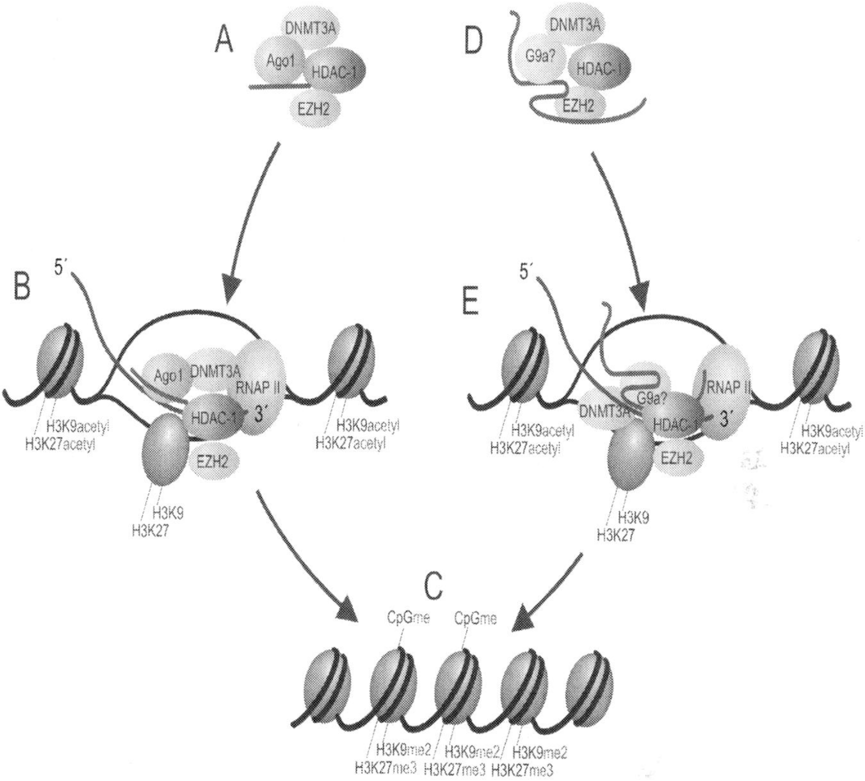

Figure 3: Antisense RNA-directed TGS. Small antisense non-coding RNAs can be (A) introduced into the nucleus and (B) interact with and recruit epigenetic silencing complexes consisting of DNMT3a, Ago1, EZH2 and HDAC1 to homology containing targeted loci by interactions with low copy promoter-associated transcripts resulting in (C) epigenetic silencing consisting of histone and DNA methylation and ultimately chromatin compaction of the targeted locus. (D) Long antisense non-coding RNAs have also been observed to interact with similar epigenetic silencing complexes (53,54) and (E) localize with these complexes at targeted loci resulting in (C) epigenetic silencing of the lncRNA targeted locus.

Early studies carried out with *S. cerevisiae* indicated that antisense non-coding RNAs function endogenously to direct epigenetic gene silencing in place of RdRP-mediated mechanisms (82,83). The parallels between *S. cerevisiae* and previous observations of small antisense RNA-directed TGS in human cells (18) have emerged, suggesting that antisense transcripts also function to direct TGS (Figure 3). Most notable are the observations that particular antisense lncRNAs, first observed in tumor suppressor genes, p15 (38) and p21(39), function to epigenetically modulate their protein-coding counterparts (Figure 1, Table 1). One interesting, and surprisingly overlooked, early study found antisense transcription was involved in DNA methylation in Thalassemia (40), and even early work linked antisense transcripts and DNA methylation in regulating HIV (51) and MYC (84,85).

Mechanistically, far less is known about how antisense lncRNAs direct epigenetic silencing in human cells. Studies carried out with the lncRNA, HOTAIR, indicate that bimodal chromatin modifying complexes can be localized to the HOX locus via the action of this lncRNA (86). A common theme is also evident with Kcnq1ot1 (45) and the p53 regulatory lincRNA p21, which indicates that the entire p53 expressed pathway is controlled by the action of this lncRNA at the p53 locus (57). Indeed, many lncRNAs have been observed to be associated with chromatin (87), but mechanistic insights into the process of lncRNA-directed gene regulation remain less clear. Interesting insights into the mechanism of action of lncRNA-directed TGS came from a recent study looking at the PTEN pseudogene. It had been reported previously that the PTEN pseudogene functions as a miRNA 'sponge' (64), similar to the CEBPA lncRNA that acts to sponge DNMT1 away from the *CEBPA* promoter (88). Studies to interrogate the PTEN pseudogene in greater detailed determined that this pseudogene also expressed an antisense lncRNA in *trans* which functions to direct TGS to the PTEN promoter and control PTEN expression epigenetically (53). Mechanistically, the PTEN pseudogene expressed antisense lncRNA modulated PTEN transcription by recruiting DNMT3a and EZH2 to the PTEN promoter. The parallels between the functions of the PTEN pseudogene and previous observations with small antisense ncRNA-directed TGS are notable, as both involved the action of DNMT3a (Figure 3). It is noteworthy that DNMT3a is the only known *de novo*DNA methyltransferase in human cells (89) and has been observed previously to be the only DNA methylatransferase to bind non-coding RNAs including small ncRNAs, both antisense and double stranded RNAs (18,77,78,90), and lncRNAs (53,91). There is an interesting connection between DNMT3a and epigenetic silencing, which including studies indicating DNTM3a co-immunoprecipitates with HDAC1 (92,93) and EZH2 (94), as well as early predictions that

DNA methylation is an active participant in X-inactivation (95), one of the first *bona fide* lncRNA regulatory pathways described. Collectively, a paradigm is emerging in human cells, which proposes that non-coding RNAs, both small and long forms (Figure (Figure3),3), function through the action of DNMT3a to modulate chromatin and epigenetic states of gene expression. While there are several other mechanisms of action described for lncRNAs in human cells, the interactions with DNMT3a and targeting of transcriptional and epigenetic states is of particular interest, as this mode of gene regulation has the potential to be long-lasting, heritable and may be of significant relevance to the development of targeted therapeutics (reviewed in (96)).

Therapeutic Applications of RNA-Directed Epigenetic Regulation of Gene Expression

The utility of small RNA-induced TGS as a therapeutic has been largely ignored, mainly due to the pervasiveness of using RNAi targeted approaches to degrade mRNAs. The main concern with RNAi and post-transcriptional mechanisms of gene silencing (Figure 2) is the duration of their therapeutic effect. The effector siRNAs required to drive RNAi must be administered continuously to repress a therapeutic target gene. This is not the case with RNA-induced TGS, where stable, long-term, silencing can be achieved following a relatively short duration of promoter targeting with the siRNAs (19,20,97–100) or small antisense RNA (14,101). This is because the mode of action for the observed gene silencing is transcriptional and driven ultimately by epigenetic silencing (79,102) and not 'slicing' of the genes messenger RNA as is the case with RNAi. One universal hurdle that both RNAi and RNA-induced TGS face with is the targeted delivery of the effector RNAs to those cells requiring treatment. One approach is to utilize synthetic antisense oligonucleotides targeted to promoters of interest. This approach has worked with regards to blocking transcription (103) but was not found to induce robust epigenetic silencing, unless the particular oligonucleotides were RNA based (104). However, it may be that better interrogation of each non-coding RNA targeted promoter is required to delineate the best promoter-associated transcripts to target and that many of the earlier studies may have neglected this notion. Indeed, establishing TGS in the absence of a target promoter RNA has not been reported and attempts by some groups, including ours, have proven fruitless. Another approach might be to deliver the effector RNAs using receptor targeted aptamers, which has shown promise for targeting HIV infected cells (105,106). While delivery remains an important concern, the notion that one needs to only target a particular gene for 2–4 days to instill stable epigenetic silencing is promising with regards to minimizing the need for sustained delivery. Recent studies suggest that small RNA-directed TGS is feasible and that stable epigenetic marks can be imposed at small RNA target loci *in vivo* (99,107).

Another area of therapeutic utility can be found in the plethora of lncRNAs that are appearing to be involved in various diseases. Emerging evidence suggest that non-coding RNAs play a wide role (108) in various disease states in humans. Genome-wide observations of diseased states, such as heart failure (109), indicate significant differential and discordant expression between protein-coding and non-coding antisense and pseudogenes is prevalent (110). To date the list of those lncRNAs involved in human diseases is expanding at an unprecedented rate. LncRNAs have been observed in disease ranging from Cancer (57,86,105,106), to HIV (111,112), to autism (113), to pluripotency and differentiation (114–116). It is worth underscoring that many of the disease relevant lncRNAs have been observed to be antisense to particular protein coding genes. A significant obstacle to using RNAi and other post-transcriptional effectors for targeting antisense lncRNAs is the fact that double stranded siRNAs have an ability to target both sense and antisense transcripts (117). The use of RNA-directed TGS avoids this issue by targeting the lncRNA promoter with single stranded antisense transcripts (52). The targeting of endogenous effector antisense lncRNAs can result in the de-repression and subsequent transcriptional activation of the lncRNA targeted locus (Figure 4). Using this mode of action, it becomes feasible to activate gene expression to affect those protein-coding genes under sustained lncRNA-directed TGS (Figure 4). This has proven an effective approach to inducing genes both *in vitro* (39,52,118–121) and *in vivo* (42,107,122), but presupposes that there are known antisense lncRNAs regulating the therapeutic target gene. Collectively, the advantages to using RNA-directed TGS as a therapeutic are many and include: (i) strand specific targeting of a gene, (ii) stable long-term epigenetic based silencing can be established to particular genes of therapeutic interest and (iii) antisense

RNA-based approaches work as well, if not better than double stranded RNAs, as the endogenous pathway of RNA-directed TGS appears to contain significant overlap with small antisense RNAs and antisense lncRNAs (Figure 3).

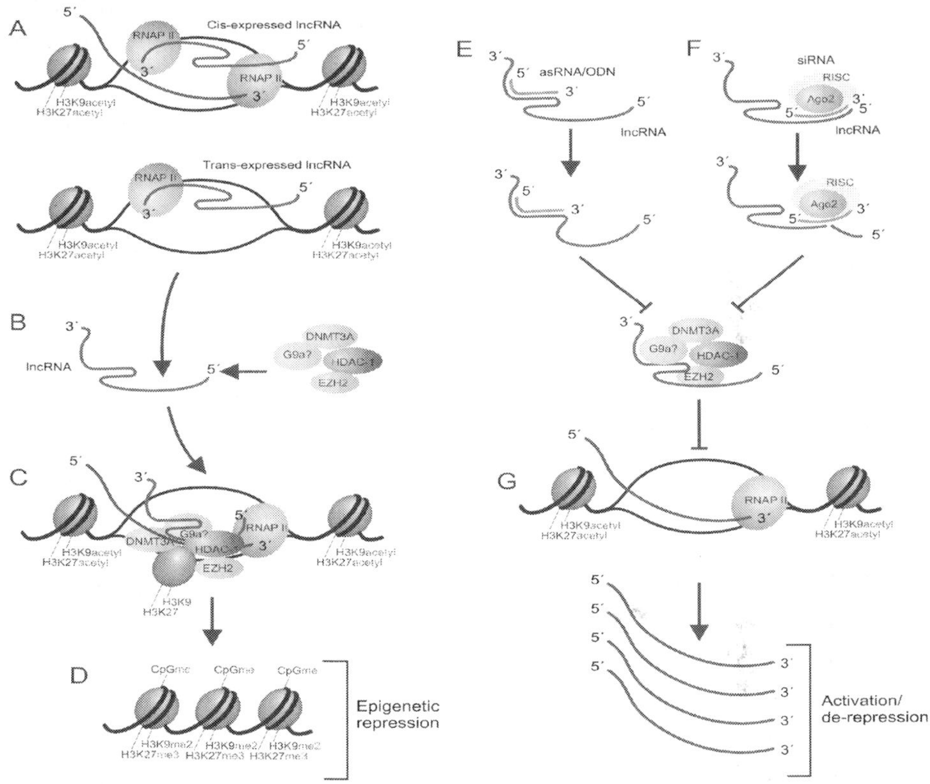

Figure 4: LncRNA pathways of transcriptional silencing and de-repression. LncRNAs can be expressed in (A) Cis or trans and can (B) interact with those proteins involved in epigenetic silencing. The lncRNAs act to (C) target and tether the epigenetic silencing complexes to homology containing loci resulting in (D) chromatin compaction and transcrtiptinal gene silencing of the targeted locus. These endogenous regulatory lncRNAs can be targeted with (E) antisense oligonucleotides or (F) siRNAs, which results in the loss of the lncRNA and activation/de-repression of those loci actively under lncRNA regulation.

CONCLUSION

It has been roughly 10 years since the first observation that promoter-directed RNAs can affect gene transcription (Figure 1 and Table 1). This seminal observation in 2004 (9) was indicative of a role for RNA in regulating gene expression, a notion proposed ~5 decades ago but largely overlooked (123,124). Possible reasons for the poor early adoption of RNA-directed TGS (Table 1) are varied but may include (i) the unfortunate retraction of a similar paper published in Nature (125), and/or (ii) the overwhelmingly positive response to PTGS and the rejection of any RNAi-related phenomena occurring in the nucleus, despite the fact that RNAi was shown to be functional in the human nucleus in 2005 (126) and confirmed in many subsequent studies (25–27,127–131).

The notion that RNA may function as the master gene regulator in the cell was something proposed by Britten and Davidson in 1969 (123), which at the time was largely neglected by the broad scientific community. With the advent of high-throughput technologies and the findings from ENCODE, that most of the human genome is transcribed and likely plays a functional role (132–139), it is becoming apparent that Britten and Davidson's theory should be reappraised. Certainly, lncRNAs are abundantly active in the nucleus, and many of them are active modulators of transcriptional and epigenetic modes of gene expression (reviewed in (37,76) and appear to share many of the mechanistic characteristics observed in small RNA-directed TGS (Figure 3). Collectively, the mounting observations that antisense non-coding RNAs, both small and long RNAs, directed to gene promoters can affect transcription by the recruitment of silent state epigenetic complexes suggests that a pervasive and underappreciated role for non-coding RNAs is part of the basic fabric of life. Knowledge of this molecular pathway may prove incredibly insightful with regards to the development of disease, including epigenetic silencing of gene expression and the development of new-targeted therapeutics aimed at specifically affecting gene expression. The next decade could prove an exciting time for our understanding of non-coding RNAs in the transcriptional gene expression and their application as novel therapeutics.

FUNDING

KVM acknowledges support from NIH [PO1 AI099783-01, RO1 CA151574-01, R01 DK104681-01] and the Australian Research Council Future Fellowship [FT130100572]. Funding for open access charge:

[PO1 AI099783-01, RO1 CA151574-01, RO1 DK104681-01]; ARC Future Fellowship [T130100572] to KVM. MSW acknowledges support from the Strategic Health Innovation Partnerships (SHIP) Unit of South African Medical Research Council (SAMRC), with funds received from the South African Department of Science and Technology (DST).

REFERENCES

1. Matzke M.A., Primig M., Trnovsky J., Matzke A.J.M. Reversible methylation and inactivation of marker genes in sequentially transformed tobacco plants. EMBO J. 1989;8:643–649.
2. Mette M.F., Aufsatz W., Van der Winden J., Matzke A.J.M., Matzke M.A. Transcriptional silencing and promoter methylation triggered by double-stranded RNA. EMBO J. 2000;19:5194–5201.
3. Wassenegger M., Heimes S., Riedel L., Sänger H.L. RNA-directed de novo methylation of genomic sequences in plants. Cell. 1994;76:567–576.
4. Lippman Z., May B., Yordan C., Singer T., Martienssen R. Distinct mechanisms determine transposon inheritance and methylation via small interfering RNA and histone modification. PLoS Biol. 2003;1:E67.
5. Waterhouse P.M., Graham M.W., Wang M.B. Virus resistance and gene silencing in plants can be induced by simultaneous expression of sense and antisense RNA. Proc. Natl. Acad. Sci. U.S.A. 1998;95:13959–13964.
6. Fire A., Xu S., Montgomery M.K., Kostas S.A., Driver S.E., Mello C.C. Potent and specific genetic interference by double-stranded RNA in *Caenorhabditis elegans*. Nature. 1998;391:806–811.
7. Green V.A., Weinberg M.S. Small RNA-induced transcriptional gene regulation in mammals mechanisms, therapeutic applications, and scope within the genome. Prog. Mol. Biol. Transl. Sci. 2011;102:11–46.
8. Morris K.V. RNA-directed transcriptional gene silencing and activation in human cells. Oligonucleotides. 2009;19:299–306.
9. Morris K.V., Chan S.W., Jacobsen S.E., Looney D.J. Small interfering RNA-induced transcriptional gene silencing in human

cells. Science. 2004;305:1289–1292.

10. Ting A.H., Schuebel K.E., Herman J.G., Baylin S.B. Short double-stranded RNA induces transcriptional gene silencing in human cancer cells in the absence of DNA methylation. Nat. Genet. 2005;37:906–910.
11. Janowski B.A., Huffman K.E., Schwartz J.C., Ram R., Hardy D., Shames D.S., Minna J.D., Corey D.R. Inhibiting gene expression at transcription start sites in chromosomal DNA with antigene RNAs. Nat. Chem. Biol. 2005;1:210–215.
12. Morris K.V. RNA-mediated transcriptional gene silencing in human cells. Curr. Top. Microbiol. Immunol. 2008;320:211–224.
13. Morris K.V. siRNA-mediated transcriptional gene silencing: the potential mechanism and a possible role in the histone code. Cell Mol. Life Sci. 2005;62:3057–3066.
14. Hawkins P.G., Santoso S., Adams C., Anest V., Morris K.V. Promoter targeted small RNAs induce long-term transcriptional gene silencing in human cells. Nucleic Acids Res. 2009:2984–2995.
15. Castanotto D., Tommasi S., Li M., Li H., Yanow S., Pfeifer G.P., Rossi J.J. Short hairpin RNA-directed cytosine (CpG) methylation of the RASSF1A gene promoter in HeLa cells. Mol. Ther. 2005;12:179–183.
16. Han J., Kim D., Morris K.V. Promoter-associated RNA is required for RNA-directed transcriptional gene silencing in human cells. Proc. Natl. Acad. Sci. U.S.A. 2007;104:12422–12427. [PMC free article]
17. Kim D.H., Villeneuve L.M., Morris K.V., Rossi J.J. Argonaute-1 directs siRNA-mediated transcriptional gene silencing in human cells. Nat. Struct. Mol. Biol. 2006;13:793–797.
18. Weinberg M.S., Villeneuve L.M., Ehsani A., Amarzguioui M., Aagaard L., Chen Z.X., Riggs A.D., Rossi J.J., Morris K.V. The antisense strand of small interfering RNAs directs histone methylation and transcriptional gene silencing in human cells. RNA. 2006;12:256–262.
19. Suzuki K., Ishida T., Yamagishi M., Ahlenstiel C., Swaminathan S., Marks K., Murray D., McCartney E.M., Beard M.R., Alexander M., et al. Transcriptional gene silencing of HIV-1 through promoter targeted RNA is highly specific. RNA Biol. 2011;8:1035–1046.

20. Suzuki K., Juelich T., Lim H., Ishida T., Watanebe T., Cooper D.A., Rao S., Kelleher A.D. Closed chromatin architecture is induced by an RNA duplex targeting the HIV-1 promoter region. J. Biol. Chem. 2008;283:23353–23363.
21. Yamagishi M., Ishida T., Miyake A., Cooper D.A., Kelleher A.D., Suzuki K., Watanabe T. Retroviral delivery of promoter-targeted shRNA induces long-term silencing of HIV-1 transcription. Microbes Infect. 2009;11:500–508.
22. Napoli S., Pastori C., Magistri M., Carbone G.M., Catapano C.V. Promoter-specific transcriptional interference and c-myc gene silencing by siRNAs in human cells. EMBO J. 2009;28:1708–1719.
23. Seila A.C., Calabrese J.M., Levine S.S., Yeo G.W., Rahl P.B., Flynn R.A., Young R.A., Sharp P.A. Divergent transcription from active promoters. Science. 2008;322:1849–1851. [PMC free article]
24. Core L.J., Waterfall J.J., Lis J.T. Nascent RNA sequencing reveals widespread pausing and divergent initiation at human promoters. Science. 2008;322:1845–1848.
25. Janowski B.A., Huffman K.E., Schwartz J.C., Ram R., Nordsell R., Shames D.S., Minna J.D., Corey D.R. Involvement of AGO1 and AGO2 in mammalian transcriptional silencing. Nat. Struct. Mol. Biol. 2006;13:787–792.
26. Ahlenstiel C.L., Lim H.G., Cooper D.A., Ishida T., Kelleher A.D., Suzuki K. Direct evidence of nuclear Argonaute distribution during transcriptional silencing links the actin cytoskeleton to nuclear RNAi machinery in human cells. Nucleic Acids Res. 2012;40:1579–1595.
27. Kim D.H., Saetrom P., Snove O., Jr, Rossi J.J. MicroRNA-directed transcriptional gene silencing in mammalian cells. Proc. Natl. Acad. Sci. U.S.A. 2008;105:16230–16235.
28. Omoto S., Fujii Y.R. Regulation of human immunodeficiency virus 1 transcription by nef microRNA. J. Gen. Virol. 2005;86:751–755.
29. Klase Z., Kale P., Winograd R., Gupta M.V., Heydarian M., Berro R., McCaffrey T., Kashanchi F. HIV-1 TAR element is processed by Dicer to yield a viral micro-RNA involved in chromatin remodeling of the viral LTR. BMC Mol. Biol. 2007;8:63.
30. Tan Y., Zhang B., Wu T., Skogerbo G., Zhu X., Guo X., He S., Chen R. Transcriptional inhibition of Hoxd4 expression by noncoding

RNAs in human breast cancer cells. BMC Mol. Biol. 2009;10:12.

31. Carninci P., Kasukawa T., Katayama S., Gough J., Frith M.C., Maeda N., Oyama R., Ravasi T., Lenhard B., Wells C., et al. The transcriptional landscape of the mammalian genome. Science. 2005;309:1559–1563.
32. Katayama S., Tomaru Y., Kasukawa T., Waki K., Nakanishi M., Nakamura M., Nishida H., Yap C.C., Suzuki M., Kawai J., et al. Antisense transcription in the mammalian transcriptome. Science. 2005;309:1564–1566.
33. Dahary D., Elroy-Stein O., Sorek R. Naturally occurring antisense: transcriptional leakage or real overlap. Genome Res. 2005;15:364–368.
34. Lavorgna G., Dahary D., Lehner B., Sorek R., Sanderson C.M., Casari G. In search of antisense. Trends Biochem. Sci. 2004;29:88–94.
35. Latos P.A., Barlow D.P. Regulation of imprinted expression by macro non-coding RNAs. RNA Biol. 2009;6:100–106.
36. Malecova B., Morris K.V. Transcriptional gene silencing through epigenetic changes mediated by non-coding RNAs. Curr. Opin. Mol. Ther. 2010;12:214–222.
37. Morris K.V. Long antisense non-coding RNAs function to direct epigenetic complexes that regulate transcription in human cells. Epigenetics. 2009;4:296–301.
38. Yu W., Gius D., Onyango P., Muldoon-Jacobs K., Karp J., Feinberg A.P., Cui H. Epigenetic silencing of tumour suppressor gene p15 by its antisense RNA. Nature. 2008;451:202–206.
39. Morris K.V., Santoso S., Turner A.M., Pastori C., Hawkins P.G. Bidirectional transcription directs both transcriptional gene activation and suppression in human cells. PLoS Genet. 2008;4:e1000258.
40. Tufarelli C., Stanley J.A., Garrick D., Sharpe J.A., Ayyub H., Wood W.G., Higgs D.R. Transcription of antisense RNA leading to gene silencing and methylation as a novel cause of human genetic disease. Nat. Genet. 2003;34:157–165.
41. Lipovich L., Dachet F., Cai J., Bagla S., Balan K., Jia H., Loeb J.A. Activity-dependent human brain coding/noncoding gene regulatory

networks. Genetics. 2012;192:1133–1148.

42. Modarresi F., Faghihi M.A., Lopez-Toledano M.A., Fatemi R.P., Magistri M., Brothers S.P., van der Brug M.P., Wahlestedt C. Inhibition of natural antisense transcripts in vivo results in gene-specific transcriptional upregulation. Nat. Biotechnol. 2012;30:453–459.
43. Vadie N., Saayman S., Lenox A., Ackley A., Clemson M., Burdach J., Hart J., Vogt P.K., Morris K.V. MYCNOS functions as an antisense RNA regulating MYCN. RNA Biol. 2015;12:893–899.
44. Li Q., Su Z., Xu X., Liu G., Song X., Wang R., Sui X., Liu T., Chang X., Huang D. AS1DHRS4, a head-to-head natural antisense transcript, silences the DHRS4 gene cluster in cis and trans. Proc. Natl. Acad. Sci. U.S.A. 2012;109:14110–14115.
45. Kanduri C. Kcnq1ot1: a chromatin regulatory RNA. Semin. Cell Dev. Biol. 2011;22:343–350.
46. Mohammad F., Pandey R.R., Nagano T., Chakalova L., Mondal T., Fraser P., Kanduri C. Kcnq1ot1/Lit1 noncoding RNA mediates transcriptional silencing by targeting to the perinucleolar region. Mol. Cell Biol. 2008;28:3713–3728.
47. Pandey R.R., Mondal T., Mohammad F., Enroth S., Redrup L., Komorowski J., Nagano T., Mancini-Dinardo D., Kanduri C. Kcnq1ot1 antisense noncoding RNA mediates lineage-specific transcriptional silencing through chromatin-level regulation. Mol. Cell. 2008;32:232–246.
48. Pandey G.K., Mitra S., Subhash S., Hertwig F., Kanduri M., Mishra K., Fransson S., Ganeshram A., Mondal T., Bandaru S., et al. The risk-associated long noncoding RNA NBAT-1 controls neuroblastoma progression by regulating cell proliferation and neuronal differentiation. Cancer Cell. 2014;26:722–737.
49. Landry S., Halin M., Lefort S., Audet B., Vaquero C., Mesnard J.M., Barbeau B. Detection, characterization and regulation of antisense transcripts in HIV-1. Retrovirology. 2007;4:71.
50. Ludwig L.B., Ambrus J.L., Jr, Krawczyk K.A., Sharma S., Brooks S., Hsiao C.B., Schwartz S.A. Human immunodeficiency virus-Type 1 LTR DNA contains an intrinsic gene producing antisense RNA and protein products. Retrovirology. 2006;3:80.

51. Miller R.H. Human immunodeficiency virus may encode a novel protein on the genomic DNA plus strand. Science. 1988;239:1420–1422.
52. Saayman S., Ackley A., Turner A.M., Famiglietti M., Bosque A., Clemson M., Planelles V., Morris K.V. An HIV-encoded antisense long noncoding RNA epigenetically regulates viral transcription. Mol. Ther. 2014;22:1164–1175.
53. Johnsson P., Ackley A., Vidarsdottir L., Lui W.O., Corcoran M., Grander D., Morris K.V. A pseudogene long-noncoding-RNA network regulates PTEN transcription and translation in human cells. Nat. Struct. Mol. Biol. 2013;20:440–446.
54. Hawkins P.G., Morris K.V. Transcriptional regulation of Oct4 by a long non-coding RNA antisense to Oct4-pseudogene 5. Transcription. 2010;1:165–175.
55. Mohammad F., Pandey G.K., Mondal T., Enroth S., Redrup L., Gyllensten U., Kanduri C. Long noncoding RNA-mediated maintenance of DNA methylation and transcriptional gene silencing. Development. 2012;139:2792–2803.
56. Bierhoff H., Schmitz K., Maass F., Ye J., Grummt I. Noncoding transcripts in sense and antisense orientation regulate the epigenetic state of ribosomal RNA genes. Cold Spring Harb. Symp. Quant. Biol. 2010;75:357–364.
57. Huarte M., Guttman M., Feldser D., Garber M., Koziol M.J., Kenzelmann-Broz D., Khalil A.M., Zuk O., Amit I., Rabani M., et al. A large intergenic noncoding RNA induced by p53 mediates global gene repression in the p53 response. Cell. 2010;142:409–419.
58. Gong C., Li Z., Ramanujan K., Clay I., Zhang Y., Lemire-Brachat S., Glass D.J. A long non-coding RNA, LncMyoD, regulates skeletal muscle differentiation by blocking IMP2-mediated mRNA translation. Dev. Cell. 2015;34:181–191.
59. Wang P., Xue Y., Han Y., Lin L., Wu C., Xu S., Jiang Z., Xu J., Liu Q., Cao X. The STAT3-binding long noncoding RNA lnc-DC controls human dendritic cell differentiation. Science. 2014;344:310–313.
60. Blewitt M.E., Gendrel A.V., Pang Z., Sparrow D.B., Whitelaw N., Craig J.M., Apedaile A., Hilton D.J., Dunwoodie S.L., Brockdorff N., et al. SmcHD1, containing a structural-maintenance-of-

chromosomes hinge domain, has a critical role in X inactivation. Nat. Genet. 2008;40:663–669.

61. Tsai M.C., Manor O., Wan Y., Mosammaparast N., Wang J.K., Lan F., Shi Y., Segal E., Chang H.Y. Long noncoding RNA as modular scaffold of histone modification complexes. Science. 2010;329:689–693.

62. Wang L., Guo Z.Y., Zhang R., Xin B., Chen R., Zhao J., Wang T., Wen W.H., Jia L.T., Yao L.B., et al. Pseudogene OCT4-pg4 functions as a natural micro RNA sponge to regulate OCT4 expression by competing for miR-145 in hepatocellular carcinoma. Carcinogenesis. 2013;34:1773–1781.

63. Wang Y., Xu Z., Jiang J., Xu C., Kang J., Xiao L., Wu M., Xiong J., Guo X., Liu H. Endogenous miRNA sponge lincRNA-RoR regulates Oct4, Nanog, and Sox2 in human embryonic stem cell self-renewal. Dev. Cell. 2013;25:69–80.

64. Poliseno L., Salmena L., Zhang J., Carver B., Haveman W.J., Pandolfi P.P. A coding-independent function of gene and pseudogene mRNAs regulates tumour biology. Nature. 2010;465:1033–1038.

65. Zhou H.L., Luo G., Wise J.A., Lou H. Regulation of alternative splicing by local histone modifications: potential roles for RNA-guided mechanisms. Nucleic Acids Res. 2014;42:701–713.

66. Liu Y., Luo F., Xu Y., Wang B., Zhao Y., Xu W., Shi L., Lu X., Liu Q. Epithelial-mesenchymal transition and cancer stem cells, mediated by a long non-coding RNA, HOTAIR, are involved in cell malignant transformation induced by cigarette smoke extract. Toxicol. Appl. Pharmacol. 2015;282:9–19.

67. Xue M., Li X., Li Z., Chen W. Urothelial carcinoma associated 1 is a hypoxia-inducible factor-1alpha-targeted long noncoding RNA that enhances hypoxic bladder cancer cell proliferation, migration, and invasion. Tumour Biol. 2014;35:6901–6912.

68. Gutschner T., Hammerle M., Eissmann M., Hsu J., Kim Y., Hung G., Revenko A., Arun G., Stentrup M., Gross M., et al. The noncoding RNA MALAT1 is a critical regulator of the metastasis phenotype of lung cancer cells. Cancer Res. 2013;73:1180–1189.

69. Ramos A.D., Andersen R.E., Liu S.J., Nowakowski T.J., Hong S.J., Gertz C.C., Salinas R.D., Zarabi H., Kriegstein A.R., Lim D.A.

The long noncoding RNA Pnky regulates neuronal differentiation of embryonic and postnatal neural stem cells. Cell Stem Cell. 2015;16:439–447.

70. Orom U.A., Derrien T., Beringer M., Gumireddy K., Gardini A., Bussotti G., Lai F., Zytnicki M., Notredame C., Huang Q., et al. Long noncoding RNAs with enhancer-like function in human cells. Cell. 2010;143:46–58.
71. Kim T.K., Hemberg M., Gray J.M., Costa A.M., Bear D.M., Wu J., Harmin D.A., Laptewicz M., Barbara-Haley K., Kuersten S., et al. Widespread transcription at neuronal activity-regulated enhancers. Nature. 2010;465:182–187.
72. Schaukowitch K., Joo J.Y., Liu X., Watts J.K., Martinez C., Kim T.K. Enhancer RNA facilitates NELF release from immediate early genes. Mol. Cell. 2014;56:29–42.
73. Pulakanti K., Pinello L., Stelloh C., Blinka S., Allred J., Milanovich S., Kiblawi S., Peterson J., Wang A., Yuan G.C., et al. Enhancer transcribed RNAs arise from hypomethylated, Tet-occupied genomic regions. Epigenetics. 2013;8:1303–1320.
74. Beckedorff F.C., Ayupe A.C., Crocci-Souza R., Amaral M.S., Nakaya H.I., Soltys D.T., Menck C.F., Reis E.M., Verjovski-Almeida S. The intronic long noncoding RNA ANRASSF1 recruits PRC2 to the RASSF1A promoter, reducing the expression of RASSF1A and increasing cell proliferation. PLoS Genet. 2013;9:e1003705.
75. Kotake Y., Nakagawa T., Kitagawa K., Suzuki S., Liu N., Kitagawa M., Xiong Y. Long non-coding RNA ANRIL is required for the PRC2 recruitment to and silencing of p15(INK4B) tumor suppressor gene. Oncogene. 2011;30:1956–1962.
76. Morris K.V., Mattick J.S. The rise of regulatory RNA. Nat. Rev. Genet. 2014;15:423–437.
77. Holz-Schietinger C., Reich N.O. RNA modulation of the human DNA methyltransferase 3A. Nucleic Acids Res. 2012;40:8550–8557.
78. Ross J.P., Suetake I., Tajima S., Molloy P.L. Recombinant mammalian DNA methyltransferase activity on model transcriptional gene silencing short RNA-DNA heteroduplex substrates. Biochem. J. 2010;432:323–332.

79. Suzuki K., Shijuuku T., Fukamachi T., Zaunders J., Guillemin G., Cooper D., Kelleher A. Prolonged transcriptional silencing and CpG methylation induced by siRNAs targeted to the HIV-1 promoter region. J. RNAi Gene Silencing. 2005;1:66–78.
80. Turner A.M., Ackley A.M., Matrone M.A., Morris K.V. Characterization of an HIV-targeted transcriptional gene-silencing RNA in primary cells. Hum. Gene Ther. 2012;23:473–483.
81. Matzke M.A., Mosher R.A. RNA-directed DNA methylation: an epigenetic pathway of increasing complexity. Nat. Rev. Genet. 2014;15:394–408.
82. Camblong J., Iglesias N., Fickentscher C., Dieppois G., Stutz F. Antisense RNA stabilization induces transcriptional gene silencing via histone deacetylation in S. cerevisiae. Cell. 2007;131:706–717.
83. Camblong J., Beyrouthy N., Guffanti E., Schlaepfer G., Steinmetz L.M., Stutz F. Trans-acting antisense RNAs mediate transcriptional gene cosuppression in S. cerevisiae. Genes Dev. 2009;23:1534–1545.
84. Celano P., Berchtold C.M., Kizer D.L., Weeraratna A., Nelkin B.D., Baylin S.B., Casero R.A., Jr Characterization of an endogenous RNA transcript with homology to the antisense strand of the human c-myc gene. J. Biol. Chem. 1992;267:15092–15096.
85. Spicer D.B., Sonenshein G.E. An antisense promoter of the murine c-myc gene is localized within intron 2. Mol. Cell. Biol. 1992;12:1324–1329.
86. Gupta R.A., Shah N., Wang K.C., Kim J., Horlings H.M., Wong D.J., Tsai M.C., Hung T., Argani P., Rinn J.L., et al. Long non-coding RNA HOTAIR reprograms chromatin state to promote cancer metastasis. Nature. 2010;464:1071–1076.
87. Khalil A.M., Guttman M., Huarte M., Garber M., Raj A., Rivea Morales D., Thomas K., Presser A., Bernstein B.E., van Oudenaarden A., et al. Many human large intergenic noncoding RNAs associate with chromatin-modifying complexes and affect gene expression. Proc. Natl. Acad. Sci. U.S.A. 2009;106:11667–11672.
88. Di Ruscio A., Ebralidze A.K., Benoukraf T., Amabile G., Goff L.A., Terragni J., Figueroa M.E., De Figureido Pontes L.L., Alberich-

Jorda M., Zhang P., et al. DNMT1-interacting RNAs block gene-specific DNA methylation. Nature. 2013;503:371–376.

89. Bestor T.H. The DNA methyltransferases of mammals. Hum. Mol. Genet. 2000;9:2395–2402.

90. Jeffery L., Nakielny S. Components of the DNA methylation system of chromatin control are RNA-binding proteins. J. Biol. Chem. 2004;279:49479–49487.

91. Wang L., Zhao Y., Bao X., Zhu X., Kwok Y.K., Sun K., Chen X., Huang Y., Jauch R., Esteban M.A., et al. LncRNA Dum interacts with Dnmts to regulate Dppa2 expression during myogenic differentiation and muscle regeneration. Cell Res. 2015;25:335–350.

92. Fuks F., Hurd P.J., Deplus R., Kouzarides T. The DNA methyltransferases associate with HP1 and the SUV39H1 histone methyltransferase. Nucleic Acids Res. 2003;31:2305–2312.

93. Fuks F., Burgers W.A., Godin N., Kasai M., Kouzarides T. Dnmt3a binds deacetylases and is recruited by a sequence-specific repressor to silence transcription. EMBO J. 2001;20:2536–2544.

94. Vire E., Brenner C., Deplus R., Blanchon L., Fraga M., Didelot C., Morey L., Van Eynde A., Bernard D., Vanderwinden J.M., et al. The Polycomb group protein EZH2 directly controls DNA methylation. Nature. 2006;439:871–874.

95. Riggs A.D. X inactivation, differentiation, and DNA methylation. Cytogenet. Cell Genet. 1975;14:9–25.

96. Morris K.V. Non-coding RNAs, epigenetic memory, and the passage of information to progeny. RNA Biol. 2009;6:242–247.

97. Mehndiratta M., Palanichamy J.K., Pal A., Bhagat M., Singh A., Sinha S., Chattopadhyay P. CpG hypermethylation of the C-myc promoter by dsRNA results in growth suppression. Mol. Pharm. 2011;8:2302–2309.

98. Esmaeili F., Bamdad T., Ghasemi S. Stable suppression of gene expression by short interfering RNAs targeted to promoter in a mouse embryonal carcinoma stem cell line. In Vitro Cell. Dev. Biol. Anim. 2010;46:834–840.

99. Suzuki K., Hattori S., Marks K., Ahlenstiel C., Maeda Y., Ishida

T., Millington M., Boyd M., Symonds G., Cooper D.A., et al. Promoter targeting shRNA suppresses HIV-1 infection in vivo through transcriptional gene silencing. Mol. Ther. Nucleic acids. 2013;2:e137.

100. Lim H.G., Suzuki K., Cooper D.A., Kelleher A.D. Promoter-targeted siRNAs induce gene silencing of simian immunodeficiency virus (SIV) infection in vitro. Mol. Ther. 2008;16:565–570.

101. Lister N.C., Clemson M., Morris K.V. RNA-directed epigenetic silencing of periostin inhibits cell motility. R. Soc. Open Sci. 2015;2:140545.

102. Kim J.W., Zhang Y.H., Zern M.A., Rossi J.J., Wu J. Short hairpin RNA causes the methylation of transforming growth factor-beta receptor II promoter and silencing of the target gene in rat hepatic stellate cells. Biochem. Biophys. Res. Commun. 2007;359:292–297.

103. Watts J.K., Yu D., Charisse K., Montaillier C., Potier P., Manoharan M., Corey D.R. Effect of chemical modifications on modulation of gene expression by duplex antigene RNAs that are complementary to non-coding transcripts at gene promoters. Nucleic Acids Res. 2010;38:5242–5259.

104. Knowling S.S.K., Turner A.M., Uhlmann E., Lehmann T., Vollmer J., Morris K.V. Chemically modified oligonucleotides modulate an epigenetically varied and transient form of transcription silencing of HIV-1 in human cells. Mol. Ther. Nucleic Acids. 2012;1:e16.

105. Zhou J., Satheesan S., Li H., Weinberg M.S., Morris K.V., Burnett J.C., Rossi J.J. Cell-specific RNA aptamer against human CCR5 specifically targets HIV-1 susceptible cells and inhibits HIV-1 infectivity. Chem. Biol. 2015;22:379–390.

106. Neff C.P., Zhou J., Remling L., Kuruvilla J., Zhang J., Li H., Smith D.D., Swiderski P., Rossi J.J., Akkina R. An aptamer-siRNA chimera suppresses HIV-1 viral loads and protects from helper CD4(+) T cell decline in humanized mice. Sci. Transl. Med. 2011;3:66ra66.

107. Turunen M.P., Lehtola T., Heinonen S.E., Assefa G.S., Korpisalo P., Girnary R., Glass C.K., Vaisanen S., Yla-Herttuala S. Efficient regulation of VEGF expression by promoter-targeted lentiviral shRNAs based on epigenetic mechanism: a novel example of

epigenetherapy. Circ. Res. 2009;105:604–609.

108. Hart J.R., Roberts T.C., Weinberg M.S., Morris K.V., Vogt P.K. MYC regulates the non-coding transcriptome. Oncotarget. 2014;5:12543–12554.

109. Di Salvo T.G., Guo Y., Su Y.R., Clark T., Brittain E., Absi T., Maltais S., Hemnes A. Right ventricular long noncoding RNA expression in human heart failure. Pulm. Circ. 2015;5:135–161.

110. Johnsson P., Lipovich L., Grander D., Morris K.V. Evolutionary conservation of long noncoding RNAs; sequence, structure, function. Biochim. Biophys. Acta. 2013;1840:1063–1071.

111. Saayman S., Ali S.A., Morris K.V., Weinberg M.S. The therapeutic application of CRISPR/Cas9 technologies for HIV. Expert Opin. Biol. Ther. 2015;15:819–830.

112. Kobayashi-Ishihara M., Yamagishi M., Hara T., Matsuda Y., Takahashi R., Miyake A., Nakano K., Yamochi T., Ishida T., Watanabe T. HIV-1-encoded antisense RNA suppresses viral replication for a prolonged period. Retrovirology. 2012;9:38.

113. Kerin T., Ramanathan A., Rivas K., Grepo N., Coetzee G.A., Campbell D.B. A noncoding RNA antisense to moesin at 5p14.1 in autism. Sci. Transl. Med. 2012;4:128ra140.

114. Guttman M., Donaghey J., Carey B.W., Garber M., Grenier J.K., Munson G., Young G., Lucas A.B., Ach R., Bruhn L., et al. lincRNAs act in the circuitry controlling pluripotency and differentiation. Nature. 2011;477:295–300.

115. Loewer S., Cabili M.N., Guttman M., Loh Y.H., Thomas K., Park I.H., Garber M., Curran M., Onder T., Agarwal S., et al. Large intergenic non-coding RNA-RoR modulates reprogramming of human induced pluripotent stem cells. Nat. Genet. 2010;42:1113–1117.

116. Kretz M., Siprashvili Z., Chu C., Webster D.E., Zehnder A., Qu K., Lee C.S., Flockhart R.J., Groff A.F., Chow J., et al. Control of somatic tissue differentiation by the long non-coding RNA TINCR. Nature. 2013;493:231–235.

117. Wei J.X., Yang J., Sun J.F., Jia L.T., Zhang Y., Zhang H.Z., Li X., Meng Y.L., Yao L.B., Yang A.G. Both strands of siRNA have potential to guide posttranscriptional gene silencing in mammalian

cells. PLoS One. 2009;4:e5382.

118. Schwartz J.C., Younger S.T., Nguyen N.B., Hardy D.B., Monia B.P., Corey D.R., Janowski B.A. Antisense transcripts are targets for activating small RNAs. Nat. Struct. Mol. Biol. 2008;15:842–848.
119. Janowski B.A., Younger S.T., Hardy D.B., Ram R., Huffman K.E., Corey D.R. Activating gene expression in mammalian cells with promoter-targeted duplex RNAs. Nat. Chem. Biol. 2007;3:166–173.
120. Li L.C., Okino S.T., Zhao H., Pookot D., Place R.F., Urakami S., EnokidaH.,DahiyaR. SmalldsRNAsinducetranscriptionalactivation in human cells. Proc. Natl. Acad. Sci. U.S.A. 2006;103:17337–17342.
121. Place R.F., Li L.C., Pookot D., Noonan E.J., Dahiya R. MicroRNA-373 induces expression of genes with complementary promoter sequences. Proc. Natl. Acad. Sci. U.S.A. 2008;105:1608–1613.
122. Turunen M.P., Husso T., Musthafa H., Laidinen S., Dragneva G., Laham-Karam N., Honkanen S., Paakinaho A., Laakkonen J.P., Gao E., et al. Epigenetic upregulation of endogenous VEGF-A reduces myocardial infarct size in mice. PLoS One. 2014;9:e89979.
123. Britten R.J., Davidson E.H. Gene regulation for higher cells: a theory. Science. 1969;165:349–357.
124. Britten R.J., Davidson E.H. Repetitive and non-repetitive DNA sequences and a speculation on the origins of evolutionary novelty. Q. Rev. Biol. 1971;66:111–138.
125. Taira K. Induction of DNA methylation and gene silencing by short interfering RNAs in human cells. Nature. 2006;441:1176.
126. Robb G.B., Brown K.M., Khurana J., Rana T.M. Specific and potent RNAi in the nucleus of human cells. Nat. Struct. Mol. Biol. 2005;12:133–137.
127. Rudel S., Flatley A., Weinmann L., Kremmer E., Meister G. A multifunctional human Argonaute2-specific monoclonal antibody. RNA. 2008;14:1244–1253.
128. Ameyar-Zazoua M., Rachez C., Souidi M., Robin P., Fritsch L., Young R., Morozova N., Fenouil R., Descostes N., Andrau J.C., et al. Argonaute proteins couple chromatin silencing to alternative

splicing. Nat. Struct. Mol. Biol. 2012;19:998–1004.

129. Allo M., Agirre E., Bessonov S., Bertucci P., Gomez Acuna L., Buggiano V., Bellora N., Singh B., Petrillo E., Blaustein M., et al. Argonaute-1 binds transcriptional enhancers and controls constitutive and alternative splicing in human cells. Proc. Natl. Acad. Sci. U.S.A. 2014;111:15622–15629.

130. Gagnon K.T., Li L., Janowski B.A., Corey D.R. Analysis of nuclear RNA interference in human cells by subcellular fractionation and Argonaute loading. Nat. Protoc. 2014;9:2045–2060.

131. Gagnon K.T., Li L., Chu Y., Janowski B.A., Corey D.R. RNAi factors are present and active in human cell nuclei. Cell Rep. 2014;6:211–221.

132. Rosenbloom K.R., Dreszer T.R., Long J.C., Malladi V.S., Sloan C.A., Raney B.J., Cline M.S., Karolchik D., Barber G.P., Clawson H., et al. ENCODE whole-genome data in the UCSC Genome Browser: update 2012. Nucleic Acids Res. 2012;40:D912–D917.

133. Pennisi E. Genomics. ENCODE project writes eulogy for junk DNA. Science. 2012;337:1159. 1161.

134. Pei B., Sisu C., Frankish A., Howald C., Habegger L., Mu X.J., Harte R., Balasubramanian S., Tanzer A., Diekhans M., et al. The GENCODE pseudogene resource. Genome Biol. 2012;13:R51.

135. Djebali S., Davis C.A., Merkel A., Dobin A., Lassmann T., Mortazavi A., Tanzer A., Lagarde J., Lin W., Schlesinger F., et al. Landscape of transcription in human cells. Nature. 2012;489:101–108.

136. Derrien T., Johnson R., Bussotti G., Tanzer A., Djebali S., Tilgner H., Guernec G., Martin D., Merkel A., Knowles D.G., et al. The GENCODE v7 catalog of human long noncoding RNAs: analysis of their gene structure, evolution, and expression. Genome Res. 2012;22:1775–1789.

137. Bernstein B.E., Birney E., Dunham I., Green E.D., Gunter C., Snyder M. An integrated encyclopedia of DNA elements in the human genome. Nature. 2012;489:57–74.

138. Consortium E.P. The ENCODE (ENCyclopedia Of DNA Elements) Project. Science. 2004;306:636–640.

139. Rinn J.L., Chang H.Y. Genome regulation by long noncoding

RNAs. Annu. Rev. Biochem. 2012;81:145–166.

140. Suzuki K., Suter S., Ward R., Cooper D.A., Kelleher A.D. 10th Conference on Retroviruses and Opportunistic Infections. Boston: 2003. Silencing gene expression by short interfering RNA targeting NF-kB binding site within 5′ LTR.

141. Turner A.M., De La Cruz J., Morris K.V. Mobilization-competent Lentiviral Vector-mediated Sustained Transcriptional Modulation of HIV-1 Expression. Mol. Ther. 2009;17:360–368.

142. Turner A.M., Morris K.V. Controlling transcription with noncoding RNAs in mammalian cells. Biotechniques. 2010;48 ix–xvi.

143. Singh A., Palanichamy J.K., Ramalingam P., Kassab M.A., Bhagat M., Andrabi R., Luthra K., Sinha S., Chattopadhyay P. Long-term suppression of HIV-1C virus production in human peripheral blood mononuclear cells by LTR heterochromatization with a short double-stranded RNA. J. Antimicrob. Chemother. 2013;69:404–415.

144. Ahlenstiel C., Mendez C., Lim S.T., Marks K., Turville S., Cooper D.A., Kelleher A.D., Suzuki K. Novel RNA duplex locks HIV-1 in a latent state via chromatin-mediated transcriptional silencing. Mol. Ther. Nucleic Acids. 2015;4:e261.

145. Zhang M.X., Zhang C., Shen Y.H., Wang J., Li X.N., Chen L., Zhang Y., Coselli J.S., Wang X.L. Effect of 27nt small RNA on endothelial nitric-oxide synthase expression. Mol. Biol. Cell. 2008;19:3997–4005.

146. Hoffman A.R., Hu J.F. Directing DNA methylation to inhibit gene expression. Cell. Mol. Neurobiol. 2006;26:425–438.

147. Allo M., Buggiano V., Fededa J.P., Petrillo E., Schor I., de la Mata M., Agirre E., Plass M., Eyras E., Elela S.A., et al. Control of alternative splicing through siRNA-mediated transcriptional gene silencing. Nat. Struct. Mol. Biol. 2009;16:717–724.

148. Park C.W., Chen Z., Kren B.T., Steer C.J. Double-stranded siRNA targeted to the huntingtin gene does not induce DNA methylation. Biochem. Biophys. Res. Commun. 2004;323:275–280.

149. Buhler M., Mohn F., Stalder L., Muhlemann O. Transcriptional silencing of nonsense codon-containing immunoglobulin minigenes. Mol. Cell. 2005;18:307–317.

150. Gonzalez S., Pisano D.G., Serrano M. Mechanistic principles of chromatin remodeling guided by siRNAs and miRNAs. Cell Cycle. 2008;7:2601–2608.

151. Wang X., Feng Y., Pan L., Wang Y., Xu X., Lu J., Huang B. The proximal GC-rich region of p16(INK4a) gene promoter plays a role in its transcriptional regulation. Mol. Cell. Biochem. 2007;301:259–266.

152. Pulukuri S.M., Rao J.S. Small interfering RNA directed reversal of urokinase plasminogen activator demethylation inhibits prostate tumor growth and metastasis. Cancer Res. 2007;67:6637–6646.

153. Yue X., Schwartz J.C., Chu Y., Younger S.T., Gagnon K.T., Elbashir S., Janowski B.A., Corey D.R. Transcriptional regulation by small RNAs at sequences downstream from 3′ gene termini. Nat. Chem. Biol. 2010;6:621–629.

154. Chu Y., Yue X., Younger S.T., Janowski B.A., Corey D.R. Involvement of argonaute proteins in gene silencing and activation by RNAs complementary to a non-coding transcript at the progesterone receptor promoter. Nucleic Acids Res. 2010;38:7736–7748.

155. Cho S., Park J.S., Kang Y.K. AGO2 and SETDB1 cooperate in promoter-targeted transcriptional silencing of the androgen receptor gene. Nucleic Acids Res. 2014;42:13545–13556.

156. Ackley A., Lenox A., Stapleton K., Knowling S., Lu T., Sabir K.S., Vogt P.K., Morris K.V. An algorithm for generating small RNAs capable of epigenetically modulating transcriptional gene silencing and activation in human cells. Mol. Ther. Nucleic Acids. 2013;2:e104.

157. Zakaria M.K., Khan I., Mani P., Chattopadhyay P., Sarkar D.P., Sinha S. Combination of hepatocyte specific delivery and transformation dependent expression of shRNA inducing transcriptional gene silencing of c-Myc promoter in hepatocellular carcinoma cells. BMC Cancer. 2014;14:582.

158. Palanichamy J.K., Mehndiratta M., Bhagat M., Ramalingam P., Das B., Das P., Sinha S., Chattopadhyay P. Silencing of integrated human papillomavirus-16 oncogenes by small interfering RNA-mediated heterochromatization. Mol. Cancer Ther. 2010;9:2114–2122.

159. Roberts T.C., Andaloussi S.E., Morris K.V., McClorey G., Wood

M.J. Small RNA-mediated epigenetic myostatin silencing. Mol. Ther. Nucleic Acids. 2012;1:e23.

160. Feng X.Z., He X.S., Zhuang Y.Z., Luo Q., Jiang J.H., Yang S., Tang X.F., Liu J.L., Chen T. Investigation of transcriptional gene silencing and mechanism induced by shRNAs targeted to RUNX3 in vitro. World J. Gastroenterol. 2008;14:3006–3014.

161. Salvi A., Arici B., Portolani N., Giulini S.M., De Petro G., Barlati S. In vitro c-met inhibition by antisense RNA and plasmid-based RNAi down-modulates migration and invasion of hepatocellular carcinoma cells. Int. J. Oncol. 2007;31:451–460.

162. Jiang G., Zheng L., Pu J., Mei H., Zhao J., Huang K., Zeng F., Tong Q. Small RNAs targeting transcription start site induce heparanase silencing through interference with transcription initiation in human cancer cells. PLoS One. 2012;7:e31379.

163. Murayama A., Sakura K., Nakama M., Yasuzawa-Tanaka K., Fujita E., Tateishi Y., Wang Y., Ushijima T., Baba T., Shibuya K., et al. A specific CpG site demethylation in the human interleukin 2 gene promoter is an epigenetic memory. EMBO J. 2006;25:1081–1092.

164. Tan Y., Zhang B., Wu T., Skogerbo G., Zhu X., Guo X., He S., Chen R. Transcriptional inhibition of Hoxd4 expression by miRNA-10a in human breast cancer cells. BMC Mol. Biol. 2009;10:12.

165. Ouellet D.L., Plante I., Landry P., Barat C., Janelle M.E., Flamand L., Tremblay M.J., Provost P. Identification of functional microRNAs released through asymmetrical processing of HIV-1 TAR element. Nucleic Acids Res. 2008;36:2353–2365.

166. Cho D.H., Thienes C.P., Mahoney S.E., Analau E., Filippova G.N., Tapscott S.J. Antisense transcription and heterochromatin at the DM1 CTG repeats are constrained by CTCF. Mol. Cell. 2005;20:483–489.

167. Wang Q.Y., Zhou C., Johnson K.E., Colgrove R.C., Coen D.M., Knipe D.M. Herpesviral latency-associated transcript gene promotes assembly of heterochromatin on viral lytic-gene promoters in latent infection. Proc. Natl. Acad. Sci. U.S.A. 2005;102:16055–16059.

168. Mohammad F., Mondal T., Guseva N., Pandey G.K., Kanduri C. Kcnq1ot1 noncoding RNA mediates transcriptional gene silencing by interacting with Dnmt1. Development. 2010;137:2493–2499.

169. Takayama K., Horie-Inoue K., Katayama S., Suzuki T., Tsutsumi

S., Ikeda K., Urano T., Fujimura T., Takagi K., Takahashi S., et al. Androgen-responsive long noncoding RNA CTBP1-AS promotes prostate cancer. EMBO J. 2013;32:1665–1680.

170. Brenner S., Jacob F., Meselson M. An unstable intermediate carrying information from genes to ribosomes for protein synthesis. Nature. 1961;190:576–581.

171. Berget S.M., Moore C., Sharp P.A. Spliced segments at the 5′ terminus of adenovirus 2 late mRNA. Proc. Natl. Acad. Sci. U.S.A. 1977;74:3171–3175.

172. Chow L.T., Gelinas R.E., Broker T.R., Roberts R.J. An amazing sequence arrangement at the 5′ ends of adenovirus 2 messenger RNA. Cell. 1977;12:1–8.

173. Brannan C.I., Dees E.C., Ingram R.S., Tilghman S.M. The product of the H19 gene may function as an RNA. Mol. Cell. Biol. 1990;10:28–36.

174. Brockdorff N., Ashworth A., Kay G.F., McCabe V.M., Norris D.P., Cooper P.J., Swift S., Rastan S. The product of the mouse Xist gene is a 15 kb inactive X-specific transcript containing no conserved ORF and located in the nucleus. Cell. 1992;71:515–526.

175. Lee R.C., Feinbaum R.L., Ambros V. The C. elegans heterochronic gene lin-4 encodes small RNAs with antisense complementarity to lin-14. Cell. 1993;75:843–854.

176. Wightman B., Ha I., Ruvkun G. Posttranscriptional regulation of the heterochronic gene lin-14 by lin- 4 mediates temporal pattern formation in *C. elegans*. Cell. 1993;75:855–862.

177. Fire A., Xu S., Montgomery M.K., Kostas S.A., Driver S.E., Mello C.C. Potent and specific genetic interference by double stranded RNA in Caenorhabditis elegans. Nature. 1998;391:806–811.

CHAPTER 5

HUMAN GENOME-WIDE RNAI SCREEN FOR HOST FACTORS THAT FACILITATE SALMONELLA INVASION REVEALS A ROLE FOR POTASSIUM SECRETION IN PROMOTING INTERNALIZATION

Joshua M. Thornbrough[1], Adarsh Gopinath[1], Tom Hundley[1], Micah J. Worley[1,2]

[1] Department of Biology, University of Louisville, Louisville, KY, 40292, United States of America

[2] Department of Microbiology and Immunology, University of Louisville, Louisville, KY, 40202, United States of America

ABSTRACT

Salmonella enterica can actively invade the gastro-intestinal epithelium. This frequently leads to diarrheal disease, and also gives the pathogen access to phagocytes that can serve as vehicles for dissemination into deeper tissue. The ability to invade host cells is also important in maintaining the carrier state. While much is known about the bacterial factors that promote invasion, relatively little is known about the host factors involved. To gain insight into how *Salmonella enterica* serovar Typhimurium is able to invade normally non-phagocytic cells, we undertook a global RNAi screen with *S.* Typhimurium-infected human epithelial cells. In all, we identified 633 genes as contributing to bacterial internalization. These genes fall into a diverse group of functional categories revealing that cytoskeletal regulators are not the only factors that modulate invasion. In fact, potassium ion transport was the most enriched molecular function category in our screen, reinforcing a link between potassium and internalization. In addition to providing new insights into the molecular mechanisms underlying the ability of pathogens to invade host cells, all 633 host factors identified are candidates for new anti-microbial targets for treating *Salmonella* infections, and may be useful in curtailing infections with other pathogens as well.

INTRODUCTION

S. Typhimurium produces diarrheal disease in humans. It is generally self-limiting but can be serious in infants and the elderly and also in immunocompromised patients. The closely related *Salmonella enterica* serovar Typhi on the other hand causes typhoid fever, a more serious, systemic disease, which is potentially fatal in otherwise healthy people. The public health threat is compounded by the emergence of multi-drug resistant strains[1]. In fact, strains of *S.* Typhi resistant to ciprofloxacin and most other antibiotics have become endemic in India and have also been reported in developed countries[2]. *Salmonella* is also studied because it is a convenient model organism for dissecting basic pathogenic properties such as the ability of microorganisms to invade host cells.

Following consumption, *Salmonella* deploys two distinct type III secretion systems harbored by *Salmonella* pathogenicity islands 1 and 2 (SPI-1 and SPI-2)[3]. These systems span the bacterial envelope and either the plasma or vacuolar membranes of host cells[3]. They function as 'molecular syringes', injecting a cocktail of effectors into host cell cytosol that manipulate various cellular functions in ways that benefit the pathogen[3]. *Salmonella* utilizes SPI-1 to invade cells and invoke an inflammatory response[4–6], important components of the gastrointestinal (GI) phase of disease. The role of SPI-1 in *Salmonella* pathogenesis in not limited to the GI phase of disease however, as numerous genes located in SPI-1 are synthesized at late stages of infection in the murine model of acute typhoid fever and are also required for persistence and shedding in a murine model of the carrier state[7–9]. After invasion of host cells, *Salmonella* utilizes SPI-2 to grow within them[10–12] and to influence phagocyte movement[13–15]. Inside cells, *Salmonella* resides within the *Salmonella* containing vacuole (SCV), which diverges from the endocytic pathway due to the action of both SPI-1 and SPI-2[16].

The host processes that are required for bacterial entry and the early establishment of a successful infection are not yet fully elucidated. With a global RNAi screen, we identified 633 human genes, which facilitate *S.* Typhimurium invasion of MCF-7 cells, including a large suite of genes involved in potassium transport. This study should contribute to the study of both GI and systemic salmonellosis and the genes identified may be able to serve as new anti-microbial targets.

RESULTS

Human Epithelial Cell Screen to Identify Host Factors required for *S.* Typhimurium Invasion

In our genome-wide screen, using a reverse transfection protocol, we targeted ≈22,000 known or predicted genes in human epithelial-like MCF-7 cells with siRNAs arrayed in 384 well plates. Four independent siRNAs, two in one plate and another two in a parallel plate targeted each gene. This 2 X 2 format reduced the likelihood that off-target effects accounted for the phenotypes. We incubated the cells for 72 hours following transfection, before infecting them with *S.* Typhimurium for thirty minutes. The 72-hour incubation allowed even long-lived proteins a chance to turn over. The strain expressed a copy of the GFP from a plasmid. Gentamicin was used to eliminate extracellular bacteria. Eighteen hours after infection, we fixed the cells, stained nuclei with Hoechst 33342 and determined the percentage of host cells that were infected by measuring GFP fluorescence within the estimated periphery of cells. The 18 hour incubation allowed us to identify host genes that modulated the intracellular growth of the bacteria, which were presented elsewhere[17], as well as those that facilitate invasion. Genes that reduced bacterial infection and had a p-value ≤ 0.02 in parallel wells were considered hits. We did not consider as hits wells that had fewer than 500 cells present as there was a positive correlation between the percentage of cells infected and the total present below this number, but not above it. The phenotypes of some of the hits may be attributable to physiology changes that allowed gentamicin to build up inside of the epithelial cells. However, this is probably not the case for the majority of the hits as a cell affected in this manner would likely be unhealthy and moreover we did not identify genes involved in drug efflux as would be expected if it was an issue.

The Host Network that Promotes *S.* Typhimurium Invasion of Epithelial Cells

In all, 633 genes were identified as facilitating *S.* Typhimurium entry into MCF-7 cells. The two most prominent string-derived subnetworks are displayed in Figs 1 and 2. Both subnetworks are involved in potassium ion transport. The molecular function categories of 633 genes, for which there was significant enrichment are show in Fig 3. The most enriched category by far, with a remarkable p-value of 9.86×10^{-11} was potassium transport. Potassium levels, but not potassium transport, have been shown to play a role in *Salmonella* pathogenesis. The bacterial potassium transporter Trk and external potassium modulate *Salmonella* type III secretion, invasion of epithelial cells and pathogenesis in mice and chickens[18,19]. In addition to genes involved in potassium transport, the molecular function categories of metal ion binding and transport were also highly enriched with p-values of 7.15×10^{-8} and 6.49×10^{-7} respectively. Iron has been shown to play a role in regulating invasion[20].

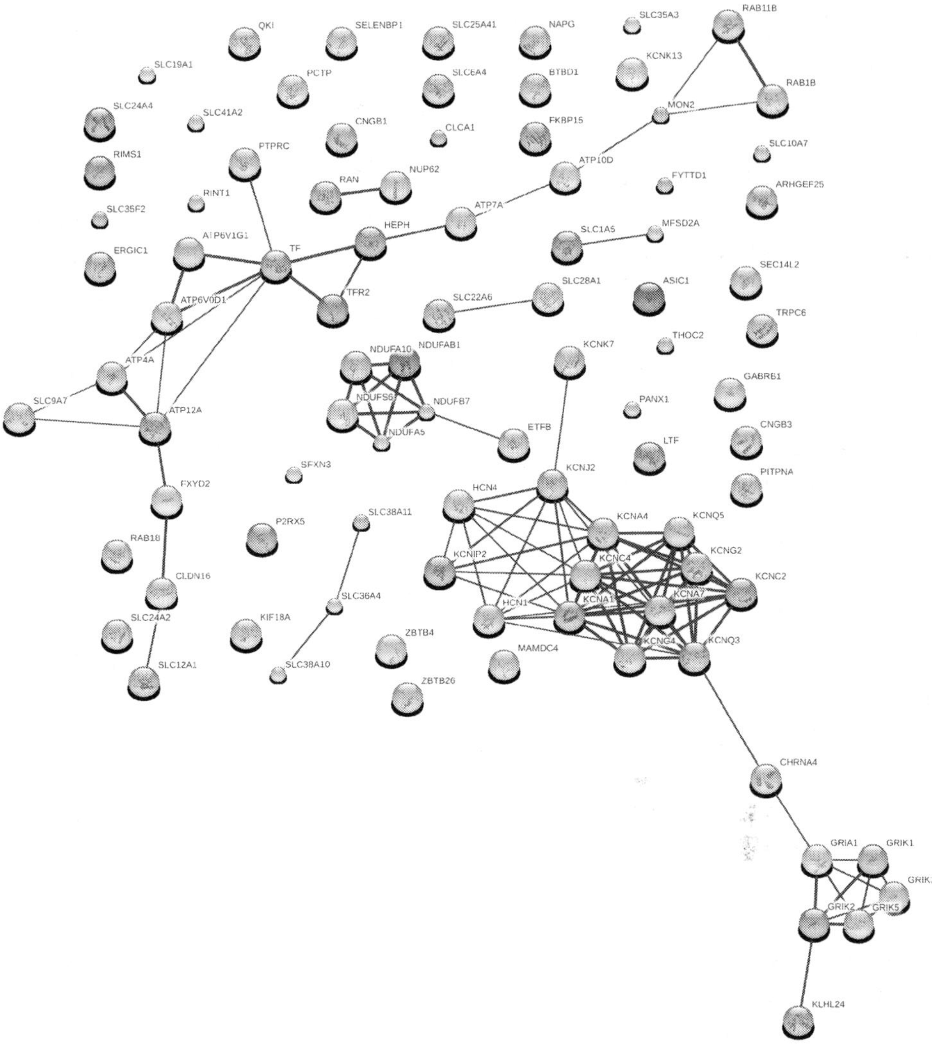

Figure 1: The most prominent Sting-derived subnetwork that promotes the invasion of epithelial cells by *S*. typhimurium.

The subnetwork is involved in potassium transport.

https://doi.org/10.1371/journal.pone.0166916.g001

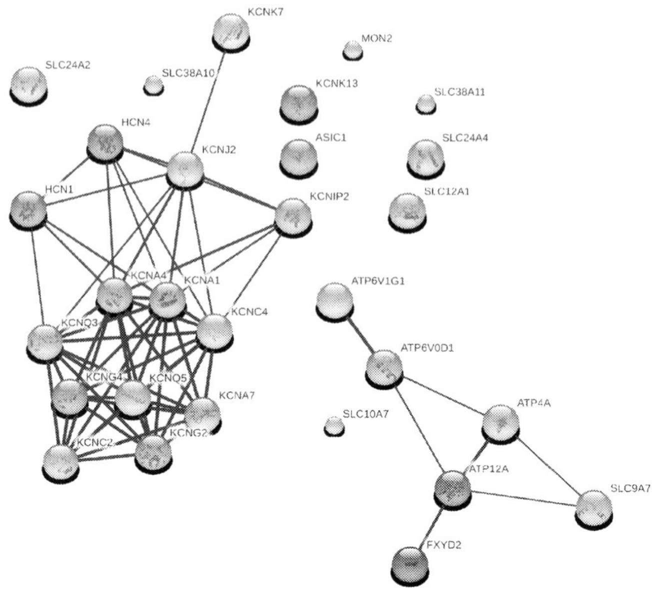

Figure 2: The second most prominent Sting-derived subnetwork that promotes the invasion of epithelial cells by *S.* typhimurium is also involved in potassium transport.

https://doi.org/10.1371/journal.pone.0166916.g002

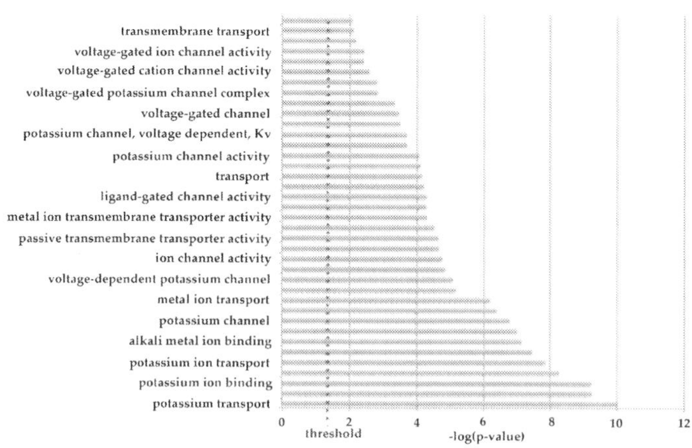

Figure 3: The overrepresented molecular function categories for the hit list.

The–log(p-value) is plotted with 0.05 being the significance threshold.

https://doi.org/10.1371/journal.pone.0166916.g003

Comparison with another RNAi Screen that Identified Genes involved in Invasion

A different group performed an RNAi screen on the 'druggable' genome with HeLa epithelial-like cells and *S.* Typhimurium to identify genes involved in invasion. The genes shared between our study (633) and Misselwitz et. al.[21] (90) are CYP26C1, TUBA4B, CYP27A1 and MUC5AC. Cytochrome P450 26C1 (CYP26C1) is inferred from electronic annotation to play a role in retinoic acid metabolism. Although it was identified in both studies, it is not clear how it would contribute to invasion. Putative tubulin-like protein alpha-4B (TUBA4B) is electronically inferred to bind GTP and to be a structural component of the cytoskeleton and accordingly, could plausibly play a role in internalization. Sterol 26-hydroxylase, mitochondrial (CYP27A1) is electronically inferred to catalyze the first step in sterol intermediate side chain oxidation. It is unclear how this would promote invasion. Mucin-5AC (MUC5AC) is an extracellular matrix structural constituent, which is a gel forming glycoprotein of gastric and respiratory tract epithelial that binds microorganisms that are then removed by the mucocilary system thereby protecting the epithelial cells from infection[22]. It seems possible that this gene was identified in both screens because the encoded protein facilitates close association of the epithelial cells and *S.* Typhimurium without a mucocilary system in the *in vitro* systems to remove the bacteria.

Identification of Genes that likely cause the SCV to Fuse with Lysosomes

We previously published the human genes we identified as promoting the growth of *S.* Typhimurium within MCF-7 cells[17] This list of 252 genes had 43 in common with the ones presented here that promote invasion. The absence of the genes likely does not affect invasion but rather causes the SCV to merge with the lysosome. Notable among this list are copine 5 (CPNE5), Ras-related protein 1B (Rab1b), Myotubularin-related protein 3 (MTMR3) and the lysosomal H^+ transporting subunit

V0D1 (ATP6V0D1). CPNE5 functions in membrane trafficking and its absence could affect the fate of the early SCV[23] as could Rab1b, as Rabs are small GTPases that are key regulators of intracellular membrane trafficking[24]. MTMR3 is a phosphatase that acts on phosphatidyl inositol (PtdIns) species that control the endocytic pathway[25]. ATP6V0D1 is a component of the vacuolar ATPase that is responsible for acidifying the SCV[26]. *S.* Typhimurium uses acid as a cue to express SPI-2 related virulence genes which are necessary for avoiding fusion of the SCV with lysosomes[27].

The Identification of Human Genes that could participate in *S.* Typhimurium Entry into Human Cells

While many of the 633 host factors identified as mediating the internalization of *S.* Typhimurium likely play novel roles in the process and fall into a diverse group of functional categories, numerous hits can be immediately rationalized as proteins that could facilitate invasion. Some of these hits are described below. The raw microscopy data for these hits is displayed in Fig 4.

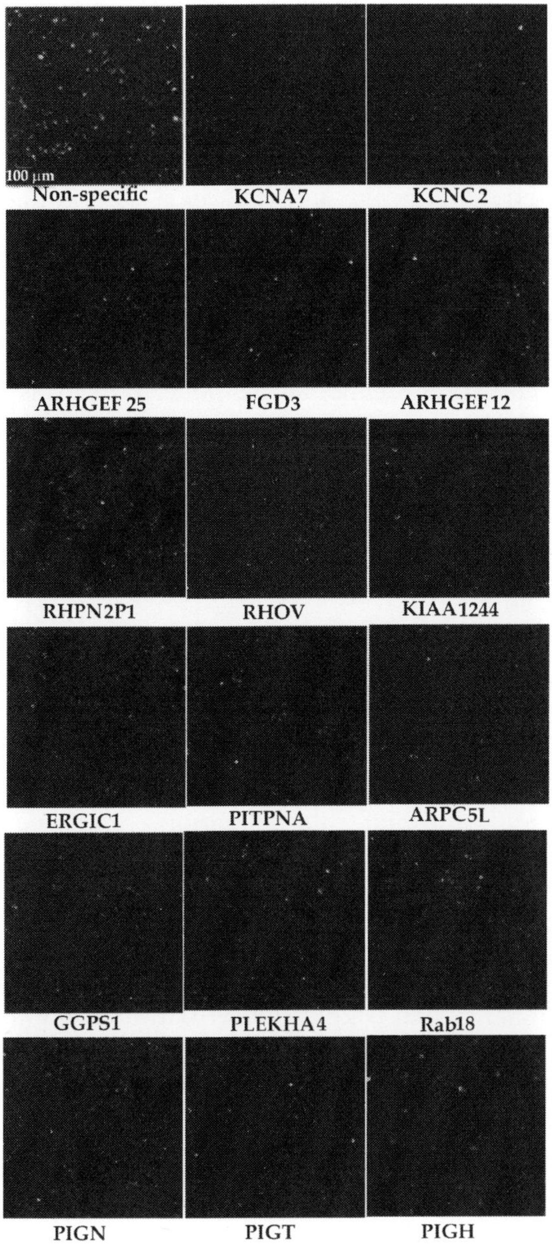

Figure 4: The microscopy data from some of the hits that can immediately be rationalized as playing a role in the invasion of epithelial cells by *S.* typhimurium.

Host cell nuclei are in blue and the bacteria are in green. The upper left image is a representative non-silencing control.

https://doi.org/10.1371/journal.pone.0166916.g004

KCNA7 is predicted to be a potassium voltage-gated channel of the shaker-related subfamily. KCNC2 is a potassium voltage-gated channel of subfamily C. In response to the voltage difference across the membrane, KCNC2 assumes an open or closed conformation, which when open forms a potassium selective channel that potassium ions can pass through[28]. KCNA7 and KCNC2 are two among at least 22 genes identified that play roles in mediating potassium transport. Potassium ions are known to activate SPI-1 and invasion[1,19].

We identified three Rho guanine nucleotide exchange factors (GEFs), one phophilin-2 like protein, which binds Rho proteins as well as one actual member of the Rho family as contributing to the invasion of epithelial cells by *S.* Typhimurium. Rho GEF 25 (ARHGEF25) and FYVE, RhoGEF and PH domain-containing protein 3 (FGD3) as well as Rho GEF 12 (ARHGEF12) are hits that can be easily rationalized as playing roles in the invasion of epithelial cells by *Salmonella* even though they have not been previously implicated in this process[29]. These proteins activate Rac, Cdc42 and Rho that are all well established to play a role in microbial internalization. *Salmonella* secretes the SPI-1 type III effectors SopE and SopE2 into epithelial cells, where they serve as nucleotide exchange factors (GEF) for the small GTP binding proteins Rac and Cdc42, catalyzing the exchange of GTP for GDP[30,31]. These proteins are active when bound to GTP and control downstream molecules that trigger the formation of highly branched and interconnected actin networks that lead to the formation of membrane ruffles, which facilitate *Salmonella* entry. While bacterially encoded GEFs have been known to facilitate entry for some time, this is to the best of our knowledge the first report of endogenous GEFs being required. It will be interesting to determine if these GEFs are directly targeted by *Salmonella* secreted proteins. We also identified putative rhophilin-2-like protein (RHPN2P1) as contributing to *S.* Typhimurium invasion, which is a member of the rhophilin family of Rho binding proteins. The protein binds both GTP- and GDP-bound RhoA and GTP-bound RhoB and may be involved in the organization of the actin cytoskeleton[32].

In addition to Rho GEFs and RHPN2P1, we also identified a Rho family member, RhoV, which may stimulate the JNK pathway to influence the actin cytoskeleton[33].

ARFGEF3 has Arf GTPase activity, which could plausibly stimulate *Salmonella* invasion of epithelial cells as it could trigger the activation of the WAVE regulatory complex, which *Salmonella* recruits to the underlying membrane where it governs actin polymerization. In fact, invading *Salmonella* are known to recruit and activate Arf1 to facilitate ruffling and Arf3 and Arf6 also promote invasion[34]. Thus, ARFGEF3 could be a new player in mediating the internalization of *Salmonella* by human cells.

The identification of ER-Golgi intermediate compartment-1 (ERGIC1) as one of the stronger hits in our screen was at first surprising. This gene encodes an ERGIC cycling membrane protein that regulates the stability of other members of this protein family. Recently, it was reported that depletion of the COPI complex resulted in a strong invasion defect for *S.* Typhimurium[21]. Like ERGIC1, the COPI complex regulates traffic between the ER and the Golgi apparatus. Depleting cells of COPI resulted in the mislocalization of cholesterol, sphingolipids, Rac1 and Cdc42 away from the plasma membrane into a large intracellular compartment, which decreased virulence factor injection and membrane ruffling[21].

ERGIC1 is a novel cycling membrane protein, which shares homology with two yeast proteins that are enriched in COPI vesicles: Erv41p and Erv46p. ERGIC1 localizes to the ERGIC and colocalizes with the human homologs of Erv41p and Erv46p. Covalent cross-linking revealed that ERGIC1 interacts with human ERGIC3 (hErv46). Depleting cells of ERGIC1 increases the turnover rate of ERGIC3 however the converse does not occur. Silencing ERGIC2 increases the turnover of ERGIC3, but not ERGIC1. It was thus proposed that ERGIC1 serves as a modulator of the ERGIC2-ERGIC3 complex through the stabilization of ERGIC3[35]. The function of these three proteins in mammalian cells remains to be fully elucidated. Depletion of ERGIC1 and ERGIC3 does not effect protein or glycoprotein secretion. Since ERGIC3 and ERGIC2 are interdependent, this double knock down is actually a triple knock down. This data suggests that these proteins are not essential for vesicular trafficking. The invasion defect in ERGIC1 depleted cells may be due to the mislocalization of cholesterol away from the plasma membrane as cholesterol in the plasma membrane is required for *S.* Typhimurium invasion[21]. Interestingly, in our screen, depleting cells of components of the COPI complex adversely affected host cell health whereas reducing the level of ERGIC1 did not.

PtdIns transfer protein (PITPNA) can transfer phospholipids between membranes *in vitro*[36]. PtdIns species are known to play roles in bacterial infection[37]. They can provide a negative charge to the inner side of the plasma membrane that attracts and anchors signaling and effector molecules that trigger phagosome formation. The loss of charge following their metabolism as phagosomes seal then causes the dissociation of proteins with the concomitant termination of signaling and cytoskeleton assembly[38].

By sequence similarity, Actin-related protein 2/3 complex subunit 5-like protein (ARPC5L) is predicted to function as a component of the Arp2/3 complex. This complex is involved in regulation of actin polymerization and together with an activating nucleation-promoting factor mediates the formation of branched actin networks. It is required for the invasion of epithelial cells by *Salmonella*[39].

Interfering with geranylgeranyl diphosphate synthase 1 (GGPS1) was also determined to significantly reduce *S.* Typhimurium invasion. This enzyme synthesizes geranylgeranyl diphosphate, which is an important precursor to geranylated proteins, from farnesyl diphosphate and isopentenyl diphosphate[40]. Another study showed that treating cells with geranylgeranyltransferase inhibitor, prevented Rho GTPases from being geranylgeranylated, dislocalizing them from the plasmid membrane thereby inhibiting the invasion of HeLa cells by *S.* Typhimurium[21], likely explaining why it came up in our screen.

Pleckstrin homology domain containing family A (phosphoinositide binding specific) member 4 (PLEKHA4) was another invasion hit. It binds specifically to phosphatidylinositol 3-phosphate (PtdIns3P) but not other PtdIns species, although the consequence of this interaction is not known[41]. It could plausibly play a role in mediating *S.* Typhimurium invasion, as PtdIns3P has been implicated in the process and can be found at the ruffling membrane and nascent phagosomes[42]. Rabs are known to regulate the endocytic pathway and several of them have been implicated in SCV biogenesis and development and been shown to be modulated by *S.* Typhimurium[43]. Rab18, one of our invasion hits, is not known to be involved in *Salmonella* pathogenesis, but it has been implicated in facilitating transport between the plasma membrane and early endosomes and in light of our results, likely plays a role[43].

GPI ethanolamine phosphate transferase 1 (PIGN), phosphatidylinositol glycan anchor biosynthesis, class T (PIGT) and phosphatidylinositol glycan anchor biosynthesis, class H (PIGH) are all three involved in attaching glycosylphosphatidyl inositol (GPI) anchors to proteins and localizing them to the apical side of the epithelial cell plasma membrane[44]. Here, these proteins could interact with microbes and facilitate invasion.

Confirmation of the Role of Potassium Secretion in Promoting Invasion

To confirm the role of potassium secretion in promoting the invasion of epithelial cells by *S.* Typhimurium, we utilized alpha dendrotoxin, a chemical inhibitor of a subset of voltage-gated potassium channels. In this experiment, cells were either untreated or treated with dendrotoxin or treated with dendrotoxin and potassium and infected with bacteria. We observed that dendrotoxin deters the invasion of epithelial cells by *S.* Typhimurium and that addition of exogenous potassium to the media alleviates the inhibition (Fig 5).

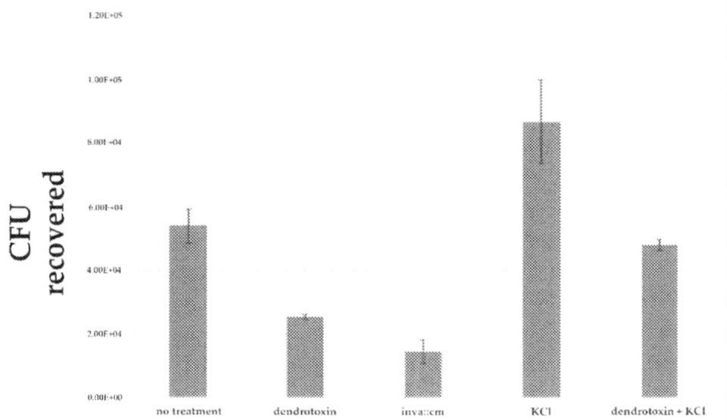

Figure 5: An invasion assay was performed with MCF-7 cells that were untreated or treated with dendrotoxin or treated with dendrotoxin plus potassium or just potassium.

An *invA::cm* strain is included as a control. The experiment was performed in duplicate on two independent occasions. Error bars depict the standard error of the mean. *p-value <0.05

https://doi.org/10.1371/journal.pone.0166916.g005

DISCUSSION

While many microbial factors that *S.* Typhimurium uses to invade normally non-phagocytic cells have been identified, relatively little is known about the host factors involved in this process. This genome-wide screen identifies the human genes exploited by *S.* Typhimurium and likely other pathogens as well to facilitate microbial entry into epithelial cells. In addition to generating much new information that will ultimately help unravel the molecular mechanisms underlying the ability of *S.* Typhimurium to invade such human cells, all of the 633 factors identified here are candidate for a new class of anti-microbial targets.

The set of genes identified in our screen had very modest overlap with an RNAi screen designed to detect host factors associated with invasion mediated by the SPI-1 type III effector SopE. There are many reasons for the lack of significant overlap. The study by Misselwitz et. al utilized a different *S.* Typhimurium strain. They used strain SL1344 whereas we employed strain 14028s. Also, their strain lacked three effectors that can affect invasion, SipA, SopE2 and SopB whereas we used wild type 14028s. Their screen was SopE driven whereas strain 14028s lacks SopE. Additionally, a different siRNA library was used, which was only composed of the 6,978 'druggable' genes whereas we screened the entire genome of apoproximately 22,000 genes. Further, different host cells were used. We employed MCF-7 cells whereas Misselwitz et. al used HeLa cells. Finally, we did not include as hits genes whose absence affected host cell health.

The most enriched molecular function category by far was potassium transport. There is precedence for the manipulation of host ion transport by *Salmonella*. *Salmonella* injects the SPI-1 type III effector SopB into epithelial cells to manipulate PtdIns-P species in order to among other things, produce PtnIns (1,4,5,6)P_4 to trigger chloride secretion and concomitant fluid entry into the GI tract, resulting in diarrhea[45]. The effect is due to the antagonism of ion channel closing. To the best of our knowledge, nobody has tested whether or not *Salmonella*stimulates the secretion of potassium as well as chloride. In light of our results, it seems likely that it may stimulate potassium secretion as external potassium is used as a cue by the bacteria to up-regulate SPI-1 and thereby promote invasion[18,19]. While our results suggest that potassium secretion by epithelial cells is a requirement for efficient *Salmonella* invasion, additional work will be required to determine if epithelial cells continuously secrete potassium, which the bacteria use as a cue to invade or it is a direct or indirect result of the presence of the bacteria.

Another study demonstrated that the CopI complex, the components of which have their stability regulated by ERGIC1, plays a role in regulating plasma membrane levels of cholesterol[21]. Caveolae are long-lived membrane microdomains composed of caveolins, cavins and a cholesterol rich membrane. These cholesterol-enriched domains can bind activated Rho GTPases via their geranylgeranyl anchors[46,47]. Misselwitz et. al. showed that Rho GTPases, which are responsible for host cell ruffling and concomitant *S.* Typhimurium entry are not localized to the plasma membrane in cells depleted for components of the CopI complex[21]. We observed that depleting cells of components of the CopI complex dramatically affected the health of the host cells whereas silencing ERGIC1 did not. Perhaps silencing ERGIC1 results in only a partial loss of CopI complex components that affect *S.* Typhimurium invasion but still allow for basic host cells processes to be carried out. The ability of ERGIC1 to increase the turnover rate of numerous ERGIC proteins, coupled with its apparent dispensability for the short-term health of host cells, may make it a unique anti-microbial target for a variety of pathogens as the requirement for cholesterol-enriched microdomains in the plasma membrane is not unique to *S.* Typhimurium. In fact, these internalization requirements are shared amongst *Staphylococcus aureus*[48,49], *Escherichia coli*[48,50,51], *Listeria monocytogenes*[52–55], *Mycobacterium fortuitum*[50], *Mycobacterium tuberculosis*[56], *Chlamydia caviae*[57], *Chlamydia trachomatis*[58,59], influenza virus[60,61], *Shigella flexneri*[62], *Brucella* abortus[63], hepatitis C virus[64,65] and vesicular stomatitis virus[21].

MTMR3, a protein whose absence likely results in the SCV merging with the lysosome is also interesting to consider further. The action of MTMR3 may change the electrostatic charge on the SCV causing Rabs that promote the fusion of endosomes with lysosomes to disassociate. The type III effector SopB, in addition to generating PtnIns(1,4,5,6) P_4 was speculated to recruit Rab5 vacuoles in part by dephosphorylating PtdIns(4,5)P_2, a tetravalent anion at physiological pH, thereby reducing the negative charge on the SCV. This may stimulate the association of some Rabs, but has been shown by Bakowski et. al. to cause other Rabs such as 8b, 13, 23 and 35, which share a polycationic prenyl membrane targeting motif in their carboxyl terminus to dissociate[66]. This dissociation halts the endocytic pathway. Smith et. al. in fact demonstrated that Rabs 23 and 35, which disassociate when SopB alters the electrostatic charge of the SCV, normally promote phagosome-lysosome fusion[43]. This likely causes the observed delay in SCV development where Rab7 containing SCVs do not fuse with late endosomes. This is reminiscent of the "pregnant pause" mechanism used by *Legionella pneumophila* and other intracellular bacterial pathogens, which temporarily halt maturation of their vacuole so they can form a replicative niche before the vacuole matures to the next state[67]. However, importantly, another study demonstrated that after a while, the PtdIns(3)P 5 kinase PIKfyve phosphorylates PtdIns(3) to PtdIns(3,5)P_2[27], which should increase the negative charge on the SCV, triggering the reassociation of Rabs 8b, 13, 23 and 35 thereby restarting the endocytic pathway. Whether it is host or the microbe that does this is not known. It may be the bacteria, as *Salmonella* cannot replicate in this intermediate SCV as acidification of the vacuole subsequent to the pregnant pause is necessary for activating a subset of intracellular virulence gene including those associated with SPI-2[27]. What then prevents the SCV from fusing with the lysosome when Rabs 23 and 35 re-associate is an open question that nobody has considered. *Salmonella* may recruit MTMR3 to the SCV at this point to prevent fusion with the lysosome, through its ability to dephosphorylate PtdIns(3,5)P_2, which should decrease the negative charge on the SCV that was increased by PIKfyve, causing Rabs 8b, 13, 23 and 35 to disassociate for a second time. PLEKHA4 could also be involved in SCV biogenesis as it is known to bind PtdIns3P[41]. While the consequences of this interaction are not known, it is possible that it facilitates phosphorylation to PtdIns(3,5)P_2, countering the effect of SopB, which is necessary for *Salmonella* virulence. It is also interesting to consider the role of Rab1b in *Salmonella* pathogenesis. Rab1b is involved in the secretory pathway between the ER and the Golgi and is required for autophagosome formation[68]. There is conflicting evidence in the literature about why we identified it. One study showed that some *Salmonella* escape the SCV and grow in the cytosol where they are targeted by autophagy and their growth is reduced[69]. Another more recent study demonstrated that autophagy facilitates *Salmonella* replication in HeLa cells[70]. In fact, knockdown of Rab1, which is required for autophagy, inhibits *Salmonella* replication[71]. It is possible that Rab1b depletion prevented *Salmonella* from growing at all inside host cells and would be more accurately described as having an intracellular growth defect than an invasion defect.

With a genome-wide RNAi-based forward genetic screen, we have identified 633 host factors that are involved in the internalization of *S.* Typhimurium by human epithelial cells. These factors fall into a group of diverse functional categories, indicating that invasion involves much more than merely cytoskeletal regulators. The most notable category was genes involved in potassium transport. In future work, it will be important to determine which if any *Salmonella*virulence factors directly stimulate the secretion of potassium ions. It will also be important to confirm and characterize some of the host factors that facilitate invasion in detail. The genome-wide results from our human RNAi screen provide us with new insights into how *Salmonella*actively invades human cells.

MATERIALS AND METHODS

Bacterial and Eukaryotic Cell Growth Conditions

As previously described[17], *S.* Typhimurium 14028s expressing the GFP from pACYC184 was grown in Luria Bertani broth (LB; Sigma-Aldrich, St. Louis, MO), supplemented with 10μg/mL tetracycline for 18 hrs at 37°C with aeration. The GFP was from *Pontellina plumata*. After 18 hours of growth, the bacteria were diluted 1:33 in fresh LB and subcultured for 3.5 hrs at 37°C with agitation to induce SPI-1 expression and then used to infect MCF-7 cells (ATCC, Manassas, VA). MCF-7 cells were cultured in Iscove's Modified Dulbecco's Medium (IMDM; Invitrogen, Carlsbad, CA) supplemented with 7.5% fetal bovine serum (FBS; Sigma-Aldrich) without antibiotics. Cell culture was maintained at

37°C in 5% CO_2 in a humidified tissue culture incubator. The cells were subcultured 8.00 x 10^4 cells/cm^2 every fourth day.

Infections and Gentamicin Protection Assay

These assays were performed as previously described[17]. Briefly, MCF-7 cells were infected with *S.* Typhimurium at an MOI of 100. An MOI of 100 was chosen because we empirically determined by testing a range of cell densities and MOIs that 3,000 host cells and an MOI of 100 resulted in the largest intracellular growth phenotype. This was important for identifying genes that normally promote intracellular growth, which were reported elsewhere[17]. In follow up experiments to confirm the role of potassium secretion in promoting invasion we used an MOI of 25. Following infection, the cells were then centrifuged at 230 rcf for 10 min at room temperature. The cells were then incubated at 37°C in 5% CO_2 for 30 min so that the bacteria could invade. The infected cell culture was then washed twice with PBS. After washing, the cells were incubated in IMDM supplemented with gentamicin (Gibco, Carlsbad, CA) at a concentration of 100 ug/mL for one hour at 37°C in 5% CO_2 to kill extracellular bacteria. The cells were washed two times with PBS after the gentamicin kill and then incubated in IMDM supplemented with gentamicin at a concentration of 10 ug/mL at 37°C in 5% CO_2 for 18 hours before imaging.

High-throughput siRNA screen of Human Genome

As previously described[17], a siRNA library was obtained from the Duke University RNAi screening facility targeting the human genome (Human whole-genome siRNA library v1.0; Qiagen, Valencia, CA). The library came in a 2 X 2, parallel plate format with four individual siRNAs targeting each gene arrayed in two sets, AB and CD. MCF-7 cells were seeded in 384 well clear flat bottom plates (Corning 3712; Corning, Lowell, MA) at a density of 3000 cells per well. Reverse transfections were carried out with 0.05uL Dharmafect 4 (Dharmacon, Lafayette, CO) and 20 nM siRNA/well in a total volume of 50uL. Transfection efficiency was optimized utilizing AllStars Hs Cell Death Control siRNA (Qiagen), a blend of siRNAs that are toxic and counting how many cells died. Efficiency was determined to be > 95% under these conditions. 72 hours after transfection, MCF-7 cells were infected with bacteria induced for SPI-1 expression and fixed eighteen hours later with 6% paraformaldehyde (Sigma-Aldrich) in 1 X PBS (Invitrogen) for 1 hour. Subsequently, the cells were washed two times with PBS. Nuclei were stained with Hoechst 33342 (Sigma-Aldrich) and automated imaging and analysis was performed on each well with the Cellomics ArrayScan VTI High Content imaging system (Thermo Scientific, Pittsburgh, PA). Images were analyzed via trained algorithm that first identified nuclei stained with Hoechst 33342. Next, the estimated cellular margins were established by extending the nuclear margin a predetermined radial distance. Only bacteria within this region were analyzed to avoid including any residual extracellular bacteria. Invasion defects were identified utilizing a Wilcoxin sum rank method for the percentage of infected cells in each well. Genes that reduced bacterial infection and had a p-value ≤ 0.02 in parallel wells were selected for follow-up. We normalized using Z-score based on mean fluorescence of plate/day batch. A p-value of 0.02 was chosen arbitrarily to allow for both enrichment and a manageable number of hits to follow-up.

Bioinformatics

Genes meeting selection criteria were manually curated with the National Center for Biotechnology Information (NCBI; http://ncbi.nlm.nih.gov) gene database. Functional clusters were analyzed with the Database for Annotation, Visualization and Integrated Discovery (DAVID) v6.7 (http://david.abcc.ncifcrf.gov/). Parameters were set to default with the following exceptions, EASE score was set for 0.05 and only clusters with an enrichment score ≥ 1.0 were selected for further analysis. For developing networks and assessing potential protein interactions the Search Tool for the Retrieval of Interacting Genes/Proteins (STRING) v9.1 (http://string-db.org/) and assessed for significant interaction enrichment within the whole genome and druggable genome categories.

Confirmation of the Role of Potassium Secretion in Promoting Invasion

MCF-7 cells were untreated or treated with 10μM alpha dendrotoxin (VWR). Twenty-four hours later, the cells were infected with *S.* Typhimurium grown under SPI-1 inducing conditions as described above at an MOI of 25. To one group of wells, KCl was added to a final concentration of 3.8mM. Bacteria were centrifuged onto the monolayers at 230g for ten minutes and then incubated for thirty minutes. After the thirty minute invasion, the infected cells were washed three times with PBS and overlaid with fresh media supplemented with 100µg/mL gentamicin and incubated for one hour. After the gentamicin kill, the infected cells were washed three times with PBS and lysed in 1% Triton X-100 and CFU recovered on agar plates

REFERENCES

1. Boyle EC, Bishop JL, Grassl GA, Finlay BB. Salmonella: from pathogenesis to therapeutics. J Bacteriol. 2007;189(5):1489–95. pmid:17189373.
2. Gaind R, Paglietti B, Murgia M, Dawar R, Uzzau S, Cappuccinelli P, et al. Molecular characterization of ciprofloxacin-resistant Salmonella enterica serovar Typhi and Paratyphi A causing enteric fever in India. J Antimicrob Chemother. 2006;58(6):1139–44. Epub 2006/10/31. dkl391 [pii] pmid:17071955.
3. Hansen-Wester I, Hensel M. Salmonella pathogenicity islands encoding type III secretion systems. Microbes Infect. 2001;3(7):549–59. pmid:11418329.
4. Galan JE, Ginocchio C, Costeas P. Molecular and functional characterization of the Salmonella invasion gene invA: homology of InvA to members of a new protein family. J Bacteriol. 1992;174(13):4338–49. pmid:1624429
5. Groisman EA, Ochman H. Cognate gene clusters govern invasion of host epithelial cells by Salmonella typhimurium and Shigella flexneri. Embo J. 1993;12(10):3779–87. pmid:8404849
6. Galyov EE, Wood MW, Rosqvist R, Mullan PB, Watson PR, Hedges S, et al. A secreted effector protein of Salmonella dublin is translocated into eukaryotic cells and mediates inflammation and fluid secretion in infected ileal mucosa. Mol Microbiol. 1997;25(5):903–12. pmid:9364916
7. Lawley TD, Bouley DM, Hoy YE, Gerke C, Relman DA, Monack DM. Host transmission of Salmonella enterica serovar Typhimurium is controlled by virulence factors and indigenous intestinal microbiota. Infect Immun. 2008;76(1):403–16. Epub 2007/10/31.

IAI. 01189-07 [pii] pmid:17967858; PubMed Central PMCID: PMC2223630.

8. Lawley TD, Chan K, Thompson LJ, Kim CC, Govoni GR, Monack DM. Genome-wide screen for Salmonella genes required for long-term systemic infection of the mouse. PLoS Pathog. 2006;2(2):e11. pmid:16518469.
9. Giacomodonato MN, Uzzau S, Bacciu D, Caccuri R, Sarnacki SH, Rubino S, et al. SipA, SopA, SopB, SopD and SopE2 effector proteins of Salmonella enterica serovar Typhimurium are synthesized at late stages of infection in mice. Microbiology. 2007;153(Pt 4):1221–8. pmid:17379731.
10. Ochman H, Soncini FC, Solomon F, Groisman EA. Identification of a pathogenicity island required for Salmonella survival in host cells. Proceedings of the National Academy of Sciences of the United States of America. 1996;93(15):7800–4. pmid:8755556
11. Cirillo DM, Valdivia RH, Monack DM, Falkow S. Macrophage-dependent induction of the Salmonella pathogenicity island 2 type III secretion system and its role in intracellular survival. Mol Microbiol. 1998;30(1):175–88. pmid:9786194
12. Hensel M, Shea JE, Waterman SR, Mundy R, Nikolaus T, Banks G, et al. Genes encoding putative effector proteins of the type III secretion system of Salmonella pathogenicity island 2 are required for bacterial virulence and proliferation in macrophages. Mol Microbiol. 1998;30(1):163–74. pmid:9786193
13. McLaughlin LM, Govoni GR, Gerke C, Gopinath S, Peng K, Laidlaw G, et al. The Salmonella SPI2 effector SseI mediates long-term systemic infection by modulating host cell migration. PLoS Pathog. 2009;5(11):e1000671. Epub 2009/12/04. pmid:19956712; PubMed Central PMCID: PMC2777311.
14. Worley MJ, Nieman GS, Geddes K, Heffron F. Salmonella typhimurium disseminates within its host by manipulating the motility of infected cells. Proc Natl Acad Sci U S A. 2006;103(47):17915–20. pmid:17095609; PubMed Central PMCID: PMCPMC1635543.
15. Thornbrough JM, Worley MJ. A naturally occurring single nucleotide polymorphism in the Salmonella SPI-2 type III effector srfH/sseI controls early extraintestinal dissemination. PLoS One.

2012;7(9):e45245. pmid:23028876; PubMed Central PMCID: PMCPMC3445477.

16. Knodler LA, Steele-Mortimer O. Taking possession: biogenesis of the Salmonella-containing vacuole. Traffic. 2003;4(9):587–99. pmid:12911813.
17. Thornbrough JM, Hundley T, Valdivia R, Worley MJ. Human genome-wide RNAi screen for host factors that modulate intracellular Salmonella growth. PLoS One. 2012;7(6):e38097. pmid:22701604; PubMed Central PMCID: PMCPMC3372477.
18. Su J, Gong H, Lai J, Main A, Lu S. The potassium transporter Trk and external potassium modulate Salmonella enterica protein secretion and virulence. Infect Immun. 2009;77(2):667–75. pmid:19001074; PubMed Central PMCID: PMC2632022.
19. Liu Y, Ho KK, Su J, Gong H, Chang AC, Lu S. Potassium transport of Salmonella is important for type III secretion and pathogenesis. Microbiology. 2013;159(Pt 8):1705–19. pmid:23728623; PubMed Central PMCID: PMC4089031.
20. Dostal A, Gagnon M, Chassard C, Zimmermann MB, O'Mahony L, Lacroix C. Salmonella adhesion, invasion and cellular immune responses are differentially affected by iron concentrations in a combined in vitro gut fermentation-cell model. PLoS One. 2014;9(3):e93549. pmid:24676135; PubMed Central PMCID: PMC3968171.
21. Misselwitz B, Dilling S, Vonaesch P, Sacher R, Snijder B, Schlumberger M, et al. RNAi screen of Salmonella invasion shows role of COPI in membrane targeting of cholesterol and Cdc42. Mol Syst Biol. 2011;7:474. pmid:21407211; PubMed Central PMCID: PMC3094068.
22. Meezaman D, Charles P, Daskal E, Polymeropoulos MH, Martin BM, Rose MC. Cloning and analysis of cDNA encoding a major airway glycoprotein, human tracheobronchial mucin (MUC5). J Biol Chem. 1994;269(17):12932–9. pmid:7513696.
23. Dorer MS, Kirton D, Bader JS, Isberg RR. RNA interference analysis of Legionella in Drosophila cells: exploitation of early secretory apparatus dynamics. PLoS Pathog. 2006;2(4):e34. pmid:16652170; PubMed Central PMCID: PMC1447669.

24. Hutagalung AH, Novick PJ. Role of Rab GTPases in membrane traffic and cell physiology. Physiological reviews. 2011;91(1):119–49. pmid:21248164.
25. Walker DM, Urbe S, Dove SK, Tenza D, Raposo G, Clague MJ. Characterization of MTMR3. an inositol lipid 3-phosphatase with novel substrate specificity. Curr Biol. 2001;11(20):1600–5. pmid:11676921.
26. van Hille B, Vanek M, Richener H, Green JR, Bilbe G. Cloning and tissue distribution of subunits C, D, and E of the human vacuolar H(+)-ATPase. Biochem Biophys Res Commun. 1993;197(1):15–21. pmid:8250920.
27. Kerr MC, Wang JT, Castro NA, Hamilton NA, Town L, Brown DL, et al. Inhibition of the PtdIns(5) kinase PIKfyve disrupts intracellular replication of Salmonella. EMBO J. 2010;29(8):1331–47. pmid:20300065; PubMed Central PMCID: PMC2868569.
28. Kasten MR, Rudy B, Anderson MP. Differential regulation of action potential firing in adult murine thalamocortical neurons by Kv3. 2, Kv1, and SK potassium and N-type calcium channels. J Physiol. 2007;584(Pt 2):565–82. pmid:17761775; PubMed Central PMCID: PMC2277158.
29. Fukuhara S, Chikumi H, Gutkind JS. Leukemia-associated Rho guanine nucleotide exchange factor (LARG) links heterotrimeric G proteins of the G(12) family to Rho. FEBS Lett. 2000;485(2–3):183–8. pmid:11094164.
30. Hardt WD, Chen LM, Schuebel KE, Bustelo XR, Galan JE. S. typhimurium encodes an activator of Rho GTPases that induces membrane ruffling and nuclear responses in host cells. Cell. 1998;93(5):815–26. pmid:9630225
31. Stender S, Friebel A, Linder S, Rohde M, Mirold S, Hardt WD. Identification of SopE2 from Salmonella typhimurium, a conserved guanine nucleotide exchange factor for Cdc42 of the host cell. Mol Microbiol. 2000;36(6):1206–21. pmid:10931274.
32. Peck JW, Oberst M, Bouker KB, Bowden E, Burbelo PD. The RhoA-binding protein, rhophilin-2, regulates actin cytoskeleton organization. J Biol Chem. 2002;277(46):43924–32. pmid:12221077.
33. Aronheim A, Broder YC, Cohen A, Fritsch A, Belisle B, Abo A.

Chp, a homologue of the GTPase Cdc42Hs, activates the JNK pathway and is implicated in reorganizing the actin cytoskeleton. Curr Biol. 1998;8(20):1125–8. pmid:9778532.

34. Humphreys D, Davidson A, Hume PJ, Koronakis V. Salmonella virulence effector SopE and Host GEF ARNO cooperate to recruit and activate WAVE to trigger bacterial invasion. Cell Host Microbe. 2012;11(2):129–39. pmid:22341462; PubMed Central PMCID: PMC3314957.
35. Breuza L, Halbeisen R, Jeno P, Otte S, Barlowe C, Hong W, et al. Proteomics of endoplasmic reticulum-Golgi intermediate compartment (ERGIC) membranes from brefeldin A-treated HepG2 cells identifies ERGIC-32, a new cycling protein that interacts with human Erv46. J Biol Chem. 2004;279(45):47242–53. pmid:15308636.
36. Wirtz KW. Phospholipid transfer proteins. Annu Rev Biochem. 1991;60:73–99. pmid:1883207.
37. Viaud J, Lagarrigue F, Ramel D, Allart S, Chicanne G, Ceccato L, et al. Phosphatidylinositol 5-phosphate regulates invasion through binding and activation of Tiam1. Nature communications. 2014;5:4080. pmid:24905281.
38. Yeung T, Grinstein S. Lipid signaling and the modulation of surface charge during phagocytosis. Immunol Rev. 2007;219:17–36. pmid:17850479.
39. Criss AK, Casanova JE. Coordinate regulation of Salmonella enterica serovar Typhimurium invasion of epithelial cells by the Arp2/3 complex and Rho GTPases. Infect Immun. 2003;71(5):2885–91. pmid:12704163; PubMed Central PMCID: PMC153244.
40. Kavanagh KL, Dunford JE, Bunkoczi G, Russell RG, Oppermann U. The crystal structure of human geranylgeranyl pyrophosphate synthase reveals a novel hexameric arrangement and inhibitory product binding. J Biol Chem. 2006;281(31):22004–12. pmid:16698791.
41. Dowler S, Currie RA, Campbell DG, Deak M, Kular G, Downes CP, et al. Identification of pleckstrin-homology-domain-containing proteins with novel phosphoinositide-binding specificities. Biochem J. 2000;351(Pt 1):19–31. pmid:11001876; PubMed Central PMCID:

PMC1221362.

42. Hernandez LD, Hueffer K, Wenk MR, Galan JE. Salmonella modulates vesicular traffic by altering phosphoinositide metabolism. Science. 2004;304(5678):1805–7. pmid:15205533.
43. Smith AC, Heo WD, Braun V, Jiang X, Macrae C, Casanova JE, et al. A network of Rab GTPases controls phagosome maturation and is modulated by Salmonella enterica serovar Typhimurium. J Cell Biol. 2007;176(3):263–8. pmid:17261845; PubMed Central PMCID: PMC2063952.
44. Brown DA, Rose JK. Sorting of GPI-anchored proteins to glycolipid-enriched membrane subdomains during transport to the apical cell surface. Cell. 1992;68(3):533–44. pmid:1531449.
45. Norris FA, Wilson MP, Wallis TS, Galyov EE, Majerus PW. SopB, a protein required for virulence of Salmonella dublin, is an inositol phosphate phosphatase. Proc Natl Acad Sci U S A. 1998;95(24):14057–9. pmid:9826652.
46. Palazzo AF, Eng CH, Schlaepfer DD, Marcantonio EE, Gundersen GG. Localized stabilization of microtubules by integrin- and FAK-facilitated Rho signaling. Science. 2004;303(5659):836–9. pmid:14764879.
47. del Pozo MA, Alderson NB, Kiosses WB, Chiang HH, Anderson RG, Schwartz MA. Integrins regulate Rac targeting by internalization of membrane domains. Science. 2004;303(5659):839–42. pmid:14764880.
48. Ramet M, Manfruelli P, Pearson A, Mathey-Prevot B, Ezekowitz RA. Functional genomic analysis of phagocytosis and identification of a Drosophila receptor for E. coli. Nature. 2002;416(6881):644–8. pmid:11912489.
49. Potrich C, Bastiani H, Colin DA, Huck S, Prevost G, Dalla Serra M. The influence of membrane lipids in Staphylococcus aureus gamma-hemolysins pore formation. J Membr Biol. 2009;227(1):13–24. pmid:19067025.
50. Philips JA, Rubin EJ, Perrimon N. Drosophila RNAi screen reveals CD36 family member required for mycobacterial infection. Science. 2005;309(5738):1251–3. pmid:16020694.
51. Riff JD, Callahan JW, Sherman PM. Cholesterol-enriched membrane

microdomains are required for inducing host cell cytoskeleton rearrangements in response to attaching-effacing Escherichia coli. Infect Immun. 2005;73(11):7113–25. pmid:16239505; PubMed Central PMCID: PMC1273830.

52. Seveau S, Bierne H, Giroux S, Prevost MC, Cossart P. Role of lipid rafts in E-cadherin—and HGF-R/Met—mediated entry of Listeria monocytogenes into host cells. J Cell Biol. 2004;166(5):743–53. pmid:15337781; PubMed Central PMCID: PMC2172418.
53. Agaisse H, Burrack LS, Philips JA, Rubin EJ, Perrimon N, Higgins DE. Genome-wide RNAi screen for host factors required for intracellular bacterial infection. Science. 2005;309(5738):1248–51. pmid:16020693.
54. Cheng LW, Viala JP, Stuurman N, Wiedemann U, Vale RD, Portnoy DA. Use of RNA interference in Drosophila S2 cells to identify host pathways controlling compartmentalization of an intracellular pathogen. Proc Natl Acad Sci U S A. 2005;102(38):13646–51. pmid:16157870; PubMed Central PMCID: PMC1224656.
55. Gekara NO, Jacobs T, Chakraborty T, Weiss S. The cholesterol-dependent cytolysin listeriolysin O aggregates rafts via oligomerization. Cell Microbiol. 2005;7(9):1345–56. pmid:16098221.
56. Munoz S, Rivas-Santiago B, Enciso JA. Mycobacterium tuberculosis entry into mast cells through cholesterol-rich membrane microdomains. Scand J Immunol. 2009;70(3):256–63. pmid:19703015.
57. Derre I, Pypaert M, Dautry-Varsat A, Agaisse H. RNAi screen in Drosophila cells reveals the involvement of the Tom complex in Chlamydia infection. PLoS Pathog. 2007;3(10):1446–58. pmid:17967059.
58. Jutras I, Abrami L, Dautry-Varsat A. Entry of the lymphogranuloma venereum strain of Chlamydia trachomatis into host cells involves cholesterol-rich membrane domains. Infect Immun. 2003;71(1):260–6. pmid:12496174; PubMed Central PMCID: PMC143347.
59. Elwell CA, Ceesay A, Kim JH, Kalman D, Engel JN. RNA interference screen identifies Abl kinase and PDGFR signaling in Chlamydia trachomatis entry. PLoS Pathog. 2008;4(3):e1000021.

pmid:18369471; PubMed Central PMCID: PMC2267011.
60. Hao L, Sakurai A, Watanabe T, Sorensen E, Nidom CA, Newton MA, et al. Drosophila RNAi screen identifies host genes important for influenza virus replication. Nature. 2008;454(7206):890–3. pmid:18615016; PubMed Central PMCID: PMC2574945.
61. Konig R, Stertz S, Zhou Y, Inoue A, Hoffmann HH, Bhattacharyya S, et al. Human host factors required for influenza virus replication. Nature. 2010;463(7282):813–7. pmid:20027183; PubMed Central PMCID: PMC2862546.
62. Lafont F, Tran Van Nhieu G, Hanada K, Sansonetti P, van der Goot FG. Initial steps of Shigella infection depend on the cholesterol/sphingolipid raft-mediated CD44-IpaB interaction. EMBO J. 2002;21(17):4449–57. pmid:12198147; PubMed Central PMCID: PMC126195.
63. Watarai M, Makino S, Fujii Y, Okamoto K, Shirahata T. Modulation of Brucella-induced macropinocytosis by lipid rafts mediates intracellular replication. Cell Microbiol. 2002;4(6):341–55. pmid:12067319.
64. Tai AW, Benita Y, Peng LF, Kim SS, Sakamoto N, Xavier RJ, et al. A functional genomic screen identifies cellular cofactors of hepatitis C virus replication. Cell Host Microbe. 2009;5(3):298–307. pmid:19286138; PubMed Central PMCID: PMC2756022.
65. Popescu CI, Dubuisson J. Role of lipid metabolism in hepatitis C virus assembly and entry. Biol Cell. 2010;102(1):63–74. pmid:19857204.
66. Bakowski MA, Braun V, Lam GY, Yeung T, Heo WD, Meyer T, et al. The phosphoinositide phosphatase SopB manipulates membrane surface charge and trafficking of the Salmonella-containing vacuole. Cell Host Microbe. 2010;7(6):453–62. pmid:20542249.
67. Swanson MS, Fernandez-Moreira E. A microbial strategy to multiply in macrophages: the pregnant pause. Traffic. 2002;3(3):170–7. pmid:11886587.
68. Zoppino FC, Militello RD, Slavin I, Alvarez C, Colombo MI. Autophagosome formation depends on the small GTPase Rab1 and functional ER exit sites. Traffic. 2010;11(9):1246–61. pmid:20545908.
69. Birmingham CL, Smith AC, Bakowski MA, Yoshimori T, Brumell

JH. Autophagy controls Salmonella infection in response to damage to the Salmonella-containing vacuole. J Biol Chem. 2006;281(16):11374–83. pmid:16495224.

70. Yu HB, Croxen MA, Marchiando AM, Ferreira RB, Cadwell K, Foster LJ, et al. Autophagy facilitates Salmonella replication in HeLa cells. mBio. 2014;5(2):e00865–14. pmid:24618251; PubMed Central PMCID: PMC3952155.

71. Huang J, Birmingham CL, Shahnazari S, Shiu J, Zheng YT, Smith AC, et al. Antibacterial autophagy occurs at PI(3)P-enriched domains of the endoplasmic reticulum and requires Rab1 GTPase. Autophagy. 2011;7(1):17–26. pmid:20980813; PubMed Central PMCID: PMC3039730.

CHAPTER 6

RNAi-BASED FUNCTIONAL GENOMICS IDENTIFIES NEW VIRULENCE DETERMINANTS IN MUCORMYCOSIS

Trung Anh Trieu[1] , Marı´a Isabel Navarro-Mendoza[1] , Carlos Pe´rez-Arques[1] , Marta Sanchis[2] , Javier Capilla[2] , Patricia Navarro-Rodriguez[2] , Loida Lopez-Fernandez[2] , Santiago Torres-Martı´nez[1] , Victoriano Garre[1] , Rosa Marı´a Ruiz-Va´zquez[1] , Francisco E. Nicola´s[1]

[1] Departmento de Gene´tica y Microbiologı´a, Facultad de Biologı´a, Universidad de Murcia, Spain

[2] Unidad de Microbiologı´a, Universitat Rovira i Virgili, IISPV, Tarragona, Spain

ABSTRACT

Mucorales are an emerging group of human pathogens that are responsible for the lethal disease mucormycosis. Unfortunately, functional studies on the genetic factors behind the virulence of these organisms are hampered by their limited genetic tractability, since they are reluctant to classical genetic tools like transposable elements or gene mapping. Here, we describe an RNAi-based functional genomic platform that allows the identification of new virulence factors through a forward genetic approach firstly described in Mucorales. This platform contains a whole-genome collection of Mucor circinelloides silenced transformants that presented a broad assortment of phenotypes related to the main physiological processes in fungi, including virulence, hyphae morphology, mycelial and yeast growth, carotenogenesis and asexual sporulation. Selection of transformants with reduced virulence allowed the identification of mcplD, which encodes a Phospholipase D, and mcmyo5, encoding a probably essential cargo transporter of the Myosin V family, as required for a fully virulent phenotype of M. circinelloides. Knock-out mutants for those genes showed reduced virulence in both Galleria mellonella and Mus musculus models, probably due to a delayed germination and polarized growth within macrophages. This study provides a robust approach to study virulence in Mucorales and as a proof of concept identified new virulence determinants in M. circinelloidesthat could represent promising targets for future antifungal therapies.

AUTHOR SUMMARY

Mucormycosis is an infectious disease caused by organisms of the order Mucorales. It is a lethal infection that is raising the alarm in the medical and scientific community due to its high mortality rates, unusual antifungal drug resistance and its emerging character. Among the reasons explaining the nescience about this disease is the lack of knowledge on the biology of the organisms that cause mucormycosis, which is encouraged by the reluctance of these species to genetic studies. In this work, we have developed an RNAi-based functional genomic platform to study virulence in Mucorales. It is a powerful tool available for the scientific community that will contribute to solve the reluctance of Mucorales to genetic studies and will help to understand the genetic basis of virulence in these organisms. Secondly, and as a proof of concept, we have used this genetic tool to identify two new virulence determinants in Mucor circinelloides. Lack of function of these determinants delays germination and growth of spores, conceding time to macrophages for the inactivation of the pathogen. The two genes identified, mcplD and mcmyo5, represent promising targets for future development of new antifungal therapies against mucormycosis.

INTRODUCTION

Mucormycosis is a fungal infection caused by species of the order Mucorales that represents the third most common angio-invasive fungal infection after candidiasis and aspergillosis. Due to the unusual antifungal drug resistance of Mucorales, mucormycosis is considered one of the most important medical complications in immunocompromised patients [1]. Among current antifungal drugs, fluconazole, voriconazole, posaconazole and itraconazole are potent agents of choice used in aspergillosis and candidiasis that, unfortunately, present poor activity against mucormycosis [2]. More specifically, amphotericin B, an old-known macrolide antifungal compound with severe adverse effects, and more recently isavuconazole are used against mucormycosis although they only achieve partial activity [3–5]. As a consequence of this lack of efficient antifungal drugs, mortality rates of mucormycosis remain higher than 50% and reach up to 90% in disseminated infections [6, 7]. Another negative aspect of mucormycosis is its emerging condition. Only a few years ago, mucormycosis was considered a rare infection limited to immunocompromised patients suffering diabetes, organ transplant or other diseases associated with immunosuppression [8]. However, the current improvement in the diagnostic techniques has revealed an alarming number of mucormycosis cases in immunocompetent patients that have severe trauma (e.g. burn patients, traumatic injuries), since it is now rarely misdiagnosed as aspergillosis [9]. Thus, the isolation of new strains that are capable of infecting healthy individuals and the increasing number of reported cases have raised the alarm on this emerging disease. Together, the lack of effective treatments and the emerging character of this devastating disease are urgently demanding new strategies to prevent and/or treat mucormycosis.

The development of therapies to treat mucormycosis is restricted by the lack of knowledge about the disease and the organisms that cause the infection. One of the main reason explaining the scarce information about mucormycosis is the high reluctance of Mucorales to modern molecular genetics techniques. Among Mucorales, Rhizopus oryzae and Mucor circinelloides are two study models in which genetic transformation is available [10, 11]. Study of pathogenesis in these two models has revealed iron uptake, spore size, spore coat proteins and dimorphism as virulence determinants in mucormycosis [12–18]. In M. circinelloides, along with genetic transformation, the application of molecular tools has allowed the dissection of its RNAi mechanism, which has become a useful tool for functional genetics in this fungus [19–22]. Besides its applications as a genetic tool, the RNAi mechanism of M. circinelloides has a regulatory role that controls complex physiological processes such as growth, sexual and asexual sporulation and death by autolysis [19, 23–27]. Moreover, the extensive study of the RNAi mechanism in M. circinelloides led to the discovery of the first link between this endogenous regulatory mechanism and the unusual antifungal drug resistance of Mucorales [28]. This novel mechanism generates spontaneous resistance to the antifungal drug FK506 by epigenetic RNAi-mediated post-transcriptional silencing of the fkbA gene encoding the protein FKBP12, which is the natural target of FK506. As a result, the lack of target blocks the action of FK506 and the fungus becomes resistant to this drug, suggesting that similar mechanisms could be behind of the exacerbated resistance to antifungal drugs in Mucorales.

The limited knowledge about Mucorales, mainly due to the phylogenetic distance and genetic differences of these basal fungi with other well-known fungi like Ascomycota and Basidiomycota [29], together with the low efficiency of the current antifungal drugs, makes it urgent the development of novel strategies to study this group of organisms and, more specifically, the finding of new virulence determinants that could become future antifungal drug targets in Mucorales. Consequently, the main purpose of this work has been the establishment of a functional genomic approach based on the RNAi mechanism of M. circinelloides to select phenotypes relevant for the biology of Mucorales and related to virulence, and subsequently to identify the genes responsible for these phenotypes. RNAi has been used as a powerful reverse genetic tool to develop functional whole-genome studies in many organisms, including worms [30], flies [31] and mammal cells [32]. In these reverse genetic approaches, a defined library is laboriously constructed by designing a silencing vector for each annotated gene of the studied organism. In these studies, the model organism requires an easy transformation method ready to be arrayed in large-scale assays in which each silencing vector/ molecule is individually delivered. Unfortunately, this is not the case of M. circinelloides or any other emerging Mucoral. These inconveniences have led us to design a different approach that uses RNAi as a forward genetic tool in which a library representing M. circinelloides whole genome was constructed in a vector that silenced the cloned inserts. Transformation with this silencing library generated a collection of silenced transformants ready for phenotype screenings in a similar way as in classic chemical or insertional mutagenesis approaches. This transformant collection and the silencing library represent a new genetic tool in Mucorales for forward genetics and functional analysis at whole genome level. Using this approach we have isolated phenotypes related to virulence leading to the identification of two new virulence determinants in M. circinelloides, the enzyme Phospholipase D (PLD) and a Myosin 5 (Myo5) motor protein, which are required for full virulence in Galleria mellonella and Mus musculus host models. Overall, this work illustrates a new approach to study virulence in Mucorales at the whole genome level.

RESULTS

A Genome Wide RNAi-based Library provides a Representative Collection of Silenced Transformants in *M. circinelloides*

A vector capable of inducing RNAi from any random DNA fragment was designed previously to the construction of the gDNA library for phenotypic screening based on RNAi. RNAi can be triggered in M. circinelloides by using self-replicative plasmids containing either complete or fragmented genes with their own promoters, obtaining silencing frequencies ranging between 3% and 30% [21]. However, highest silencing frequencies (nearly 95%) can be achieved when the plasmid contains a strong promoter and hairpin structures that directly transcribe dsRNA [22]. To circumvent the limitations of constructing a hairpin producing vector for each gene of M. circinelloides genome, we designed a high-throughput silencing vector (pMAT1700) with two convergent promoters and no terminator sequences that are flanking a multiple cloning site (MCS) in which random gDNA fragments can be cloned (Fig 1A). In addition, a sequence of 0.5 kb of carB gene was cloned next to the MCS to be used as a reporter of silencing (Fig 1A). This gene encodes a phytoene dehydrogenase involved in the production of β-carotene, a pigment responsible for the typical yellow color of M. circinelloides [33]. Triggering of RNAi after transformation by plasmids containing this reporter produces albino transformants that are easily detectable and also signalize silencing of any other sequence cloned next to it. Fragments of 0.5–4 kb were isolated from M. circinelloides gDNA partially digested with Sau3Aand filled in with dGTP-dATP to avoid self-ligation. The genomic fragments thus obtained were ligated with vector pMAT1700 digested with XhoI and filled in with dCTP-dTTP to make their ends compatible with Sau3A filled fragments and avoid self-ligation, thus favoring the frequency of recombinant clones in the library [34]. Ligation mixtures were introduced into Escherichia colicells to generate the genome-wide RNAi library (Fig 1B) consisting in roughly 83,000 clones, which determined a confidence level higher than 99%. Pooled plasmids were directly purified from the E. coli colonies and used to transform M. circinelloides MU402 (pyrG$^-$ and leuA$^-$) strain. The empty vector pMAT1700 and a version of this vector lacking the carB fragment (pMAT1701) were used as controls to monitor silencing efficiency. Up to sixty transformations following this approach were required to obtain a collection of 51,657 silenced transformants, which ensured a 95% confidence level. Comparison of silencing frequencies obtained with the empty vector and the high-throughput silencing library showed a pronounced increase of carB silenced transformants among those obtained with the library (87%) relative to the empty plasmid pMAT1700 (43%). As expected, the plasmid pMAT1701 did not trigger silencing in any of the transformants. The increase in silencing frequency among transformants obtained with the library could be explained if the 0.5 kb fragment of carBgene that is cloned between the two promoters was not long enough in the empty plasmid to allow efficient convergent transcription from both promoters. Nevertheless, once the convergent cassette assimilates new fragments in the library, the silencing efficiency increases close to the maximum previously observed with hairpin triggering molecules [22]. These results demonstrated a high silencing efficiency of our high-throughput library in M. circinelloides.

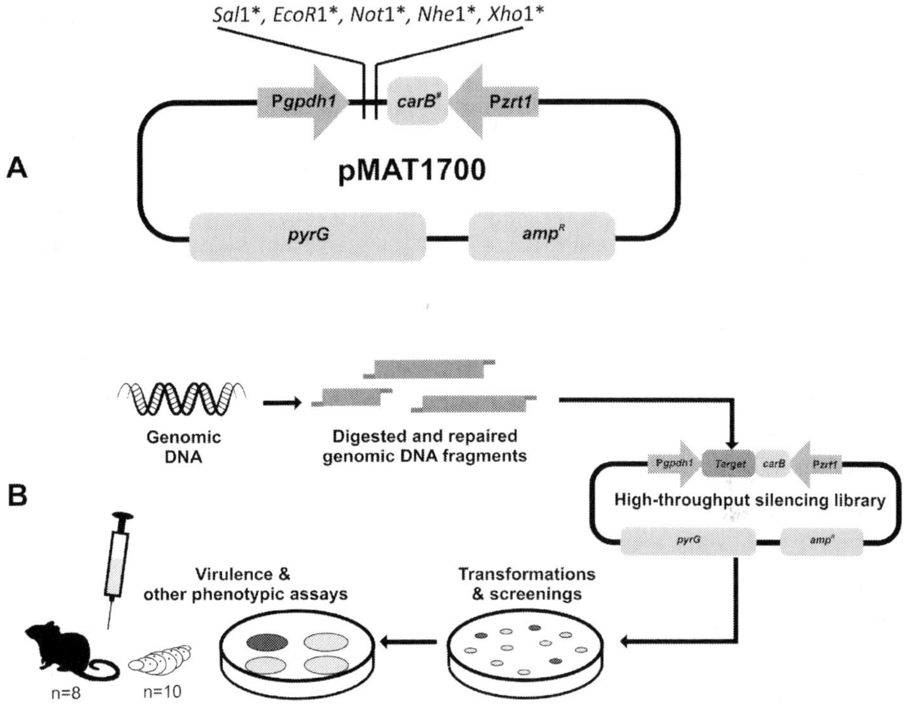

Figure 1: Schematic diagram of the RNA silencing vector pMAT1700 (A) and the platform for functional genomics in M. circinelloides **(B).**

Pzrt1, M. circinelloides zrt1 promoter; Pgpd1, M. circinelloides gpd1 promoter; carB#, 0.5 kb fragment of M circinelloides carB gene; pyrG, M. circinelloides orotidine 5›-phosphate decarboxylase gene; ampR, ampicillin-resistant gene. Asterisks indicate unique restriction sites in the vector.

https://doi.org/10.1371/journal.ppat.1006150.g001

Selection of Avirulent Phenotypes

The main purpose of generating a high-throughput functional genomic tool in M. circinelloideswas to use it as a new approach to find unknown virulence determinants in Mucorales (Fig 1B). In order to find candidate genes involved in M. circinelloides pathogenesis, we focused the screening of silenced strains on abnormal growth and morphology, since those aspects of fungal physiology have been related to pathogenesis in M. circinelloides and other fungi [14, 35]. Special attention was paid to transformants growing as yeast-like colonies and showing altered dimorphism and strong reduction of the growth rate, as it is one of the few processes previously associated with virulence in M. circinelloides [13]. Plates from the transformations with the silencing library were directly screened for abnormal phenotypes (51,657 silenced transformants), resulting in the selection of fifteen transformants with different abnormalities. Later, the transformation plates were further incubated and vegetative spores were pooled together to obtain the collection of silenced spores harboring the high-throughput silencing library described in the previous section. A total of $1x10^4$ viable spores from this collection were grown in new plates for a second screening, resulting in the selection of eleven abnormal candidates. In addition, the second screening confirmed that silencing is maintained in the collection of silenced spores, since the frequency of silenced colonies (79±4% of albino colonies) was similar to the previously described in the original transformants. Growth rate and sporulation efficiency of the twenty six isolated candidates from the two screenings were quantified and classified into five categories based on the different morphological abnormalities that they presented (Fig 2 and Table 1). The first category, the most abundant with 16 isolates, presented a reduced growth (RG1-16) compared with control transformants, but wild-type sporulation. Two transformants presenting a highly reduced growth (HRG1 and HRG2) were included in the second category, as they showed clear differences with the first category, including a reduced vegetative sporulation. The third category comprised five transformants that showed a strong lack of vegetative sporulation (LVS1-5). In addition to the lack of sporulation, some of these five transformants also presented a reduced growth similar to the first category (Table 1). The fourth category contained only one transformant that showed a yeast-like growth (YLG1). This transformant presented the slowest growth, forming small colonies similar to yeasts rather than mycelial colonies. The morphology of YLG1 under the optical microscope also showed strongly deformed cell walls incapable of forming regular filaments (Fig 2). These filaments appeared to be septated, although one of the main characteristics of Mucorales is their coenocytic mycelium. This contradictory observation could be explained if this transformant is immersed in a hyphae-yeast transition state in which the tip of the hyphae produces yeast cells that resemble a septated structure before the yeast cells are liberated (Fig 2, yeast-like growth). The last category included two transformants showing a satellite growth phenotype (SG1 and SG2). These two transformants grew slower than control transformants, producing long sporangia that bent to the media to form new colonies, acquiring this unusual satellite phenotype in transformation plates at pH 3.2 (Fig 2). When SG1 and SG2 were grown in MMC medium at pH 4.5 (a rich medium but selective for uracil auxotrophy, [36]), they showed reduced growth and sporulation, but not the satellite phenotype. HRG and SG transformants also presented abnormal mycelia under the optical microscope, showing swollen hyphae with abnormal branching (Fig 2). Accordingly with the mechanism of silencing previously described in M. circinelloides [21], the twenty six transformants showed a reversible phenotype and they lost the abnormalities when they were grown in a non-selective medium for several vegetative cycles, confirming that the phenotypes were caused by the silencing of some genes harbored in the plasmids.

Table 1: Phenotypic characterization of silenced transformants with abnormal phenotypes.

https://doi.org/10.1371/journal.ppat.1006150.t001

Strain	Growth rate*	Sporulation#	Pathogenicity
WILD TYPE			
MU402+pMAT1700	5.98 ±0.12	5.86±0.40	VIRULENT
REDUCED GROWTH (RG)			
RG1	**4.09±0.13**	5.53±0.15	VIRULENT
RG2	**4.45±0.27**	5.61±0.17	VIRULENT
RG3	**4.26±0.22**	5.62±0.21	VIRULENT
RG4	**5.01±0.19**	5.53±0.31	VIRULENT
RG5	**4.27±0.20**	5.63±0.26	VIRULENT
RG6	**4.05±0.14**	5.43±0.35	VIRULENT
RG7	**4.94±0.22**	5.56±0.11	VIRULENT
RG8	**5.27±0.10**	5.46±0.31	VIRULENT
RG9	**4.15±0.22**	5.41±0.26	VIRULENT
RG10	**3.94±0.17**	5.63±0.25	VIRULENT
RG11	**4.03±0.14**	5.71±0.21	VIRULENT
RG12	**5.01±0.20**	5.75±0.17	VIRULENT
RG13	**4.38±0.23**	5.55±0.45	VIRULENT
RG14	**4.25±0.13**	5.26±0.37	VIRULENT
RG15	**5.01±0.19**	5.43±0.41	VIRULENT
RG16	**4.07±0.33**	5.22±0.26	VIRULENT
HIGHLY REDUCED GROWTH (HRG)			
HRG1	**2.03±0.33**	**4.06±0.20**	**AVIRULENT**
HRG2	**2.04±0.27**	**4.01±0.15**	**AVIRULENT**
REDUCED SPORULATION			
RS1	6.04±0.31	**3.13±0.35**	VIRULENT
RS2	**4.74±0.19**	**4.21±0.31**	VIRULENT
RS3	6.00±0.17	**3.21±0.41**	VIRULENT
RS4	**4.23±0.25**	**4.21±0.26**	VIRULENT
RS5	**7.11±0.26**	**3.46±0.35**	VIRULENT
YEAST LIKE GROWTH (YLG)			
YLG1	**0.48±0.06**	**0.04±0.01**	**AVIRULENT**
SATELLITE GROWTH (SG)			
SG1	**3.19±0.23**	**2.51±0.41**	VIRULENT
SG2	**3.19±0.26**	**2.53±0.25**	VIRULENT

Data are represented as mean ± SEM.

* Diameter (cm).

Spores x10^6/(cm^2). Bold numbers represent significant differences (P<0.01, t-test).

doi:10.1371/journal.ppat.1006150.t001

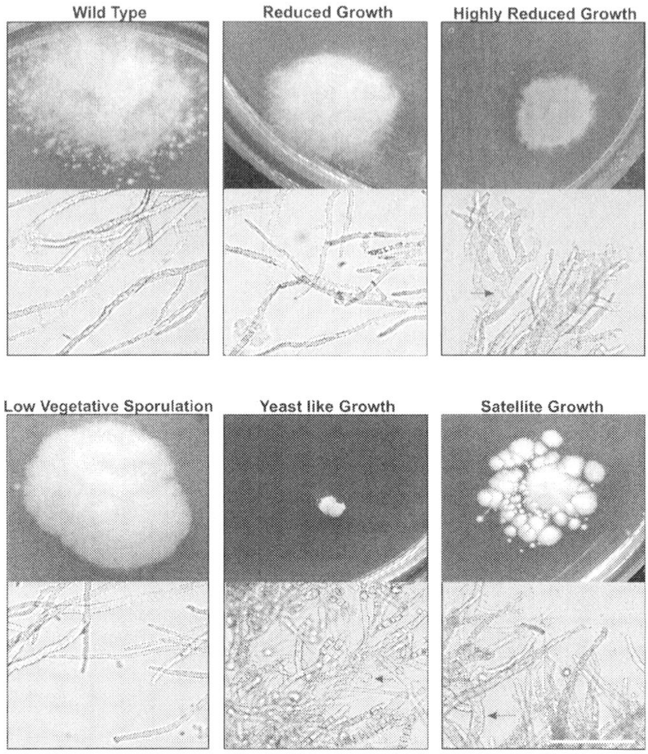

Figure 2: Representative images of wild type strain and the five groups of abnormal phenotypes selected from screenings.

Colonial mycelia were grown in MMC medium at pH 3.2 during 72 hours. For the micrographs below each colony, a piece of mycelium from the edge of each colony was observed under the light microscope. Red arrows point to swollen hyphae and green arrow points to yeast cells. Scale bars represent 50 μm.

https://doi.org/10.1371/journal.ppat.1006150.g002

As this work focuses on the identification of new virulence determinants in M. circinelloides, we performed virulence tests with all the selected transformants in a heterologous host, Galleria mellonella, which was previously established as a host model for M. circinelloides [14]. The viability of larvae infected with two thousand spores was monitored at one day intervals for all the transformants except YLG1, which was unable to produce spores for this assay and, therefore, yeast-like cells were used for the infection assay [13] (Fig 3). Among the twenty six transformants, only three isolates were significantly less virulent than the virulent control strains, the two HRG transformants (Fig 3A, HRG1 and HRG2) and the single YLG transformant (Fig 3B, YLG1). The isolation of these three transformants with reduced virulence confirmed that our RNAi-based functional genomics strategy can be used to select phenotypes related to virulence and pathogenesis in M. circinelloides.

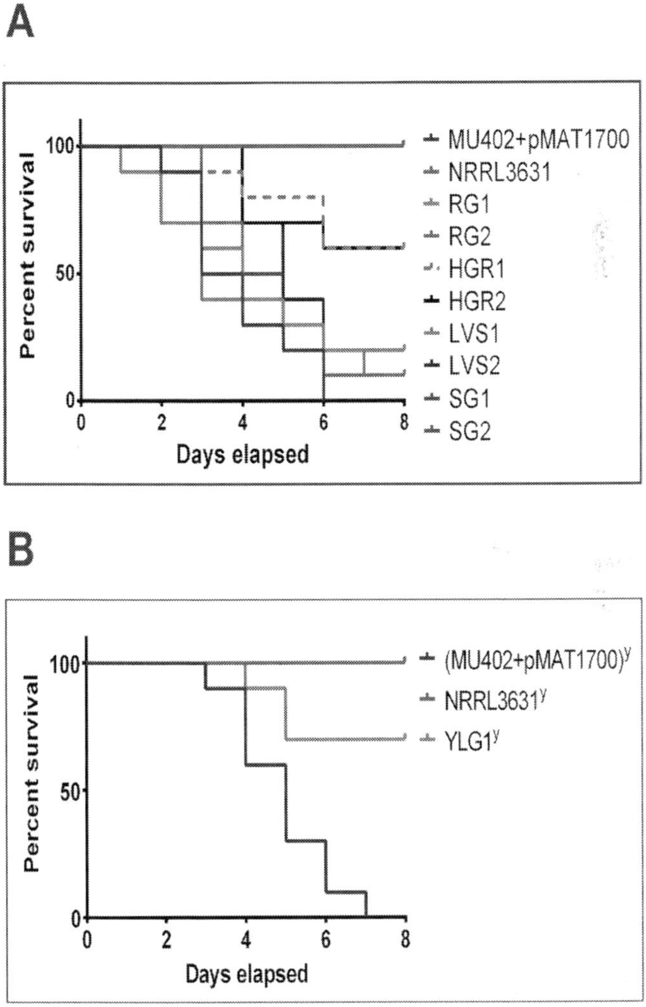

Figure 3: Virulence assays of selected transformants with abnormal phenotypes.

(A) Virulence assays of all highly reduced growth (HRG) and satellite growth (SG) transformants, and two representative transformants from reduced growth (RG) and low vegetative sporulation (LVS) phenotypes. Ten G. mellonella larvae were injected with 2,000 spores of each tested strain. Only transformants HRG1 and HRG2 showed a significant reduced virulence ($p = 0.0014$ and $p = 0.0016$, respectively) (B) Virulence assays of wild type strains and the yeast-like growth (YLG) transformant using an inoculum of 20,000 yeast cells in the infections ([y]). The transformants YLG1 showed a significant reduced virulence ($p = 0.0017$). NRRL3631 and MU402+pMAT1700 were used as avirulent and virulent control strains, respectively. These data are representative of three independent experiments.

https://doi.org/10.1371/journal.ppat.1006150.g003

Candidate Gene Identification and Validation of the Silencing Phenotypes

The self-replicative nature of M. circinelloides plasmids used to construct the RNAi libraries facilitates the identification of the silencing sequences responsible for the phenotypes in the selected transformants, since library plasmids are maintained as episomes. Thus, the gDNA sequence present in the silencing plasmids can be identified by PCR amplification and sequenced using oligonucleotides flanking the cloning site. Alternatively, silencing plasmid can be re-cloned in E. coli and sequenced. In order to validate this hypothetical forward genetic approach in Mucorales, five independent transformants (HGR1, HGR2, YLG1, SG1 and SG2) were selected for gene identification and validation of the silencing phenotype. Three transformants (HGR1, HGR2 and YLG1) were selected due to their avirulent phenotype, whereas the transformants presenting the satellite growing phenotype (SG1 and SG2) were selected to demonstrate that genes involved in other physiological processes can also be identified following this approach. Amplifications from gDNA of these five transformants generated PCR products only in YLG1, SG1 and SG2. After purification and sequencing of these PCR products, the DNA sequences were analyzed and compared to the genome database of M. circinelloides v1.0 and v2.0 (http://genome.jgi-psf.org/Mucci1/Mucci1.home.htmland http://genome.jgi-psf.org/Mucci2/Mucci2.home.html, respectively). The two strains sharing the satellite growth phenotype, SG1 and SG2, exhibited both equal size PCR products and DNA sequences, indicating that these two transformants harbored the same plasmid. The analysis of the sequence amplified from this plasmid revealed the presence of three different ORFs: ID 84675 (CLIP-associated proteins (CLASPs), v1.0), ID 156742 (intracellular protein transport, v2.0) and ID 145873 (DNA repair protein RAD51/RHP55, v2.0) (Table 2). The analysis of the sequence obtained from transformant YLG1 also unveiled a DNA insert containing two different ORFs: ID 51513 (myosin class V heavy chain, v1.0) and ID 166338 (no description in either v1.0 or v2.0). For the analysis of transformants HGR1 and HGR2, plasmid re-cloning in E. coli was required, revealing that both transformants shared a plasmid with the same insert sequence (pMAT1726). In this case, the sequence of the plasmid insert harbored only one ORF, ID 134906, which encoded a Phospholipase D like protein (v2.0). The analysis of the sequences found in the plasmids of the five selected transformants has been summarized in Table 2, which shows that six different candidate genes could be responsible for three selected phenotypes. In order to identify which genes are behind the phenotypes, a silencing validation experiment was performed for each of the six candidate genes. Five new silencing validation plasmids were engineered by cloning a 1 kb fragment of each candidate gene in the MCS of pMAT1700 (Table 2). After transformation of the recipient wild type strain with these five plasmids and pMAT1726, only three plasmids reproduced the three phenotypes previously observed in the original transformants obtained with the high-throughput silencing libraries (Table 2). Silencing of gene ID 84678 resulted in the satellite growing phenotype previously observed in the transformants SG1 and SG2, whereas the yeast like growth phenotype of YLG1 was reproduced only by silencing of gene ID 51513. As expected, silencing of the only gene (ID 134906) found in the transformants HRG1 and HRG2 resulted in the highly reduced growth phenotype.

Table 2: Gene identification and phenotypic validations

https://doi.org/10.1371/journal.ppat.1006150.t002

Strains	Candidate Genes	Description	Validation Plasmids	Phenotype
HIGHLY REDUCED GROWTH (HRG)				
HRG1 & HRG2	ID 134906[2]	Phospholipase D	pMAT1726	Highly reduced growth
YEAST LIKE GROWTH (YLG)				
YLG1	ID 51513[1]	Myosin class V heavy chain	pMAT828	Yeast like growth
	ID 166338[2]	(Unknown)	pMAT798	Wild type
SATELLITE GROWTH (SG)				
SG1 & SG2	ID 84675[1]	CLIP-associated proteins (CLASPs)	pMAT823	Satellite growth
	ID 156742[2]	Intracellular protein transport	pMAT824	Wild type
	ID 145873[2]	DNA repair protein RAD51/RHP55	pMAT825	Wild type

(1) Version 1.0 of *M. circinelloides* genome (http://genome.jgi-psf.org/Mucci1/Mucci1.home.html).
(2) Version 2.0 of *M. circinelloides* genome (http://genome.jgi-psf.org/Mucci2/Mucci2.home.html)

doi:10.1371/journal.ppat.1006150.t002

To confirm that the phenotypes obtained after the introduction of plasmids harboring sequences of genes IDs 84675, 51513 and 134906 are due to the lack of function of these genes through a canonical RNAi mechanism, we checked the mRNA levels and the production of siRNAs for the three candidate genes in both the original transformants containing the plasmids from the high-throughput RNAi libraries and the transformants obtained with the validation plasmids (Fig 4). All transformants for the three genes showed a reduction of mRNA levels and a production of siRNA for the corresponding gene, confirming the expected mechanism of action of the RNAi high-throughput library. These results demonstrated that the RNAi high-throughput library can be used as a new means to perform forward genetics and functional genomics in the study of virulence of Mucorales.

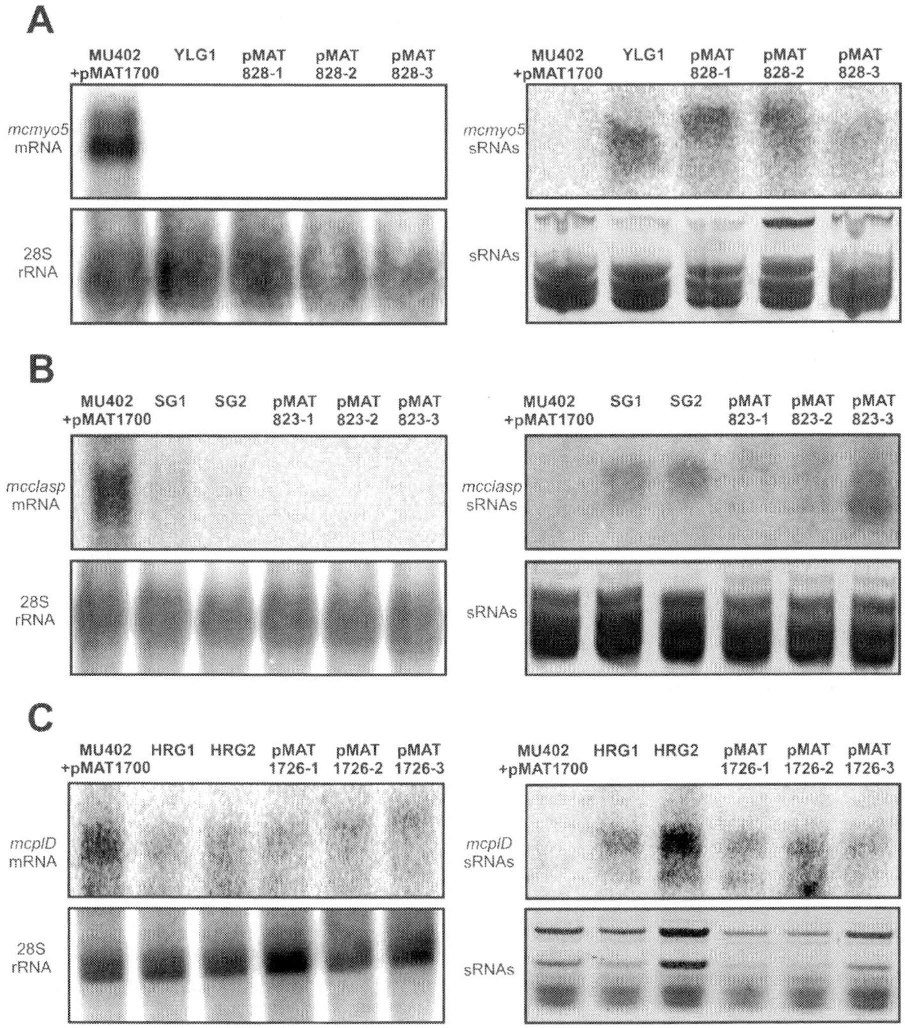

Figure 4: Validation of RNAi-mediated silencing in transformants with selected phenotypes.

(A) Levels of mcmyo5 mRNA and siRNAs in YLG1 original transformant and three transformants carrying a plasmid designed to specifically silence mcmyo5 (pMAT828). (B) Levels of mcclasp mRNA and siRNAs in SG1 and SG2 transformants, and three transformants carrying a plasmid designed to specifically silence mcclasp (pMAT823). (C) Levels of mcplD mRNA and siRNAs in HRG1 and HRG2 transformants, and three transformants carrying a plasmid designed to specifically silence mcplD (pMAT1726). The mRNA loading controls were performed by re-probing the membranes with a 28S rRNA probe, whereas siRNAs loading controls were obtained by Ethidium Bromide staining [21].

https://doi.org/10.1371/journal.ppat.1006150.g004

Disruption of Genes *mcmyo5* and *mcplD* induces Morphological Abnormalities and Avirulent Phenotypes in *M. circinelloides*

The approach of RNAi-based functional genomics in M. circinelloides resulted in the identification of two genes that could be new virulence determinants in Mucorales (Fig 3). The role of these two genes was confirmed through the generation of the corresponding knockout strains and the study of their phenotype and virulence in a heterologous host model. The gene ID 51513 (v2.0) encodes a Myosin class V protein that contains the three characteristic domains of this protein family: a motor domain, the IQ motifs and the cargo-binding globular tail. Thus, the gene encoding M. circinelloides Myosin 5 was denominated mcmyo5. In order of adding more evidence to the identity of mcmyo5 gene, we performed a detailed phylogenetic analysis that included myosin proteins identified in other fungi. This analysis revealed that gene mcmyo5 encodes a myosin protein that is perfectly clustered among other fungal myosin 5 proteins. The second gene, ID 134906 (v2.0), encodes a Phospholipase D like protein (accordingly denominated mcplD), that contains the characteristic domains (C2, PX, PH), the active site and other functionally important parts of the enzyme [37]. Similarly to mcmyo5, we performed a detailed phylogenetic analysis that included phospholipase proteins identified in other fungi. This analysis revealed that gene mcplD encodes a phospholipase protein that is perfectly clustered among other fungal phospholipases type D. In addition, the gene ID 84675 (v1.0) was also mutated, as mentioned above, to prove that the strategy presented in this work is also valid to study other fungal processes different than virulence, and also as a control to prove that not all growth defects are related to reduced virulence. This gene, named as mcclasp, encodes a CLIP-associated protein like (CLASPs), as the CLASP N-terminal domain is the main conserved region, which shares an 87% identity with a hypothetical CLASP protein of Mucor ambiguous. The disruptions of these three genes were carried out through the construction of knockout vectors designed to replace each candidate gene with pyrG gene, which was used as a selective marker. These knockout vectors contained an engineered cassette with adjacent regions of the target genes flanking the pyrG gene and were used to transform MU402 strain (pyrG$^-$, leuA$^-$). After transformations, candidates presenting the phenotype previously associated with silencing of each gene were isolated and the disruptions analyzed by Southern blot analysis. The two transformants selected from mcplDdisruption (MU466 and MU467) and the transformant selected from mcclasp disruption (MU464) only showed the DNA fragments corresponding to the correct integration of the disruption fragment at the corresponding loci, indicating that they were homokaryons for the mutant allele. In order to confirm the identity of the product encoded by the gene mcplD, activity of the enzyme PLD was measured in the mutant ΔmcplD and compared to the wild type strain ("Phospholipase D Assay Kit", from Sigma-Aldrich). This assay showed a significant reduction ($p = 0.0017$) of PLD activity of almost 30% in the mutant strain, but not a total lack of PLD activity. These results could be explained if there are other proteins with similar activity in the crude extracts of this mutant. Regarding the deletion of mcmyo5 gene, two transformants showing the yeast-like growth phenotype were selected after transformation with a replacement cassette for the gene mcmyo5. One of these transformants, MU468, probably harbored a chromosomal rearrangement at the mcmyo5 locus, whereas the second one, MU465, showed the correct pyrG insertion of 4.1 kb replacing mcmyo5 gene. However, it was impossible to obtain a homokaryotic knockout strain for the gene mcmyo5, as the transformant containing the mutant allele maintained some wild type nuclei even after ten vegetative cycles on selective media. These results suggested that mcmyo5 gene may play an essential role in the viability of M. circinelloides and a homokaryotic state of the mutant nuclei might be lethal. The heterokaryotic strain containing mcmyo5 mutant nuclei was named Δmcmyo5$^{(-)(+)}$.

The phenotype of each knockout strain was equivalent to those observed in the silencing transformants (Fig 5A). The three mutants showed a reduction of growth and sporulation rates, as well as an increase in the production of β-carotene (Fig 5C, 5D and 5E, respectively). The accumulation of β-carotene in the three mutants might be due to the growth stress present in these strains, since diverse stress factors have been previously linked to the production of β-carotene in other organisms [38]. Regarding virulence, infection assays in G. mellonella larvae with sporangiospores from mutants ΔmcplD and Δmcclasp, and yeast cells from mutant Δmcmyo5$^{(-)(+)}$, revealed a significant reduction in virulence of mutants Δmcmyo5$^{(-)(+)}$ and ΔmcplD but not in Δmcclasp ($p = 0.0007$, $p = 0.0002$ and $p = 0.4420$, respectively) (Fig 5B), as expected from the results previously obtained with the strains containing silencing vectors (Fig 3). The moth G. mellonella is a convenient model to study virulence when numerous candidates have to be tested. However, the immune system of this invertebrate model presents several differences compared to vertebrates, especially with warm blooded animals like mammals. Thus, the avirulent phenotype of mutants Δmcmyo5$^{(-)(+)}$ and ΔmcplD was tested in a mouse model, where temperature and immune system components and action mechanisms are similar to humans. Yeast cells and spores of mutants Δmcmyo5$^{(-)(+)}$ and ΔmcplD (respectively) were injected in immunodepressed mice and survival was daily monitored during twenty days after the infection with inocula of both $1x10^5$ and $1x10^6$ (Fig 6). Both inocula generated similar results, confirming the avirulent phenotype of mutant ΔmcplD in the murine model with a strong statistical significance ($p = 0.0081$ and $p = 0.0065$ in Fig 6A). However, although a reduced virulence was also observed for the heterokaryotic strain Δmcmyo5$^{(-)(+)}$compared to the wild type R7B strain, difference was not statistically significant ($p = 0.1595$ and $p = 0.0526$ in Fig 6B). This was probably due to the long time-course of the virulence assay in mice, since heterokaryotic Δmcmyo5$^{(-)(+)}$ cells might segregate to a wild type phenotype by losing mutant nuclei when grown under non-selective conditions. In fact, growing of mutant Δmcmyo5$^{(-)(+)}$ in non-selective culture medium gave rise to patches of wild-type phenotypes after few days of incubation. To test this hypothesis, retrieved CFUs from infected organs of both agonizing mice that showed signs of an imminent death and apparently healthy mice were analyzed in a Southern blot assay that distinguishes between the mutant and wild type genotypes. Quantification of the proportion of wild type and mutant nuclei in these retrieved CFUs showed correlation between the segregation to wild type genotype and the restitutions of virulence, which supported the role of the gene mcmyo5 in the pathogenesis of M. circinelloides. In order of acquiring more insights about the virulence of the strains ΔmcplD and Δmcmyo5, the fungal burden was quantified in the relevant organs of mice infected with wild type R7B and both mutant strains. Quantification of fungal gDNA on relevant target organs (brain and lung) revealed prevalence of R7B in tissues from mice infected with both yeast and spore forms, showing a more significant presence in lung tissues (Fig 6C and 6D).

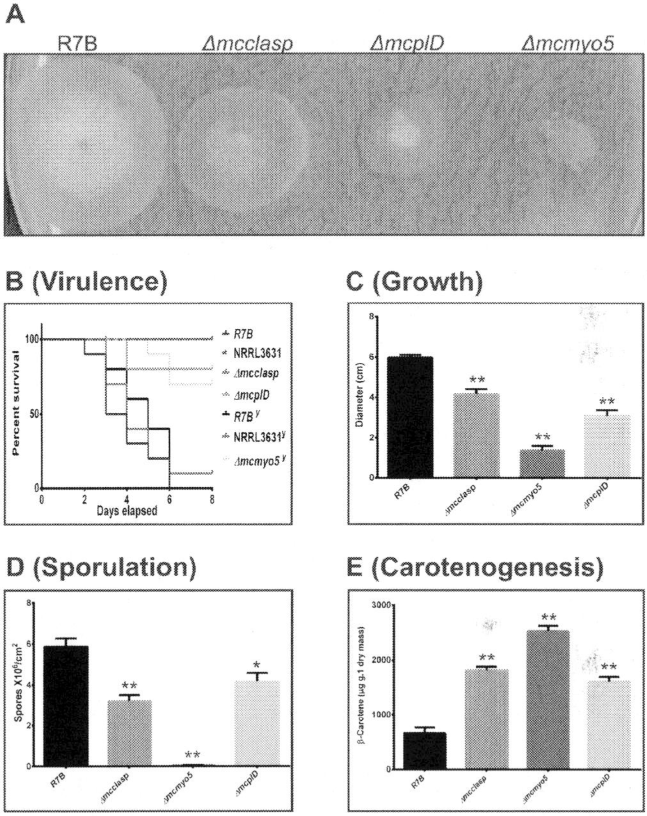

Figure 5: Phenotypic analyses of mcclasp, mcmyo5 **and** mcplD **knockout mutants.**

(A) Colonies of the three mutants and a wild type strain (R7B) cultivated in rich medium MMC at pH 4.5 for 48 h. (B) Virulence assays using spores of mcclasp and mcplDknockout mutants, and yeast cells from the mcmyo5 (y) heterokaryon mutant in the heterologous host model G. mellonella. NRRL3631 and R7B (spores and yeast cells) were used as avirulent and virulent control strains, respectively. Quantification of growth rate (C), vegetative sporulation (D) and β-carotene production (E) was performed in the same growth conditions as in (A). Values are means and standard errors of 10 independent experiments.

https://doi.org/10.1371/journal.ppat.1006150.g005

In particular, the presence of R7B in lung tissue was higher at day 2 post infection (Fig 6C) than at five days (Fig 6D), indicating a decrease of fungal biomass in mice over time. Such fungal burden decrease was less accentuated on infection with R7B yeasts, suggesting that this fungal form is more persistent in mice during the infection progression. Despite these differences in fungal load, symptoms and mortality rates were similar after infection with R7B spores and yeasts (Fig 6A and 6B). Conversely, a low amount of fungal DNA in mice infected with NRRL3631 and mutant strains ΔmcplD and Δmcmyo5 was detected, even below the limit of detection (0.005 ng) after five days of infection (Fig 6C and 6D). These results, together with survival outcomes, indicated a greater capacity of R7B strain to infect and invade mice tissues, causing higher mortality rates than the mutants ΔmcplD and Δmcmyo.

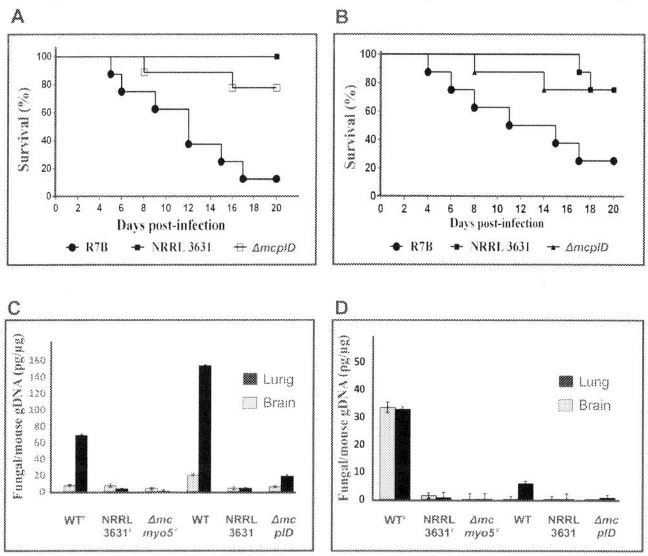

Figure 6: Virulence tests of the ***ΔmcplD* and *Δmcmyo5*$^{(-)(+)}$ mutants of** M. circinelloides **in a murine host model.**

(A) Virulence assays using spores of wild type strains and the mutant ΔmcplD. Mice were injected with 1×10^6 sporangiospores. (B) Virulence assays using yeast cells of wild type strains and the mutant Δmcmyo5$^{(-)(+)}$. Injections contained 1×10^6 yeast cells. (C) Quantification analyses of fungal DNA in mice target organs after two days post infection. (D) Quantification analyses of fungal DNA in mice target organs after five days post infection. Infections were carried out with yeast (y) or spore inocula forms of the different strains of M. circinelloides.

https://doi.org/10.1371/journal.ppat.1006150.g006

Delayed Germination and reduced Polar Growth as New Virulence Determinants in Mucormycosis

The determinants of virulence in M. circinelloides have been studied during the initial interaction of spores with macrophages, in which the main factor distinguishing virulent and avirulent strains was the size of the spores [14]. Therefore, spore and yeast cell sizes of virulent and avirulent strains were determined. The size of yeast cells produced by the avirulent control strain NRRL3631(+) was pronouncedly reduced compared with yeasts produced by virulent control strain R7B(-), in the same manner as occurred with the size of the spores [14]. However, the sizes of the spores or yeast cells of the mutant strains ΔmcplD, Δmcmyo5$^{(-)(+)}$ and Δmcclasp were not significantly reduced when compared to the virulent strain R7B. These results suggested that the reduction of virulence in the strains ΔmcplD and Δmcmyo5$^{(-)(+)}$ might be due to other factors that are independent of the initial size of the fungal spore or yeast cells. In order to find these factors, the interaction between macrophages and the mutant strains ΔmcplD and Δmcmyo5$^{(-)(+)}$ was also studied. Spores and yeast cells (from ΔmcplD and Δmcmyo5$^{(-)(+)}$, respectively) were co-cultured with the mouse macrophage cell line J774A.1 (ATCC, TIB-67), during four hours. At this time of interaction, all the spore/yeast cells have been phagocytized by macrophages and virulent strains initiate germination and polar growth trying to escape before being inactivated [14]. Thus, we quantified the germination rate and polarity index (a quotient between cell length and cell width [39]) as a measure of virulence of the different strains tested here. A germination delay and a reduced polarity index were observed in mutants ΔmcplD and Δmcmyo5$^{(-)(+)}$ relative to wild type (Fig 7). Mutant ΔmcplD presented a germination delay and polarity index similar to the avirulent strain NRRL3631, whereas mutant Δmcmyo5$^{(-)(+)}$ showed a reduction of the polarity index even more pronounced than the avirulent control strain, as well as a similar germination delay. Mutants grown in absence of macrophages also presented the same delay in germination and polar growth. The knockout strain in the mcclasp gene showed a non-significant reduction of the polarity index and no changes in the germination rate when compared to the wild type. As mutant Δmcclasp is not affected in virulence, these results highlight the relevance of spore germination and hyphal growth rates within macrophages for M. circinelloides virulence.

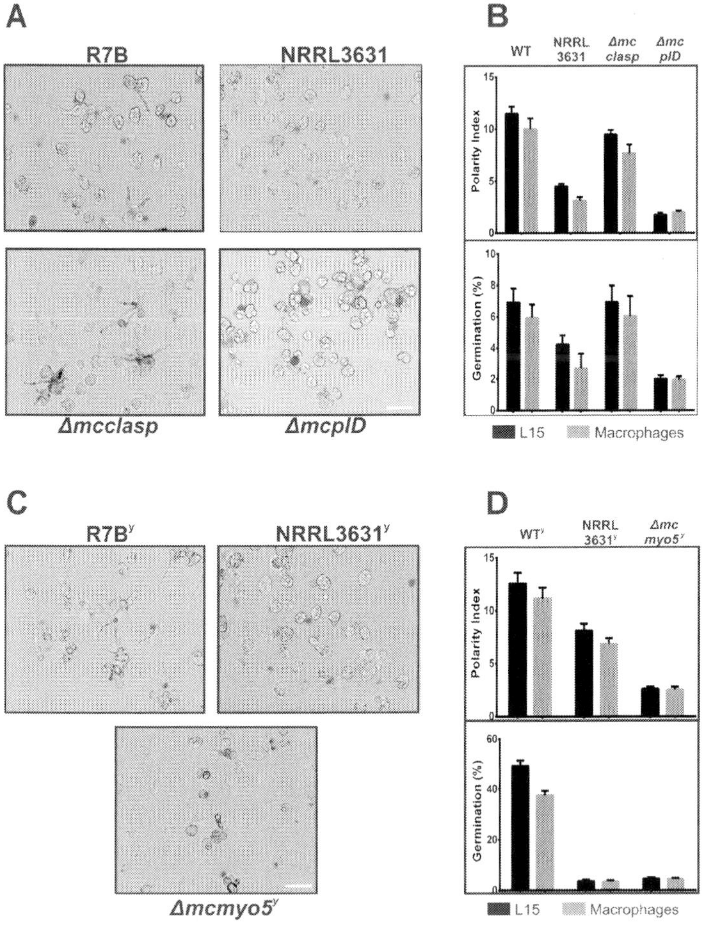

Figure 7: Interaction of macrophage cells and mcclasp, mcmyo5 **and** mcplD **knockout mutants.**

(A) Representative optical microscopic images of the interaction between macrophage cells (J774A.1) and spores of mcclasp and mcplD mutants. Scale bars represent 50μm. (B) Quantification of germination and polar index of mcclasp and mcplD mutant spores after four hours of interaction in the presence or absence of macrophage cells. (C) Representative optical microscopic images of the interaction between macrophage cells and yeast cells of the mcmyo5 knockout mutant. (D) Quantification of germination and polarity index of mc-myo5 mutant yeast cells after four hours of interaction with macrophage cells. NRRL3631 and R7B were used as avirulent and virulent control strains, respectively. Values are means and standard errors of 500 independent measurements.

https://doi.org/10.1371/journal.ppat.1006150.g007

DISCUSSION

Here, we have developed a new approach based on RNAi high-throughput libraries that allows the identification of genes responsible for virulence in M. circinelloides. This work represents the first application of a functional genomic approach to identify virulence determinants in Mucorales. RNAi-based reverse genetics has allowed successful whole-genome functional studies in animals (see Introduction), but this technology presents several restrictions in fungi that have prevented its use at whole genome level. Following this reverse genomic approach, the only known study carried out in fungi was firstly reported in a plant pathogen, Magnaporthe oryzae, in which the function of a calcium-signaling family of 37 genes was studied using silencing plasmids for each gene of this family [40]. Our platform presents an opposite approach in which a collection of phenotypes are generated by transformation with a whole-genome RNAi library and afterward the genes responsible for a particular phenotype are identified. This approach represents the first forward genetic strategy to study gene function at the whole genome level in Mucorales. Following this strategy, a general screening of a collection of silenced M. circinelloides transformants led us to the isolation of twenty-six strains showing a wide range of distinct phenotypes in fungal processes such as virulence, growth and sporulation. One advantage of this approach versus traditional chemical or insertional mutagenesis is that essential genes could be isolated, since RNAi usually reduces the expression of the target gene rather than a total inhibition. Another advantage of this RNAi-based functional genomic platform is the possibility of designing conditional screenings. The screenings shown here were performed under non-specific growth conditions, as it was intended to prove the general utility of the platform to apply functional genomics and to show how genes related to virulence can be identified from general screenings. However, the collection of silenced transformants offers the opportunity to carry out specific screenings under particular conditions to select phenotypes related to virulence, such as yeast growth, thermotolerance, protease over-production, etc. Moreover, another attractive advantage of this approach is its exportability to other Mucorales or fungi from different groups like Ascomycota and Basidiomycota, since the only conditions required are the existence of a functional RNAi mechanism and an efficient transformation method, which are both present in many fungal groups [41].

The application of the RNAi-based functional genomic platform has facilitated the identification of the genes mcplD as an essential factor to maintain full virulence in M. circinelloides, both in a heterologous model like G. mellonella and a murine host model. Knockout mutants in mcplDshowed deficient growth accompanied with sporulation reduction and increased production of β-carotene, and more importantly, reduced virulence in the host models G. mellonella and M. musculus. The gene mcplD codes for a Phospholipase D enzyme (PLD), a well-known protein that is highly conserved in different organisms. This enzyme catalyzes the hydrolysis of the phosphodiester bond of glycerophospholipids to generate phosphatidic acid and a free headgroup [37]. Phosphatidic acid functions as an intracellular lipid messenger that activates different target kinases, which in turn activate a broad range of cellular processes such as receptor signaling, control of intracellular membrane transport, and reorganization of the actin cytoskeleton [37]. This pleiotropic function of PLD could explain the complex phenotype observed in the M. circinelloides mutant for the mcplD gene. In addition, PLD has been described as a major virulence factor in Corynebacterium pseudotuberculosis, being involved in macrophage death and systemic dissemination of this pathogen [42]. In fungi, Aspergillus fumigatus PLD regulates its internalization into lung epithelial cells, and the pld gene of Purpureocillium lilacinum is significantly up-regulated during infection of Meloidogyne incognitaeggs [43, 44]. In M. circinelloides, mcplD may regulate some signaling pathways involved in germination and hyphal growth, since mutants in this gene showed delayed germination and reduction of the polarity index. The other gene related to virulence that has been found in this study, mcmyo5 encodes a processive cargo transporter belonging to the Myosin V Class (Myo5). Myosins play important roles in morphogenesis of filamentous fungi, since they are involved in the establishment and/or maintenance of polarity [45]. Among Myosins, Myo5 supplies a constant transport of organelles, membranous cargo, secretory vesicles, mRNA, lipid and protein vesicles on actin tracks [46]. In M. circinelloides, Myo5 might play an essential role in viability, since the knockout mutant was viable only as a heterokaryon containing a small proportion of wild type nuclei. This heterokaryotic strain was unable to produce regular hyphae and presented a yeast-like phenotype with no polar growth. Likely, the lack of a continuous transport mediated by Myo5 impairs the correct formation of the mycelium. Since filamentous growth is a major determinant of virulence in M. circinelloides [13], the yeast-like knockout strain Δmcmyo5$^{(-)(+)}$ presented a pronounced reduced virulence in G. mellonella. Similarly, in the dimorphic plant pathogenic fungus Ustilago maydis, a single Myosin class V protein encoded by myo5 was involved in hyphal growth and pathogenicity [47]. However, the reduction of virulence shown by the Δmcmyo5$^{(-)(+)}$ heterokaryotic mutant in a murine host model did not reach the significance level stablished, although the reduction in the fungal burden was similar to the mutant mcplD. These partially contradictory results obtained from the two host models could be precisely due to the heterokaryotic state of mutant Δmcmyo5$^{(-)(+)}$. The virulence assays in G. mellonella are performed during eight days, whereas in M. musculusthe assays are prolonged until the twentieth day, which could be time enough for the heterokayotic strain under non selective conditions to segregate and lose mutant nuclei, reverting to the wild type phenotype. According to this hypothesis, wild type patches segregating from Δmcmyo5$^{(-)(+)}$ mutant cells growing under non selective culture conditions are easily observed after several days of incubation. In addition, a correlation between the segregation to wild type genotype and the restitutions of virulence was observed after genotyping the nuclei proportion of several retrieved CFUs, which further supported a role of gene mcmyo5 in the virulence of M. circinelloides. Similar results were obtained in Rhizopus oryzae when FTR1 gene was disrupted by double cross-over homologous recombination, but multinucleated R. oryzae could not be forced to segregate to a homokaryotic null allele [48]. The heterokaryotic strain Δmcmyo5$^{(-)(+)}$ showed the strongest reduction of polar growth and germination rates when phagocyted by macrophages, since it was tested during only four hours in non-selective medium, which is not time enough for segregation. Our results from the in vitroanalysis and the intravenous infection model showed the potential role of mcmyo5 in virulence of M. circinelloides. Although Myo5 is a highly conserved protein, which disqualifies it as a specific antifungal target, the fungal cargo domain and the proteins that interact with this domain could represent a promising target for future antifungal developments.

The analysis of germination and polar growth of Δmcmyo5$^{(-)(+)}$ and ΔmcplD mutants and their interaction with mouse macrophages revealed a delayed germination and reduced polarity index of those strains relative to the wild type strain, although the size of the infecting particles (spores or yeast cells) was not reduced in these mutants. These results suggested that a big size of the infecting particle (spore or yeast cells) is not enough to counteract a delayed germination and reduced polarity index. Time of germination and polarity index are two values that measure the velocity of the pathogen growing inside the macrophage and escaping from it. Thus, a possible explanation of the reduced virulence observed in ΔmcplD and Δmcmyo5$^{(-)(+)}$mutants might be that delayed germination and reduced polarity index concede macrophages time enough to inactivate the pathogen before it escapes from its cytoplasm. Along with the size of the spore previously described [14], our work demonstrated that the time required for germination and the hyphal elongation rate, measured as polarity index, are two new factors to be considered in the analysis of M. circinelloides virulence.

Besides genes related to virulence, a third gene named mcclasp was selected from the RNAi-based functional genomic screenings for further studies. Mutant Δmcclasp presented a complex phenotype affecting several fungal processes like vegetative growth, carotene production and sporulation in a similar manner to the Δmcmyo5$^{(-)(+)}$ and ΔmcplD mutants, although the Δmcclasp strain was not affected in virulence. Besides that, the main differences between Δmcclasp and those mutants were the formation of sporangiophores, which were normal in length in Δmcclasp mutants, allowing the formation of satellite colonies in low pH media, along with a less pronounced reduction of growth and sporulation. The gene product of mcclasp is similar to CLASP proteins that are involved in the attachment of microtubules to the cell cortex in animals and plants, thereby contributing to self-organization of cortical microtubules [49]. During mitosis, CLASP proteins control the interactions of astral microtubules with the cell cortex, helping the proper positioning and orientation of the spindle [50]. This important role of CLASP proteins during cell division might be behind the reduced growth and decreased sporulation observed in the Δmcclasp mutants. In addition, the role of CLASP proteins in the stability of microtubules is essential for the motility of motor proteins, such as kinesins and dyneins. Kinesins participate in the maintenance of the polarity of filamentous fungi (reviewed by Harris, 2006), which could explain the reduction in the polarity index of the Δmcclasp mutant relative to the wild type strain. However, unlike Myo5, kinesins does not seem to be involved in polarity establishment in M. circinelloides, since strains without CLASP protein are still able to generate hyphal growth, although their elongation rate is lower than the wild type strain. Mutant Δmcclasp showed a reduced growth rate and vegetative sporulation and an increase in β-carotene production compared to control strain R7B (Fig 5C, 5D and 5E, respectively), similarly to the phenotypes observed in ΔmcplD and Δmcmyo5 mutants. However these phenotypes are not associated with reduced virulence in mutant Δmcclasp, indicating that growth defects are not necessarily linked to attenuated virulence and suggesting a possible specific role of mcplDand mcmyo5 genes in pathogenesis. The identification and analysis of mcclasp gene demonstrated that along with virulence, other fungal processes can be studied and genetically dissected with the RNAi-based functional genomic platform developed in this work.

In summary, the absence of classic molecular genetic tools and the scarce information about virulence and pathogenesis in Mucorales encouraged us to develop a robust system for RNAi-based functional genomics in M. circinelloides. It is a new genetic tool that can be used in the study of a wide range of biological processes, including the identification and study of genes related to virulence and pathogenesis. As a result of its implementation, we have identified new virulence determinants in Mucorales that could represent new targets for future antifungal therapies.

MATERIALS AND METHODS

Strains, Growth and Transformation Conditions

The leucine auxotroph R7B, derived from the (-) mating type M. circinelloides f. lusitanicus CBS 277.49 (syn. Mucor racemosus ATCC 1216b), was used as the wild type strain. Strain MU402 is a uracil and leucine auxotroph derived from R7B used as recipient strain of the silencing library [36]. The M. circinelloides f. lusitanicus strain of the (+) mating type NRRL3631 was used in virulence assays as an avirulent control.

M. circinelloides cultures were grown at 26°C in complete YPG medium or in MMC medium as described previously [36]. Media were supplemented with uridine (200 μg/ml) when required. The pH was adjusted to 4.5 and 3.2 for mycelial and colonial growth, respectively. Transformation was carried out as described previously [10]. Macrophage cells, J774A.1 (ATCC, TIB-67), were cultured in L15 medium (Capricorn Scientific GmbH) supplemented with 10% FBS at 37°C and without CO_2 supplementation.

Plasmids and Genomic Libraries

Plasmid pMAT1726 was recovered from gDNA of HRG1 and HRG2 transformants. In order to construct dsRNA-expressing vectors with the target candidate genes, plasmid pMAT1700 was used as cloning vector. Insert fragments corresponding to the 5' end of each candidate gene (0.5–2 kb) were amplified with primers containing NotI and XhoI restriction sites to facilitate cloning into pMAT1700. Plasmid pMAT828 harbors a 2 kb fragment of gene ID 51513 which was PCR-amplified using primer pairs FYL1 and RYL1. Plasmid pMAT798 contains a 0.9 kb fragment of gene ID 166338 amplified by PCR reactions using primer pairs FYL1.2 and RYL1.2. To identify the candidate gene responsible for the satellite growth phenotype, three different plasmids were constructed: pMAT823, pMAT824 and pMAT825. These plasmids contain 1.2 kb, 0.9 kb and 0.5 kb fragments corresponding to genes ID 84675, ID 156742 and ID 145873, which were amplified using primer pairs FYL10.1/RYL10.1, FYL10.2/RYL10.2 and FYL10.3/RYL10.3, respectively. To calculate the coverage of the genomic libraries in the genome of M. circinelloides and confidence levels, we followed the formula: N = ln(1-P)/ln(1-f); where N is the necessary number of recombinants, P is the desired probability that any fragment of the genome is represented in the library at least one time, and f is the fractional proportion of the genome in a single recombinant. "f" can be further shown to be f = i/g, where i is the insert size and g is the genome size [51].

To disrupt the candidate genes, a pyrG selective marker (2 kb fragment amplified from gDNA using primers F-pyrG and R-pyrG was fused with adjacent sequences of the candidate coding regions using fusion PCRs, generating a gene replacement fragment. This fragment was cloned into pGEMT-easy vector (Promega) and used to disrupt the candidate genes via homologous recombination. Plasmid pMAT833 was constructed to disrupt the mcclasp gene. It contains a 4.2 kb fragment that includes the pyrG gene flanked by 1.3 kb of upstream and downstream sequences of mcclasp gene, amplified with primers FYL10U/RYL10-pyrG and FYL10-pyrG/RYL10D, respectively. The 4.2 kb fusion fragment was amplified with internal primers FYL10 and RYL10. Plasmid pMAT832 was constructed to disrupt mcmyo5 gene, following the same strategy described for pMAT833, but using primers FYL1U/RYL1-pyrG and FYL1-pyrG/RYL1D to amplify 1.3 kb of upstream and downstream sequences of the mcmyo5 gene, respectively. In the case of gene mcplD, plasmid pMAT1733 was constructed also following the same fusion strategy, but with the specific primers FPLD/RPLD-pyrG and RPLD/FPLD-pyrG, for two fragments of 0.95 kb from upstream and downstream regions of mcplD, respectively.

Nucleic Acid Manipulation and Analysis

Genomic DNA from M. circinelloides mycelia was extracted as previously described [36]. Recombinant DNA manipulations were performed as reported [52]. Total RNA was extracted from mycelia grown during 48 hours at 26°C in liquid MMC pH 4.5 medium under light conditions using RNeasy Plant Mini Kit following the supplier's recommendation (Qiagen). Southern blot and Northern blot hybridizations were carried out under stringent conditions [22]. DNA probes were labeled with [α-32P] dCTP using Ready-To-Go Labeling Beads (GE Healthcare Life Science). For Southern and Northern blot experiments, DNA probes were directly amplified from genomic DNA using the primer pairs FPLD/RPLD-pyrG, FYL10/RYL10-pyrG and FYL1N/RYL1-pyrG for genes mcplD, mcclasp and mcmyo5, respectively. For siRNA analysis, small RNA samples were extracted from mycelia grown 72 hours in liquid MMC medium pH 4.5 at 26°C using the miRVana kit (Ambion), following the instructions of the supplier. Northern blots for siRNAs were performed as previously described using antisense specific riboprobes generated by in vitro transcription of the DNA probes described above (MAXIscriptsT7, Ambion) [21]. Quantifications of signal intensities were estimated from autoradiograms using a Shimadzu CS-9000 densitometer and the ImageJ application (rsbweb.nih.gov/ij/). Computational phylogenetic analyses were performed using Phylogeny software (http://phylogeny.lirmm.fr) [53]. Multiple protein sequence alignments were conducted with ClustalW program and phylogenetic trees were inferred by maximum likelihood statistical methods using a bootstrapping procedure of 1000 iterations.

Phenotypic Analysis

Vegetative sporulation, growth rate, carotene production and virulence measurements were carried out as previously described [14, 19, 36, 54]. Interactions between different strains of M. circinelloides and J774A.1 macrophage cells were carried out in L15 medium, during 4 hours at 37°C. Regarding polarity index, ten images were taken from each interaction and a total of fifty germinating spores were measured from each image with ImageJ [55]. From the same images, germination was calculated considering germinated spores all those that presented a protuberant bud from the spherical spore. For the phospholipase D activity measurements, spores of the wild type strain and mutant mcplD were grown in MMC medium at pH = 4.5 during six hours. The mycelia from five biological replicates were filtrated, washed and weighted to perform the assay with exactly 100 mg of biomass from each strain.

The virulence assays in G. mellonella were performed by injection of 5 μl of phosphate buffered saline (PBS) containing 2000 spores or 20,000 yeast cells into the wax moth larvae (10 per strain). For the murine host model, groups of 8 four-week-old OF1 male mice (Charles River, Criffa S.A., Barcelona, Spain) weighing 30 g were used. Mice were immunosuppressed 2 days prior to the infection by intraperitoneal (i.p.) administration of 200 mg/kg of body weight of cyclophosphamide and once every 5 days thereafter. Animals were housed under standard conditions with free access to food and water. Mice were challenged intravenously (i.v.) via the lateral tail vein with a suspension consisting on 1×10^5 sporangiospores or 1×10^5 yeast cells per animal. Animals were checked twice daily for 20 days. Surviving animals at the end of the experimental period or those meeting criteria for discomfort were euthanized by CO_2 inhalation. Significance of mortality rate data was evaluated by using the Kaplan-Meier (Graph Pad Prism 4.0 for Windows; GraphPad Software, San Diego California USA). Differences were considered statistically significant at a P value of <0.05.

Fungal Burden Quantification and Genotyping

For absolute DNA quantification and genotyping, organs were ground up on liquid nitrogen and gDNA was extracted as previously described [56]. For DNA quantification by real-time PCR (qRT-PCR) specific primers of M. circinelloides chitin synthase gene (ID153118) and mice β2 microglobulin gene (ID12010) were used. Samples analyses were carried out in triplicate in 15 μl PCR reactions containing 180 ng of test sample gDNA form three individuals using SybrGreen kit (Fast SYBR Green Master Mix -ABI) in a StepOne Real-Time PCR System (ABI). gDNA from non-infected mice was used as negative control. Relative amount of fungal and mice gDNA was quantified on the basis of their standard curves, elaborated with known fungal DNA concentrations (0.005 ng—10 ng) in a background of 150 ng of non-infected mice gDNA and mice DNA concentrations (1 ng—200 ng) and their corresponding amplification cycle threshold (Ct).

ETHICS STATEMENT

Animal care procedures were supervised and approved by the Universitat Rovira i Virgili Animal Welfare and Ethics Committee. The experimental animal facilities are registered under reference T9900003 of the Generalitat de Catalunya in compliance with the regulations of Real Decreto 53/2013, of February 1st (BOE of 8 February). Procedures included into the project number 280 were supervised and approved by L. Loriente Sanz (ID 39671243) of the Veterinary and Animal Welfare

Advisory of the Universitat Rovira i Virgili Animal Welfare and Ethics Committee (Reus, Spain).

ACKNOWLEDGMENTS

We thank the Joint Genome Institute for early access to the M. circinelloides genome sequence.

REFERENCES

1. Kennedy KJ, Daveson K, Slavin MA, van Hal SJ, Sorrell TC, Lee A, et al. Mucormycosis in Australia: contemporary epidemiology and outcomes. Clin Microbiol Infect. 2016.
2. Salas V, Pastor FJ, Calvo E, Alvarez E, Sutton DA, Mayayo E, et al. In vitro and in vivo activities of posaconazole and amphotericin B in a murine invasive infection by Mucor circinelloides: poor efficacy of posaconazole. Antimicrob Agents Chemother. 2012;56(5):2246–50. pmid:22290952
3. Biswas D, Kotwal A, Kakati B, Ahmad S. Amphotericin B Resistant Apophysomyces elegans Causing Rhino-oculo-Cerebral Mucormycosis in an Immunocompetent Host. J Clin Diagn Res. 2015;9(8):DD01–2. pmid:26435947
4. Gebremariam T, Wiederhold NP, Alqarihi A, Uppuluri P, Azie N, Edwards JE Jr. , et al. Monotherapy or combination therapy of isavuconazole and micafungin for treating murine mucormycosis. J Antimicrob Chemother. 2016.
5. Marty FM, Ostrosky-Zeichner L, Cornely OA, Mullane KM, Perfect JR, Thompson GR 3rd, et al. Isavuconazole treatment for mucormycosis: a single-arm open-label trial and case-control analysis. Lancet Infect Dis. 2016;16(7):828–37. pmid:26969258
6. Bal AM. Importance of identification of zygomycetes in the era of newer antifungal agents. Transpl Infect Dis. 2006;8(2):122–3. pmid:16734636
7. Kontoyiannis DP, Lewis RE. Invasive zygomycosis: update on pathogenesis, clinical manifestations, and management. Infect Dis Clin North Am. 2006;20(3):581–607, vi. pmid:16984870
8. Petrikkos G, Skiada A, Lortholary O, Roilides E, Walsh TJ,

Kontoyiannis DP. Epidemiology and clinical manifestations of mucormycosis. Clin Infect Dis. 2012;54 Suppl 1:S23–34.

9. Mignogna MD, Fortuna G, Leuci S, Adamo D, Ruoppo E, Siano M, et al. Mucormycosis in immunocompetent patients: a case-series of patients with maxillary sinus involvement and a critical review of the literature. Int J Infect Dis. 2011;15(8):e533–40. pmid:21764345
10. Gutierrez A, Lopez-Garcia S, Garre V. High reliability transformation of the basal fungus Mucor circinelloides by electroporation. J Microbiol Methods. 2011;84(3):442–6. pmid:21256886
11. Skory CD. Homologous recombination and double-strand break repair in the transformation of Rhizopus oryzae. Mol Genet Genomics. 2002;268(3):397–406. pmid:12436261
12. Ibrahim AS, Spellberg B, Walsh TJ, Kontoyiannis DP. Pathogenesis of mucormycosis. Clin Infect Dis. 2012;54 Suppl 1:S16–22.
13. Lee SC, Li A, Calo S, Heitman J. Calcineurin plays key roles in the dimorphic transition and virulence of the human pathogenic zygomycete Mucor circinelloides. PLoS Pathog. 2013;9(9):e1003625. pmid:24039585
14. Li CH, Cervantes M, Springer DJ, Boekhout T, Ruiz-Vázquez RM, Torres-Martínez SR, et al. Sporangiospore size dimorphism is linked to virulence of Mucor circinelloides. PLoS Pathog. 2011;7(6):e1002086. pmid:21698218
15. Liu M, Lin L, Gebremariam T, Luo G, Skory CD, French SW, et al. Fob1 and Fob2 Proteins Are Virulence Determinants of Rhizopus oryzae via Facilitating Iron Uptake from Ferrioxamine. PLoS Pathog. 2015;11(5):e1004842. pmid:25974051
16. Gebremariam T, Liu M, Luo G, Bruno V, Phan QT, Waring AJ, et al. CotH3 mediates fungal invasion of host cells during mucormycosis. J Clin Invest. 2014;124(1):237–50. pmid:24355926
17. Chibucos MC, Soliman S, Gebremariam T, Lee H, Daugherty S, Orvis J, et al. An integrated genomic and transcriptomic survey of mucormycosis-causing fungi. Nat Commun. 2016;7:12218. pmid:27447865
18. Lee SC, Li A, Calo S, Inoue M, Tonthat NK, Bain JM, et al. Calcineurin orchestrates dimorphic transitions, antifungal drug responses and host-pathogen interactions of the pathogenic mucoralean

fungus Mucor circinelloides. Mol Microbiol. 2015;97(5):844–65. pmid:26010100

19. Cervantes M, Vila A, Nicolas FE, Moxon S, de Haro JP, Dalmay T, et al. A single argonaute gene participates in exogenous and endogenous RNAi and controls cellular functions in the basal fungus Mucor circinelloides. PLoS One. 2013;8(7):e69283. pmid:23935973
20. de Haro JP, Calo S, Cervantes M, Nicolas FE, Torres-Martinez S, Ruiz-Vazquez RM. A single dicer gene is required for efficient gene silencing associated with two classes of small antisense RNAs in Mucor circinelloides. Eukaryot Cell. 2009;8(10):1486–97. pmid:19666782
21. Nicolas FE, Torres-Martinez S, Ruiz-Vazquez RM. Two classes of small antisense RNAs in fungal RNA silencing triggered by non-integrative transgenes. EMBO J. 2003;22(15):3983–91. pmid:12881432
22. Calo S, Nicolas FE, Vila A, Torres-Martinez S, Ruiz-Vazquez RM. Two distinct RNA-dependent RNA polymerases are required for initiation and amplification of RNA silencing in the basal fungus Mucor circinelloides. Mol Microbiol. 2012;83(2):379–94. pmid:22141923
23. Ruiz-Vazquez RM, Nicolas FE, Torres-Martinez S, Garre V. Distinct RNAi Pathways in the Regulation of Physiology and Development in the Fungus Mucor circinelloides. Adv Genet. 2015;91:55–102. pmid:26410030
24. Nicolas FE, Vila A, Moxon S, Cascales MD, Torres-Martinez S, Ruiz-Vazquez RM, et al. The RNAi machinery controls distinct responses to environmental signals in the basal fungus Mucor circinelloides. BMC Genomics. 2015;16:237. pmid:25880254
25. Nicolas FE, Moxon S, de Haro JP, Calo S, Grigoriev IV, Torres-Martinez S, et al. Endogenous short RNAs generated by Dicer 2 and RNA-dependent RNA polymerase 1 regulate mRNAs in the basal fungus Mucor circinelloides. Nucleic Acids Res. 2010;38(16):5535–41. pmid:20427422
26. Nicolas FE, Ruiz-Vazquez RM. Functional diversity of RNAi-associated sRNAs in fungi. Int J Mol Sci. 2013;14(8):15348–60. pmid:23887655

27. Trieu TA, Calo S, Nicolas FE, Vila A, Moxon S, Dalmay T, et al. A non-canonical RNA silencing pathway promotes mRNA degradation in basal Fungi. PLoS Genet. 2015;11(4):e1005168. pmid:25875805
28. Calo S, Shertz-Wall C, Lee SC, Bastidas RJ, Nicolás FE, Granek JA, et al. Antifungal drug resistance evoked via RNAi-dependent epimutations. Nature. 2014;513(7519):555–8. pmid:25079329
29. Hoffmann K, Pawlowska J, Walther G, Wrzosek M, de Hoog GS, Benny GL, et al. The family structure of the Mucorales: a synoptic revision based on comprehensive multigene-genealogies. Persoonia. 2013;30:57–76. pmid:24027347
30. Kamath RS, Fraser AG, Dong Y, Poulin G, Durbin R, Gotta M, et al. Systematic functional analysis of the Caenorhabditis elegans genome using RNAi. Nature. 2003;421(6920):231–7. pmid:12529635
31. DasGupta R, Kaykas A, Moon RT, Perrimon N. Functional genomic analysis of the Wnt-wingless signaling pathway. Science. 2005;308(5723):826–33. pmid:15817814
32. Berns K, Hijmans EM, Mullenders J, Brummelkamp TR, Velds A, Heimerikx M, et al. A large-scale RNAi screen in human cells identifies new components of the p53 pathway. Nature. 2004;428(6981):431–7. pmid:15042092
33. Velayos A, Blasco JL, Alvarez MI, Iturriaga EA, Eslava AP. Blue-light regulation of phytoene dehydrogenase (carB) gene expression in Mucor circinelloides. Planta. 2000;210(6):938–46. pmid:10872226
34. Werner E, Patel K, Holder AA. Construction of a library for sequencing long regions of malarial genomic DNA. Biotechniques. 1997;23(1):20, 2, 4. pmid:9232219
35. Kashem SW, Igyarto BZ, Gerami-Nejad M, Kumamoto Y, Mohammed J, Jarrett E, et al. Candida albicans morphology and dendritic cell subsets determine T helper cell differentiation. Immunity. 2015;42(2):356–66. pmid:25680275
36. Nicolas FE, de Haro JP, Torres-Martinez S, Ruiz-Vazquez RM. Mutants defective in a Mucor circinelloides dicer-like gene are not compromised in siRNA silencing but display developmental defects. Fungal Genet Biol. 2007;44(6):504–16. pmid:17074518
37. McDermott M, Wakelam MJ, Morris AJ. Phospholipase D. Biochem Cell Biol. 2004;82(1):225–53. pmid:15052340

38. Raja R, Hemaiswarya S, Rengasamy R. Exploitation of Dunaliella for beta-carotene production. Appl Microbiol Biotechnol. 2007;74(3):517–23. pmid:17225103
39. Schuster M, Treitschke S, Kilaru S, Molloy J, Harmer NJ, Steinberg G. Myosin-5, kinesin-1 and myosin-17 cooperate in secretion of fungal chitin synthase. EMBO J. 2012;31(1):214–27. pmid:22027862
40. Nguyen QB, Kadotani N, Kasahara S, Tosa Y, Mayama S, Nakayashiki H. Systematic functional analysis of calcium-signalling proteins in the genome of the rice-blast fungus, Magnaporthe oryzae, using a high-throughput RNA-silencing system. Mol Microbiol. 2008;68(6):1348–65. pmid:18433453
41. Choi J, Kim KT, Jeon J, Wu J, Song H, Asiegbu FO, et al. funRNA: a fungi-centered genomics platform for genes encoding key components of RNAi. BMC Genomics. 2014;15 Suppl 9:S14.
42. McKean SC, Davies JK, Moore RJ. Expression of phospholipase D, the major virulence factor of Corynebacterium pseudotuberculosis, is regulated by multiple environmental factors and plays a role in macrophage death. Microbiology. 2007;153(Pt 7):2203–11. pmid:17600064
43. Li X, Gao M, Han X, Tao S, Zheng D, Cheng Y, et al. Disruption of the phospholipase D gene attenuates the virulence of Aspergillus fumigatus. Infect Immun. 2012;80(1):429–40. pmid:22083709
44. Yang F, Abdelnabby H, Xiao Y. The role of a phospholipase (PLD) in virulence of Purpureocillium lilacinum (Paecilomyces lilacinum). Microb Pathog. 2015;85:11–20. pmid:26026833
45. Harris SD. Cell polarity in filamentous fungi: shaping the mold. Int Rev Cytol. 2006;251:41–77. pmid:16939777
46. Reck-Peterson SL, Provance DW Jr. , Mooseker MS, Mercer JA. Class V myosins. Biochim Biophys Acta. 2000;1496(1):36–51. pmid:10722875
47. Weber I, Gruber C, Steinberg G. A class-V myosin required for mating, hyphal growth, and pathogenicity in the dimorphic plant pathogen Ustilago maydis. Plant Cell. 2003;15(12):2826–42. pmid:14615599
48. Ibrahim AS, Gebremariam T, Lin L, Luo G, Husseiny MI, Skory CD, et al. The high affinity iron permease is a key virulence

factor required for Rhizopus oryzae pathogenesis. Mol Microbiol. 2010;77(3):587–604. pmid:20545847

49. Ambrose JC, Wasteneys GO. CLASP modulates microtubule-cortex interaction during self-organization of acentrosomal microtubules. Mol Biol Cell. 2008;19(11):4730–7. pmid:18716054
50. Espiritu EB, Krueger LE, Ye A, Rose LS. CLASPs function redundantly to regulate astral microtubules in the C. elegans embryo. Dev Biol. 2012;368(2):242–54. pmid:22613359
51. Clarke L, Carbon J. A colony bank containing synthetic Col El hybrid plasmids representative of the entire E. coli genome. Cell. 1976;9(1):91–9. pmid:788919
52. Sambrook J and Russell D. W. . Molecular Cloning: A Laboratory Manual: Cold Spring Harbor Laboratory Press; 2001.
53. Dereeper A, Guignon V, Blanc G, Audic S, Buffet S, Chevenet F, et al. Phylogeny. fr: robust phylogenetic analysis for the non-specialist. Nucleic Acids Res. 2008;36(Web Server issue):W465–9. pmid:18424797
54. Nicolas FE, Calo S, Murcia-Flores L, Garre V, Ruiz-Vazquez RM, Torres-Martinez S. A RING-finger photocarotenogenic repressor involved in asexual sporulation in Mucor circinelloides. FEMS Microbiol Lett. 2008;280(1):81–8. pmid:18194338
55. Schneider CA, Rasband WS, Eliceiri KW. NIH Image to ImageJ: 25 years of image analysis. Nat Methods. 2012;9(7):671–5. pmid:22930834
56. Fernandez-Molina JV, Abad-Diaz-de-Cerio A, Sueiro-Olivares M, Pellon A, Ramirez-Garcia A, Garaizar J, et al. Rapid and specific detection of section Fumigati and Aspergillus fumigatus in human samples using a new multiplex real-time PCR. Diagn Microbiol Infect Dis. 2014;80(2):111–8. pmid:25063549

CHAPTER 7

RNAi IN PLANTS

Prerna Pandey

Dr Prerna Pandey, a biotechnologist with several years of wet lab research experience, worked at International Center for Genetic Engineering and Biotechnology, New Delhi. Her field of research in PhD included isolation and molecular characterization of geminiviruses, genome sequencing, gene annotation, and gene silencing using the RNA interference technology. She has also worked at Transasia Biomedicals and Advance Enzyme Technologies as a scientist. Prerna has published papers in peer-reviewed journals, and has submitted a number of annotated Geminiviral genome sequences in the GenBank, including two novel ones. She has also completed her editing and proof-reading courses from Society for the promotion of Editors and Proof-readers, UK and now works as a freelance scientific editor and writer. When Prerna is not busy with her assignments, she loves to explore historical places.

INTRODUCTION

Since nearly two decades, RNAi has been used as a functional genomics tool, and we are now starting to see the use of this technology for commercially focused applications in plants, from stress tolerance, disease resistance, flower color, improved crop quality, to plant-produced medical therapeutics and fine-tuning of metabolic pathways.

RNAi mechanism has the potential in identification and functional assessment of thousands of genes by effectively knocking down the expression of any particular gene through siRNA molecules in any target cell and determining the gene loss-of-function phenotype which leads to gene functional analysis in absence of cognate mutant alleles. The first time this mechanism found expression in plants was with a view to elevate anothocyanin pigment production in *Petunia hybrida L.* The gene chalcone synthase *(chsA)* that codes for an enzyme of the same namesake involved in anthocyanin pigment production (Napoli et al., 1990). However, contrary to the expected result of darker flowers, an unexpected color pattern was observed that was subsequently attributed to RNA silencing.

The approach of RNAi-mediated gene suppression opens new avenues in the development of eco-friendly techniques for incorporating desirable traits in various plant species including resistance to pathogens, crop improvement for better stress tolerance and improved nutritional status. The approach of RNAi is highly elegant and advanced; that shall be summarized in the following sections.

BIOTIC STRESSES IN PLANTS

When a pathogen attack on plants it can induce priming, a defense response that can pass on to the offspring, also known as "Transgenerational defense", which has been observed in response to insect herbivores too (Slaughter et al., 2012; Luna et al., 2012). Transgenerational defense is not limited to the pathogen stress, but also to insect herbivores. A few successful examples are:

- Using induced transgenerational defense mechanism, radish (*Raphanusraphanistrum*) was produced that possessed higher resistance than the parents (Agrawal et al., 1999)
- Increased trichome production in the leaf-damaged yellow monkey flower offsprings (Holeski 2007; Scoville et al., 2011).

The transgenerational resistance triggered by pathogen/insect attack is passed on to the offspring through DNA methylation, sRNA accumulation or histone modification (Luna et al., 2012). In this type of resistance, siRNAs appear to play a role, as was observed byabsence of transgenerational resistance to caterpillar in *Arabidopsis* mutants, which are deficient in hc-siRNA synthesis and processing (Rasmann et al., 2012).

According toHoleski et al., (2012), who hypothesized a transgenerational induction model: chemical and physical defenses are induced in response to environmental challenges, and sRNAs move from vegetative tissue via phloem to developing seeds as a record of stored information to be passed on to the next generation.

The role of miRNAs in plants infected by pathogens has been reported (Ruiz and Voinnet, 2009). The regulatory roles of miRNAs have also been reported especially in host-microbe associations. Gor instance, the formation of tumor in *Agrobacterium* and also fixation of Nitrogen by *Rhizobium* (Katiyar-Agarwal, 2010). The level of GC was elevated; an important point with miRNAs regulating stress (Mishra, 2009).

RNAI FOR PLANT DISEASE RESISTANCE

One of the first successful applications of gene silencing technique has been protection of plants from viral infection - with the discovery of PTGS. The first few commercial releases were the transgenic papaya (*Carica papaya*), resistant to *Papaya ringspot virus* (PRSV; Fuchs and Gonsalves, 2007); the NewLeaf Plus and NewLeaf Y potatoes (*Solanum tuberosum*) with resistance to *Potato leafroll virus* and PVY (www.research.cip.cgiar.org/). The papaya industry has seen a panacea in the form of PSRV-resistant papaya that has shown tremendous success since its release in Hawaii.

This launched the development of several plant species with the use of PTGS to gain resistance for viral outbreaks (Fuchs and Gonsalves, 2007). A few examples include:

- Wang et al., 2000 have employed RNA silencing for protecting an important crop species of barley: *Hordeum vulgare* against BYDV: Barley yellow dwarf virus. The technique involved transformation of a hpRNA construct that targeted the 5' of the virus into barley in order to produce crops that possessed complete immunity against the virus.
- Viral RNA inside a plant cell can be targeted by artificial mRNAs (amiRNAs) in order to control viruses. Resistance against turnip mosaic virus was achieved by the use of miR159a pre-miRNA transcript to target genes involved ion suppression such as P69.
- Viral suppressor 2b of *Cucumber mosaic virus* (CMV) 2b was

targeted by the use of amiRNA vector. The approach resulted in tobacco plants that were resistant to the virus (Qu et al., 2007).

Bacteria

Apart from protecting a plant from invading viruses, RNAi can confer protection from other pathogens. But, research in RNAi and its role to control bacterial and fungal infections has seen very few publications. Escobar et al., 2001 employed hpRNA constructs against *iaaM* and *ipt* oncogenes (play key roles in formation of tumors in crown gall disease mediated by *Agrobacterium*) into tomatoes. RNAi against the oncogenes: *iaaM* and *ipt* was responsible for the silencing of the target mRNAs that in turn led to development of resistance.

Fungi

The agricultural industry is constantly on the quest for robust strategies against fungi as pathogenic fungi display genetic adjustments in developing resistance to fungicides. (Yang et al., 2008). Karthikeyan et al., in 2013 and Wani et al., 2010 have reviewed that gene silencing offers an ideal strategy for control of fungal diseases and RNAi offers an emerging strategy by silencing the vital genes that help the pathogens sustain in the host. Hypermorphic mechanism is displayed by RNAi: it refers to the principle that the technology can find application for polyploidy and poly karyotic pathogenic plant fungi. For instance:

- Nowara et al., 2010 have expressed dsRNA and antisense transcripts specific against powdery mildew fungi *Blumeria graminis* in wheat and barley.
- Gene fragments of rust fungi *Puccinia striiformis* f. sp. Tritici were introduced into plant cells in order to reduce the expression of the genes in rust fungi during infection using the VIGS method. (Yin et al., 2011)
- An RNAi construct against the Syntaxin-related 1 (StSYR1) gene in potato conferred increased resistance against *Phytophthora infestans* (Eschen-Lippold et al., 2012).

Insect, Nematode -Pest Resistance

RNAi is sequence specific; this property aids in the design of strategies to eliminate harmful pests but without harming useful species. The use of conventional pesticides has unintended targets that extend across the food chain: this is circumvented in the use of RNAi. While *Bacillus thuringiensis* finds extensive use in the field of plant transgenics with minimum reports of development of resistance. However, the potential of resistance in insects propels research to explore other options of combating disease. Hence, the use of RNAi along with *Bacillus thuringiensis* based technologies can serve to boost the efficiency of the system and prevent the threat of resistance (Goldstein et al., 2009).

Certain nematodes such as *Globodera* spp, *Meloidogyne* spp and *Heterodera* cause havoc globally on significant crops. The studies on *C. elegans* facilitated the use of RNAi to target nematodes. The use of RNAi can either target the host genes of infection or functional nematode genes.

- Hoffman et al., 2008 used RNA silencing by targeting Suc transporter genes that down regulated the genes that resulted in impairment in development of female nematode *Heterodera schachtii.*
- Since in *C. elegans,* RNA silencing could be induced by feeding it dsRNA, the same principle was applied and hpRNAs containing sequences of vital nematode genes were expressed in the host plant to incapacitate or kill the parasite. hpRNA constructs against two root knot nematode genes have resulted in the substantial reduction of the target mRNAs of the plant parasitic nematode along with conferring an effective resistance against the parasite (Fairbairn et al., 2007).

A protection method entailing the use of intimate feeding associations against insect pests that are herbivores.

- Baum et al., (2007) fed western corn rootworm larvae on artificial diets supplemented with specific dsRNAs, and identified 14 genes whose knockdown by the dsRNA was lethal and ultimately killed the larvae. For instance, the use of hpRNA that targeted ATPase of the vacuoles of midgut of western corn rootworm resulted in resistance. This resistance was equivalent to that of the use of transgenics of *Bacillus thuringiensis*. Such an approach can be applied for *B. thuringiensis* resistance.

CROP IMPROVEMENT

Barley and Rice

The technique has seen application in the development of BYDV (barley yellow dwarf virus) resistant varieties of barley (Wang et al., 2000).

Kusaba et al., (2003) have applied RNAi to improve rice plants by significantly reducing glutenin levels and produced a rice variety called LGC-1 (low glutenin content 1). This rice could be consumed by kidney patients who are incapacitated to assimilate glutenin in a stable manner and inherited.

Banana

Resistance has been engineered against Banana Bract Mosaic Virus (BBrMV) (Rodoni et al., 1999). This virus has been responsible for the destruction of the bracts resulting in significant blows to economies. The virus is spread by aphids, through infected plant materials, and in fields with previous diseased crop. The use of RNAi sees few challenges as Coat Protein shows minimum variation; suggestive of the potential of RNAi.

Fiber producing plants: Cotton and Jute

Separate lignin from cellulose and hemicellulose during chemical pulping of wood is one of the most expensive and environmentally hazardous processes (Pilate et al.,2002). Plant material with lower lignin content would mean a significant reduction of processing cost and pollution.

- In order to cause a decrease in the levels of lignin; several enzymes involved in the metabolism of lignin have been targeted such as:
    - cynnamyl alcohol deshydrogenase (Baucher et al., 1996),
    - coenzyme A ligase (Hu et al., 1999),
    - caffeate/ 5-hydroxyferulate O-methyltransferase (Lapierre et al., 1999).

RNAi technology has been applied to down regulate the key enzymes, and transgenic jute variety could be created that down regulate the quantity of 4-Cl mRNA thereby reducing the lignin production, allowing for environment friendly and cost-effective processing of fiber for high quality paper and cloth (Williams et al., 2004).

- Cotton is a key crop considering its use. Although the cottonseeds that remain after fiber extraction could be as sources of protein and calories, they contain a toxic gossypol terpenoid and that confers protection against pathogens but however prevents their use as nutrition. This calls for a down regulation of gossypol in cottonseeds.

  An approach involved the use of a seed promoter to *d-cadinene synthase* gene involved in biosynthesis of gossypol. The resultant progeny had decreased amounts of gossypol (Sunilkumar et al., 2006). It is predicted that while these plants have resistance profiles similar to wild type; the decreased level of toxic metabolites can enhance their use in nutrition (Tang et al., 2007).

Tomato

Tomatoes contain high levels of molecules such as lycopene with antioxidant properties and other vitamins, carotenoids. The latter molecules are produced from chlorophyll pathway. The amounts of these key flavonoids or carotenoids have been increased through the use of RNAi-coupled to promoters specific for the fruit without compromising the growth or properties of tomatoes (Davuluri et al., 2005).

Flower Color and Flowering Time Modification of Plants

Gentians are often used as ornamental flowers. Blue-flowered carnations have been developed by using genetic engineering techniques (Tanaka et al., 2005). Nishihara et al., (2006) have reported the use of antisense technology to generate white-flowered transgenic gentians by suppressing the chalcone synthase (*CHS*) gene. It was observed that out of seventeen transgenics, three had the suppressed CHS on account of their white flowers. The application of RNAi to silence genes for color of flowers has been applied to other flora of species such as tobacco, torenia and petunia (Nakatsuka et al., 2008).

RNAi can also be used to manipulate flowering time (anthesis), beside colors and fragrances; an essential component of agriculture. As an example, the flowering of a cereal earlier can have a dipped production of grain due to insufficient levels of reserves. On the other hand, a delay in flowering can also adversely affect the produce. This calls for the control of the time of flowering.

- Flowering Locus T 1 (*FT1*) gene of wheat has been reported as a protein instrumental in regulation of flowering pathways Brachypodium and wheat leaves. When the *FT1* gene was silenced, many FT-like genes were expressed to a lower extent in *Brachypodium* and wheat: with phenotypes manifesting as late flowering in *Brachypodium* and absence of flowers in wheat (Lv et al., 2014).
- Trusov and Botella (2006) produced late flowering phenotype pineapples by silencing the A1-amino-cyclopropane-1-carboxylate synthase (*ACACS2*) gene, a trait that generates reproductive parts after generating enough vegetative tissues to support producing fruits, also providing enough adaptation to the surrounding environment before flowering.
- The generation of plants displaying early flowering has been achieved by RNAi. For instance, *EARLI1* gene of *Arabidopsis*, that codes for a proline rich protein was targeted by RNAi for downregulation and resulted in early flowering along with reduced content of lignin (Shi et al., 2011).
- In trees, Szankowski et al., (2009) developed transgenic apple trees with early flowering phenotype through RNAi by silencing the *MdTFL1* gene and showed that transgenic plants were able to generate flower buds and axillary shoots within few months of planting.
- Yamagishi et al., (2013) applied VIGS strategy to generate early flowering apple trees using apple latent spherical virus (ALSV) with a silencing trigger for the *MdTFL1* gene to accelerate flowering time in apple trees. Early flowering pear trees showed stable and inherited trait, an interesting breeding

source which accelerates the breeding period in trees (Freiman et al., 2012).

Blue Rose

Flower color is controlled by genes: a red color is attributed to *cyanidin* gene while a blue color is attributed to *Delphinidin* gene. A team of researchers of Florigene (Australia) and Suntory (Japan) applied RNAi technology to eliminate the *cyanidin* genes in rose and carnation. The *Delphinidin* gene was exogenously supplied as these flowers do not possess blue color. The outcome of this research was blue flowers (Van Uyen, 2006).

RNAi and Quality of Fruits

The breeding of fruits faces a major challenge due to self-incompatibility of certain woody plants. RNAi has been extended to alter the quality of fruits; For example, by expressing an intron spliced hairpin RNA containing *Mal d 1*-specific inverted repeat sequences, transgenic apples containing reduced amounts of the Mal d 1 allergen were obtained (Gilissen et al., 2005). The outcome of this experiment is the possibility to manufacture apples utilizing RNAi with minimum levels of allergens.

Allergens and Toxins

Allergens and toxins are found in various plant species. The RNAi technology can help in inactivating specific individual genes, allergens or toxins, which are not involved in the germination or maturation of these plants.

- Soybean (*Glycine max*) is known to possess various quantities of P34 proteins, causing allergic reaction in individuals sensitive to soybean. Herman et al., (2003) have reported gene silencing technique to successfully prevent the accumulation of such allergenic proteins, without affecting the normal physiology (germination or maturation) of the transgenic plant.
- Wakasa et al., (2011) have reported the silencing of genes encoding allergens in rice, β-glyoxalase and α-globulin, using RNAi constructs in a mutated α-amylase/trypsin inhibitor rice line. The transgenic rice produced seeds which are free from

the three potential allergens.

- In peanuts, Ara h2 protein (Dodo et al., 2008); apple: Mal d 1 allergen (Gilissen et al., 2005); tomato: Lyc e 1.01 and Lyc e 1.02 allergens (Le et al., 2006) have been significantly reduced by silencing experiments.
- Bhalla et al., (1999) have reported the silencing of a ryegrass pollen allergen, the Lol P 5 protein, by using the antisense strand of the gene without affecting the fertility of the pollen grains.

People consuming a leafy vegetable *Lathyrus sativus* suffer from a paralytic disease called, lathyrism because it contains a neurotoxin called β-oxalylaminoalanine-L-alanine (BOAA) (Spencer et al., 1986). The amount of this BOAA can be reduced to harmless or low levels by applying RNAi technology to target the gene (Williams et al., 2004).

METABOLIC PATHWAYS

Plants produce a variety of products that are plant species specific. This happens because plants possess a diverse network of metabolic pathways of producing primary and secondary metabolites (Zorrilla-López et al., 2013). Extensive study is required to understand the biochemical pathway networks for metabolic engineering. Gene silencing through reverse genetics may be one of the ways to study metabolic pathways in a plant (Alhagdow et al., 2007). Through this technique, the loss of function of a specific enzyme leads to understanding the effects on the whole plant physiology and to engineer plant metabolic pathways to overproduce desired secondary metabolites (Flores et al., 2008). A few secondary metabolites include aromatic fragrances, sugars, drugs, steroids and fatty acids.

- *SlGalLDH*genes of tomato was targeted by Alhagdow et al., (2007), by RNAi in order to study L -Galactono-1,4-lactone dehydrogenase, an enzyme in the mitochondria that is involved in catalysis of the last step of biosynthesis of Vitamin C. Following reduction in GALDH, the expression of several stress induced secondary metabolites showed a change manifesting as compromised growth, the physiology that reflected on plant structure.

- Studying the *Ipa1* mutant maize has inspired scientists to silence the expression of ATP-binding cassette (ABC) transporter, which resulted in reduction of phytic acid and an increase of phosphate in transgenic maize, without affecting the germination or growth of the plant (Shi et al., 2007).
- With increased content in amylose, encouraging easy digestibility in humans, food quality was enhanced in transgenic wheat. RNAi was used to inactivate *SBEIIa* genes to modify the synthesis of amylopectin (Sestili et al., 2010).
- Decaffeinated coffee is essential to consumers who are caffeine sensitive. Traditionally, decaf is extracted from natural coffee by using water or solvents. Through RNAi technology, Coffee that produces natural coffee can be generated with low or very low caffeine content, overcoming the need of extraction. The theobromine synthase (*CaMXMT1*) gene that codes for caffeine, was targeted and repressed in coffee beans by using RNAi technology, resulting in 70 % decrease in its levels in leaves (Ogita et al., 2003).
- Mohanpuria et al., (2011) have reported the production of low caffeine tea by silencing the caffeine synthase (*CS*).

Amino Acid Content

Transgenesis programs are racing to provide crops with enhanced levels of essential amino acids and storage proteins because the amino acid content of food and feed is considered one of the most important agronomical traits. For example, high lysine content can affect the plant phenotype and the seed germination (Tang and Galili, 2004). Successful experiments of lysine modification through RNAi technology have been done in maize by expressing a lysine synthesis enzyme and silencing a lysine degradation enzyme (Frizzi et al., 2008); the resulting transgenic maize exhibited a significant increase in lysine amino acid content. There are other reports of using VIGS to manipulate the levels of aspartate, glutamate and lysine amino acids through the suppression of the plastid NbAsp5 and NbPAT genes, respectively (de la Torre et al., 2014). Using RNAi technology to suppress the regulation and degradation of amino acids biosynthetic pathways helps in expanding the genetic pool and is considered an advantage for plant breeders.

Edible Oil Improved

Seed oils show a vast spectrum of properties such as the health benefits of certain oils or heat stability of others. This is due to the fatty acid content of the oil. For instance, palm oil contains an elevated level of palmitic acid that confers stability from heat but makes it unsuitable for cooking by virtue of its increased content of LDL. However, olive oil contains higher proportion of linoleic acid that lacks the stability to heat but is a healthy alternative.

RNAi has found application to elevate the oleic acid levels in cotton. The gene coding for the enzyme catalyzing oleic acid to another conversion was targeted that raised the level of oleic acid from 10 to 75 percent (Waterhouse et al., 1998).

Increasing Amylose Content of Grains

The RNAi technology was applied to boost the level of amylase in grains such as wheat. A study involved a combination of a wheat seed specific promoter and RNAi to silence genes coding for branching enzymes involved in the synthesis of aamylopectin. The arrangement caused down regulation of isozymes involved in starch branching (Regina et al., 2006) that raised the amount of amylose in wheat to 70% (Tang et al., 2007).

NON- FOOD APPLICATIONS

RNA silencing technology has been used for several nonfood applications.

- An increased performance in a bioreactor was achieved by altering photosynthetic pathways in algae (Mussgnug et al., 2007).
- The yield of pharmaceutically relevant compounds of poppy (*Papaver somniferum*) was the outcome of altering the morphine pathway (Allen et al., 2004).
- The demand for antibodies produced from plant is increasing as monoclonal antibodies find use in several therapeutic applications. A challenge here is a compromise of post translational modifications (PTMs) of these heterologous proteins. For example, glycosylation of Fc region of antibody by plant may have a compromised functioning. RNAi

was applied to cause silencing of two endogenous glycan-transferases of the algae *Lemna*. The strategy successfully resulted in the synthesis of antibodies that were products of homogenous glycosylation; suitable for use in therapy on account of safety and functionality (Cox et al., 2006).

- nat-siRNAATGB2 was the first endogenous siRNA associated with plant biotic stress and is a regulator of effector triggered R-gene mediated immunity (Katiyar-Agarwal, 2006).
- lsiRNAs (between thirty and forty nucleotides) classified a new group of endogenous siRNA were produced in response to conditions of growth (suspension) or infection by bacteria (Katiyar-Agarwal, 2007).
- As a response to *H. schachtii* , the levels of siRNAs showed a graded content; the level of *siRNA32* saw a dip while those of *siRNA46* , *siRNA9*, *siRNA41*were elevated (Hewezi, 2008).

IMPROVED ABIOTIC STRESS TOLERANCE

Abiotic stress tolerance in plants is developed as a responsive behavior to overcome sudden changes in the environment, temperature adaptation mechanisms, drought, light and soil nutrients stress. Plants show several physiological, biochemical and metabolic strategies to combat abiotic stress. Complete success has not been yet attributed to classical plant breeding techniques to develop improved resistance to abiotic stress (Tester, 2005).

The following approaches assist crop improvement in today's world: (Younis et al., 2014):

- Methods involving transgenics
- Functional genomics
- Molecular mapping that includes scanning Quantitative Trait Locus (QTL)
- Analyzing networks of pathways through genomics, transcriptome and proteomics

A network of biochemical events with several stress genes is involved in tolerance to abiotic stress as revealed by tolls of molecular biology (Senthil-Kumar, 2010), This suggests a role of RNAi to discover and characterize genes involved in abiotic stress tolerance.

Small noncoding RNAs were studied in response to abiotic stress in *Triticumaestivum*. A marked change in expression pattern of four siRNAs were reported in response to heat, cold, salt, and drought (Yao et al., 2010).

miRNAs Involved in ABA-Mediated Stress Responses

Abscisic acid (ABA): a phytohormone is involved in environmental stress responses. Treatment with ABA or Gibberellin has been correlated with micro RNA mediated control of development of organs. For example, the control of floral organ development and expression of *miR159* (Achard, 2004). Treatment with ABA caused an increase of miR159 in germinating *Arabidopsis* seeds (Reyes and Chua, 2007).

miRNAs in Response to Drought and Salt Stress

The role of miRNAs in response to drought and salt has been examined using *Arabidopsis*. The sequencing of small RNAs library isolated from *Arabidopsis* seedlings exposed to various stresses has identified several stress-related miRNAs (Sunkar and Zhu, 2004). The profiling and analysis of miRNAs in a genome wide genome profiling was performed in rice exposed to drought. Studies performed along several stages of development suggest the upregulation and downregulation of several miRNAs.

It was observed that miRNA gene families, such as *miR171*, *miR319*, and *miR896*, were both up and down regulated while the plants exposed to drought showed opposite results (Zhou et al., 2010).

In addition, several genes and associated pathways in plants are affected by salt stress. Several differentially regulated miRNAs have been identified in salt-stressed plants like *Arabidopsis* (Liu et al., 2008), and *Populustrichocarpa*(Zhou et al., 2008).the authors found ABA-responsive element (ABRE), a *cis*-acting element, upstream of *miR169n*, suggesting that *miR169n* may be ABA-regulated. miRNA profile through microarray experiments in salt-tolerant and -sensitive lines of maize (*Zea mays*) (Ding et al., 2009) indicate that members of the *miR156*, *miR164*, *miR167*, and *miR396* families were downregulated, while *miR162*, *miR168*, *miR395*, and *miR474* families were upregulated in salt-shocked maize roots.

Cold, Heat Stress Tolerance

On examining the miRNAs expression in cold stress, *miR172* was seen upregulated in *Arabidopsis* and *Brachypodium*. In addition, several miRNAs were induced under cold stress in *Arabidopsis*, while other miRNAs showed either transient or mild regulation. Using high-throughput sequencing, wheat miRNAs isolated from leaves showed differential expression in response to heat stress, as reported by, Xin et al., (2010). Of the 32 families of miRNAs unearthed in wheat, 9 were classified as putatively responsive to heat as their expression was altered upon heat stress.

miRNAs in Response to Hypoxia and Oxidative Stress

The respiration in mitochondria is compromised upon exposure to hypoxia or anaerobic stress (Agarwal and Grover, 2006). Changes in the transcriptome in hypoxia (Bailey-Serres and Voesenek, 2008) followed by differential regulation of several miRNAs in seedlings that were submerged.

Reactive oxygen species (ROS) are produced constantly as a result of aerobic pathways in cell organelles like mitochondria, chloroplasts and peroxisomes as an outcome of stress (Mittler, 2004). Superoxide dismutases (SODs) converts superoxide in plants into molecular oxygen and hydrogen peroxide. in *Arabidopsis* Cu-Zn SODs are encoded by *CSD1, CSD2*, and *CSD3* genes (Sunkar et al., 2006). Sunkar et al confirmed that *miR398* targets *CSD1* and *CSD2* (Jones-Rhoades and Bartel, 2004; Bonnet et al., 2004), and that *miR398* is downregulated under oxidative stress.

miRNAs in Response to UV-B Radiation

Computational analysis showed that certain miRNAs were produced as a result of exposure to UV-B. The miRNAs included: *miR393, miR169*, and *miR159*: their levels were higher in *Arabidopsis* and decreased in *Populustremula*; this is suggestive of different responses dependant on the species (Jia et al., 2009).

MECHANICAL STRESS TOLERANCE

The role of miRNAs has also been predicted to confer fitness in a mechanical context to increase the tolerance of plants to mechanical stress (Lu et al., 2005).

RNAI AND PLANT FUNCTIONAL GENOMICS

The characterization of all the genes of the plant genome is a huge trial in the post genomic era. The technique of RNAi presents a specific and efficient tool for silencing a gene or a gene family. Additionally, the use of dsRNA whose expression can be induced can facilitate manipulation of gene silencing to target specific genes at a particular stage or part (Matthew, 2004; Sato, 2005).

The activation of RNAi in plants can be done by several means. The different methods each possess merits and demerits in relation to their persistence of their effects along with the class spectrum of plant species on which they manifest. Several methods are transformation using *Agrobacterium*, particle bombardment, virus-induced gene silencing (VIGS) and hairpin RNA-expressing vectors, (VIGS) (Burch-Smith, 2004).

CONCLUSION

This chapter highlights the use of gene silencing by RNAi to combat the attack by bacteria, fungi or insects especially for crops of economic significance. However, added research is warranted to ascertain the significance of RNAi silencing in immunity of plants against pathogens. The upregulation or down expression of factors of disease resistance associated with gene silencing stands as a possible route. Another approach is over expression or a knockdown of sRNAs associated with plant defense circuits to develop resistance against non-viral pathogens. Initially, the researchers who worked on this concept probably would not have analyzed the mechanism of the resistance to virus as a tool of transcriptional control.

The mechanisms behind the observations in relation to resistance to virus, co-suppression along with RdDM are salient features of silencing. The discovery of dsRNA based gene control ushered in a new era of research towards RNA silencing across all kingdoms of life. This, along with the discoveries that followed brought in a paradigm shift in the way regulation of genes and control of development in organisms is evaluated. Despite the need for deeper comprehension; RNA silencing in plants has emerged as a candidate for the development of analysis of genes as well as crop improvement.

The imparting of viral resistance along with alterations in metabolism to achieve desirable traits in plants have been possible by techniques such as hpRNA or artificial miRNA. Further research in the elucidation of RNAi for instance, as virus-induced gene silencing can provide an added boost to augment newer strategies. As recent discoveries have unearthed that RNA silencing is a player in responses of plants to biotic and abiotic stress, the potential of RNA based silencing in meeting the roadblocks of the climate change and hence, global food security deserves mention.

REFERENCES

1. Achard P, Herr A, Baulcombe DC, Harberd NP. (2004) Modulation of floral development by a gibberellin-regulated microRNA. Development 31:3357–3365.
2. Agarwal S, Grover A. (2006) Molecular Biology. Biotechnology and Genomics of Flooding-Associated Low O2 Stress Response in Plants. Crit Rev Plant Sci 25:1–21.
3. Agrawal AA, Laforsch C, Tollrian R. (1999) Transgenerational induction of defences in animals and plants. Nature 401: 60–63.
4. Alhagdow M, Mounet F, Gilbert L et al. (2007) Silencing of mitochondrial ascorbate synthesizing enzyme L -Galactono-1.4Lactone dehydrogenase affects plant and fruit development in tomato. Plant Physiol 145:1408–1422.
5. Allen RS, Millgate AG, Chitty JA, Thisleton J, Miller JA, Fist AJ, Gerlach WL, Larkin PJ. (2004) RNAi-mediated replacement of morphine with the nonnarcotic alkaloid reticuline in opium poppy. Nat Biotechnol 22(12):1559-66.
6. Bailey-Serres J, Voesenek LACJ. (2008) Flooding Stress: Acclimations and Genetic Diversity. Annu. Rev. Plant Biol 59:313–339.

7. Baucher M, Chabber B, Pilate G, Vandoorsselaere J, Tollier MT, PetitConil M, et al. (1996) Red xylem and higher lignin extractability by down-regulating a cinnamyl alcohol dehydrogenase in Poplar. Plant Physiol 112: 1479-1490.
8. Baum JA, Bogaert T, Clinton W, Heck GR, Feldmann P, Ilagan O, Johnson S, Plaetinck G, Munyikwa T, Pleau M, Vaughn T, Roberts J. (2007) Control of coleopteran insect pests through RNA interference.Nat Biotechnol 25(11):1322-6.
9. Bhalla PL, Swoboda I, Singh MB. (1999) Antisense-mediated silencing of a gene encoding a major ryegrass pollen allergen. Proc Natl Acad Sci USA 96:11676–11680.
10. Bonnet E, Wuyts J, Rouze P, Van de PY. (2004) Detection of 91 potential conserved plant microRNAs in Arabidopsis thaliana and Oryza sativa identifies important target genes. Proc Natl Acad Sci USA 101:11511–11516.
11. Burch-Smith T.M. et al. (2004) The Plant J. 39:734-746.
12. Cox KM, Sterling JD, Regan JT, Gasdaska JR, Frantz KK, Peele CG, et al. (2006) Glycan optimization of a human monoclonal antibody in the aquatic plant Lemna minor. Nat Biotechnol 24(12):1591-1597.
13. Davuluri GR, van Tuinen A, Mustilli AC, Manfredonia A, Newman R, Burgess D, et al. (2004) Manipulation of DET1 expression in tomato results in photomorphogenic phenotypes caused by posttranscriptional gene silencing. Plant J 40: 344–354.
14. de la Torre F, El-Azaz J, Avila C, Canovas F. (2014) Deciphering the role of aspartate and prephenate aminotransferase activities in plastid nitrogen metabolism. Plant Physiol 164:92–104.
15. Ding D, Zhang L, Wang H, Liu Z, Zhang Z, Zheng Y. (2009) Differential expression of miRNAs in response to salt stress in maize roots. Ann Bot 103:29–38.
16. Dodo HW, Konan KN, Chen FC. (2008) Alleviating peanut allergy using genetic engineering: the silencing of the immunodominant allergen Ara h2 leads to its signifi cant reduction and a decrease in peanut allergenicity. Plant Biotechnol J 6(2):135–145.
17. Eschen-Lippold L, Lübken T, Smolka U, Rosahl S (2012) Characterization of potato plants with reduced StSYR1 expression.

Plant Signal Behav 7:559–562.

18. Escobar MA, Civerolo EL, Summerfelt KR, Dandekar AM. (2001) RNAi-mediated oncogene silencing confers resistance to crown gall tumorigenesis. Proc Natl Acad Sci U S A 98(23):13437-13442.

19. Fairbairn DJ, Cavallaro AS, Bernard M, Mahalinga-Iyer J, Graham MW, Botella JR. (2007) Host-delivered RNAi: an effective strategy to silence genes in plant parasitic nematodes. Planta 226(6):152533.

20. Flores T, Karpova O, Su X. (2008) Silencing of GmFAD3 gene by siRNA leads to low alphalinolenic acids (18:3) of fad3 -mutant phenotype in soybean [Glycine max (Merr.)]. Transgenic Res 17:839–850.

21. Freiman A, Shlizerman L, Golobovitch S et al. (2012) Development of a transgenic early flowering pear (Pyruscommunis L.) genotype by RNAi silencing of PcTFL1-1 and PcTFL1-2. Planta 235(6):1239–1251.

22. Frizzi A, Huang S, Gilbertson LA. (2008) Modifying lysine biosynthesis and catabolism in corn with a single bifunctional expression/silencing transgene cassette. Plant Biotechnol J 6:13–21.

23. Fuchs, M., and Gonsalves, D. (1995). Resistance of transgenic hybrid squash ZW-20 expressing the coat protein genes of Zucchini yellow mosaic virus and Watermelon mosaic virus 2 to mixed infections by both potyviruses. Bio/Technology 13:1466-1473.

24. Giliseen LJWJ, Bolhaar STH, Matos CI, Rouwendal GJA, Boone MJ, Krens FA, et al. (2005) Silencing of the major apple allergen Mal 1 by using RNA interference approach. J Allergy Clin Immunol 115: 369-384.

25. Goldstein DA, Songstad D, Sachs E, Petrick J. (2009) RNA interference in plants. Monsanto. http://www.monsanto.com

26. Herman E, Helm R, Jung R, Kinney A. (2003) Genetic modifi cation removes an immunodominant allergen from soybean. Plant Physiol 132(1):36–43.

27. Hewezi T, Howe P, Maier TR, Baum TJ. (2008) Arabidopsis small RNAs and their targets during cyst nematode parasitism. Mol Plant Microbe Interact 21(12):1622-34.

28. Hoffman Y, Aflalo C, Zarka A, Gutman J, James TY, Boussiba

S. (2008) Isolation and characterization of a novel chytrid species (phylum Blastocladiomycota), parasitic on the green alga Haematococcus. Mycol Res 112(Pt 1):70-81.

29. Holeski LM (2007) Within and between generation phenotypic plasticity in trichome density of Mimulusguttatus. J Evol Biol 20: 2092–2100.
30. Hu W, Harding SA, Lung J, Popko JL, Ralph J, Stokke DD, et al. (1999) Repression of lignin biosynthesis promotes cellulose accumulation and growth in transgenic trees. Nature Biotechnol 17: 808-812.
31. Jia X, Ren L, Chen QJ, Li R, Tang G. (2009) UV-B-responsive microRNAs in Populustremula. J Plant Physiol 166(18):2046-57.
32. Jones-Rhoades MW, Bartel DP. (2004) Computational Identification of Plant MicroRNAs Their Targets, Including a Stress-Induced miRNA. Mol cell 14:787–799.
33. Karthikeyan A, Deivamani M, Shobhana VG et al. (2013) RNA interference: evolutions and applications in plant disease management. Arch Phytopathol Plant Protect 46(12):1430–1441.
34. Katiyar-Agarwal S, Gao S, Vivian-Smith A, Jin H. (2007) A novel class of bacteria-induced small RNAs in Arabidopsis. Genes Dev 21(23):3123-34.
35. Katiyar-Agarwal S, Jin H.(2010)Role of small RNAs in hostmicrobe interactions. Annu Rev Phytopathol 48():225-46.
36. Katiyar-Agarwal S, Morgan R, Dahlbeck D, Borsani O, Villegas A Jr, Zhu JK, Staskawicz BJ, Jin H. (2006)A pathogen-inducible endogenous siRNA in plant immunity. Proc Natl Acad Sci USA 103(47):18002-7.
37. Kusaba M, Miyahara K, Lida S, Fukuoka H, Takario T, Sassa H, Nishimura M, Nishio T. (2003) Low glutenin content 1: a dominant mutation that suppresses the glutenin multigene family via RNA silencing in rice. Plant Cell 15: 1455-1467.
38. Lapierre C, Pollet B, Petit-Conil M, Toval G, Romero J, Pilate G, et al. (1999) Structural alterations of lignins in transgenic poplars with depressed cinnamyl alcohol dehydrogenase or caffeic acid O-methyltransferase activity have an opposite impact on the efficiency of industrial kraft pulping. Plant Physiology 119: 153164.

39. Le LQ, Mahler V, Lorenz Y et al. (2006) Reduced allergenicity of tomato fruits harvested from Lyc e 1-silenced transgenic tomato plants. J Allergy Clin Immunol 118:1176–1183.
40. Liu H-H, Tian X, Li Y-J, Wu C-A, Zheng C-C. (2008) Microarray based analysis of stress-regulated microRNAs in Arabidopsis thaliana. RNA 14:836–843.
41. Lu S, Sun YH, Shi R, Clark C, Li L, Chiang VL. (2005) Novel and mechanical stress-responsive MicroRNAs in Populus trichocarpa that are absent from Arabidopsis. Plant Cell 17(8):2186-203.
42. Luna E, Bruce TJ, Roberts MR, Flors V, Ton J. (2012) Next generation systemic acquired resistance. Plant Physiol 158: 844–853.
43. Lv B, Nitcher R, Han X et al. (2014) Characterization of flowering locus T1 (FT1) gene in Brachypodium and wheat. PLoS One 9(4):e94171.
44. Matthew L. (2004) Comparative and Functional Genomics 5: 240244.
45. Mishra AK, Agarwal S, Jain CK, Rani V. (2009) High GC content: critical parameter for predicting stress regulated miRNAs in Arabidopsis thaliana. Bioinformation 4(4):151-154.
46. Mittler R, Vanderauwera S, Gollery M, Van Breusegem F. (2004) Reactive oxygen gene network of plants. Trends Plant Sci 9:490–498.
47. Mohanpuria P, Kumar V, Ahuja PS, Yadav SK. (2011) Producing low-caffeine tea through posttranscriptional silencing of caffeine synthase mRNA. Plant MolBiol 76(6):523–524.
48. Mussgnug JH, Thomas-Hall S, Rupprecht J, Foo A, Klassen V, McDowall A, Schenk PM, Kruse O, Hankamer B. (2007) Engineering photosynthetic light capture: impacts on improved solar energy to biomass conversion. Plant Biotechnol J 5(6):802-814.
49. Nakatsuka T, MishibaaKI,Abe Y, Kubota A, Kakizaki Y, Yamamura S, Nishihara M. (2008) Plant Biotechnology, Flower color modification of gentian plants by RNAi mediated gene silencing. Plant Biotechnology 25:61-68.
50. Napoli C, Lemieux C, Jorgensen R.(1990) Introduction of chimeric

chalcone synthase gene into Petunia results in reversible co-suppression of homologous genes in trans. Plant Cell 2:279–289.

51. Nishihara M, Nakatsuka T, Hosokawa K, Yokoi T, Abe Y, Mishiba K, Yamamura S. (2006) Dominant inheritance of white-flowered and herbicide-resistant traits in transgenic gentian plants. Plant Biotechol 23: 25–31.
52. Niu QW, Lin SS, Reyes JL, Chen KC, Wu HW, Yeh SD, Chua NH. (2006) Expression of artificial microRNAs in transgenic Arabidopsis thaliana confers virus resistance. Nat Biotechnol 24(11):1420-1428.
53. Nowara D, Gay A, Lacomme C et al. (2010) HIGS: hostinduced gene silencing in the obligate biotrophic fungal pathogen Blumeriagraminis. Plant Cell 22:3130–3141.
54. Ogita S, Uefuji H, Yamaguchi Y. (2003) Producing decaffeinated coffee plants. Nature 423:823.
55. Pilate G, Guiney E, Holt K, Petit-Conil M, Lapierre C, Leple J, et al (2002) Field and pulping performances of transgenic trees with altered lignification. Nat Biotechnol 20: 607-612.
56. Qu J, Ye J, Fang R. (2007) Artificial microRNA-mediated virus resistance in plants. J Virol 81(12):6690-6699.
57. Rasmann S, De Vos M, Casteel CL, Tian D, Halitschke R, Sun JY, Agrawal AA, Felton GW, Jander G. (2012) Herbivory in the previous generation primes plants for enhanced insect resistance. Plant Physiol 158: 854–863.
58. Regina A, Bird A, Topping D, Bowden S, Freeman J, Barsby Tet al. (2006) High amylose wheat generated by RNA interference improves indices of large-bowel health in rats. Proc Natl Acad Sci USA 103: 3546– 3551.
59. Reyes JL, Chua NH. (2007) ABA induction of miR159 controls transcript levels of two MYB factors during Arabidopsis seed germination. Plant J 49:592–606.
60. Rodoni BC, Dale JL. (1999) Harding RM Characterization and expression of the coat protein-coding region of the banana bract mosaic potyvirus, development of diagnostic assays and detection of the virus in banana plants from five countries in Southeast Asia. Arch Virol 144: 1725-1737.

61. Ruiz-Ferrer V, Voinnet O. Roles of Plant Small RNAs in Biotic Stress Responses. (2009) Annu Rev Plant Biol 60:485–510.
62. Sato F. (2005) Plant Biotechnol 22: 431-442.
63. Scoville AG, Barnett LL, Bodbyl-Roels S, Kelly JK, Hileman LC. (2011) Differential regulation of a MYB transcription factor is correlated with transgenerational epigenetic inheritance of trichome density in Mimulus guttatus. New Phytol 191: 251–263.
64. Senthil-Kumar M, Mysore KS. (2010) RNAi in Plants: recent developments and applications in agriculture; Gene Silencing: Theory, Techniques and Applications: 183-199.
65. Sestili F, Janni M, Doherty A et al. (2010) Increasing the amylose content of durum wheat through silencing the SBEIIa genes. BMC Plant Biol 10:144–155.
66. Shi J, Wang H, Schellin K et al. (2007) Embryo-specifi c silencing of a transporter reduces phytic acid content of maize and soybean seeds. Nat Biotechnol 25(8):930–937.
67. Shi Y, Zhang X, Xu Z-Y et al. (2011) Influence of EARLI1-like genes on flowering time and lignin synthesis of Arabidopsis thaliana. Plant Biol 13:731–739.
68. Slaughter A, Daniel X, Flors V, Luna E, Hohn B, Mauch-Mani B (2012) Descendants of primed Arabidopsis plants exhibit resistance to biotic stress. Plant Physiol 158: 835–843.
69. Spencer PS, Roy DN, Ludolph A, Hugon J, Dwivedi MP, Schaumburg HH. (1986) Lathyrism: evidence for role of the neuroexcitatory aminoacid BOAA. Lancet 2(8515): 1066-7.
70. Sunilkumar G, Campbell LM, Puckhaber L, Stipanovic RD, Rathore KS. (2006) Engineering cottonseed for use in human nutrition by tissue-specific reduction of toxic gossypol. Proc Natl Acad Sci USA 103: 18054–18059.
71. Sunkar R, Kapoor A, Zhu JK. (2006) Posttranscriptional induction of two Cu/Zn superoxide dismutase genes in Arabidopsis is mediated by downregulation of miR398 and important for oxidative stress tolerance. Plant Cell 18:2051–2065.
72. Sunkar R, Zhu JK. (2004) Novel and stress-regulated microRNAs and other small RNAs from Arabidopsis. Plant Cell 16:2001–2019.

73. Szankowski I, Waidmann S, El-Din Saad O et al. (2009) RNAisilencing of MdTFL1 induces early flowering in apple. Acta Horticult 839:633–636.

74. Tanaka Y, Katsumoto Y, Brugliera F, Mason J. (2005) Genetic engineering in floriculture. Plant Cell Tiss Org Cult 80: 1–24.

75. Tang G, Galili G, Zhuang X. (2007) RNAi and microRNA: breakthrough technologies for the improvement of plant nutritional value and metabolic engineering. Metabolomics 3:357–369.

76. Tang G, Galili G. (2004) Using RNAi to improve plant nutritional value: from mechanism to application. Trends Biotechnol 22(9):463–469.

77. Tester M, Bacic A. (2005) Abiotic stress tolerance in grasses. From model plants to crop plants. Plant Physiol 137(3):791-3.

78. Trusov Y, Botella JR. (2006) Silencing of the ACC synthase gene ACACS2 causes delayed flowering in pineapple [Ananascomosus (L.) Merr.]. J Exp Bot 57(14):3953–3960.

79. Van Uyen N. (2006) Novel approaches in plants breeding RNAi technology. Proceedings of International Workshop on Biotechnology in Agriculture 12-16.

80. Wakasa Y, Hirano K, Uriso A et al. (2011) Generation of transgenic rice lines with reduced contents of multiple potential allergens using a null mutant in combination with an RNA silencing method. Plant Cell Physiol 52(12):2190–2199.

81. Wang M, Abbott D, Waterhouse P M. (2000) A single copy of a virus derived transgene encoding hairpin RNA gives immunity to barley yellow dwarf virus. Mol Plant Pathol 1: 401-410.

82. Wani SH, Sanghera GS, Singh NB. (2010) Biotechnology and plant disease control-role of RNA interference. Am J Plant Sci. 1:55–68.

83. Waterhouse PM, Graham MW, Wang MB. (1998) Virus resistance and gene silencing in plants can be induced by simultaneous expression of sense and antisense RNA. Proc Natl Acad Sci USA 95:13959–13964.

84. Williams M, Clark G, Sathasivan K, Islam AS. (2004) RNA Interference and its Application in Crop Improvement. Plant tissue culture and Biotechnology 1-18.

85. Xin M, Wang Y, Yao Y, Xie C, Peng H, Ni Z, Sun Q. (2010) Diverse set of microRNAs are responsive to powdery mildew infection and heat stress in wheat (Triticumaestivum L.) BMC Plant Biol. 10: 123.
86. Yamagishi N, Kishigama R, Yoshikawa N. (2013) Reduced generation time of apple seedlings to within a year by means of plant virus vector: a new plant breeding technique with no transmission of genetic modifi cation to the next generation. Plant Biotechnol J. doi:10.1111/pbi.12116
87. Yang L, Liu Z, Lu F, Dong A, Huang H. (2006) SERRATE is a novel nuclear regulator in primary micro RNA processing in Arabidopsis. Plant J 47(6):841-50.
88. Yao Y, Ni Z, Peng H, Sun F, Xin M, Sunkar R, Zhu JK, Sun Q. (2010) Non-coding small RNAs responsive to abiotic stress in wheat (Triticumaestivum L.). Funct Integr Genomics 10(2):187-190.
89. Yin C, Jurgenson JE, Hulbert SH. (2011) Development of a host-induced RNAi system in the wheat stripe rust fungus Pucciniastriiformis f. sp. tritici. Mol Plant Microbe Interact 24:554–561.
90. Younis A, Siddique MI, Kim C-K, Lim K-B. (2014) RNA Interference (RNAi) Induced Gene Silencing: A Promising Approach of HiTech Plant Breeding. Int J Biol Sci 10(10):1150-1158.
91. Zhou L, Liu Y, Liu Z, Kong D, Duan M, Luo L. (2010) Genomewide identification and analysis of drought-responsive microRNAs in Oryza sativa. J Exp Bot 61:4157–4168.
92. Zhou X, Sunkar R, Jin H, Zhu JK, Zhang W. (2009) Genome-wide identification and analysis of small RNAs originated from natural antisense transcripts in Oryza sativa. Genome Res 19:70–78.
93. Zorrilla-López U, Masip G, Arjo G, Bai C, Banakar R et al. (2013) Engineering metabolic pathways in plants by multigene transformation. Int J Dev Biol 57:565–576.

CHAPTER 8

SMALL RNAS IN PLANTS: RECENT DEVELOPMENT AND APPLICATION FOR CROP IMPROVEMENT

Ayushi Kamthan[1], Abira Chaudhuri[1], Mohan Kamthan[2] and Asis Datta[1]

[1]National Institute of Plant Genome Research, New Delhi, India

[2]Indian Institute of Toxicology Research, Lucknow, India

Keywords: miRNA, siRNA, RNAi, gene silencing, crop improvement

The phenomenon of RNA interference (RNAi) which involves sequence-specific gene regulation by small non-coding RNAs, i.e., small interfering RNA (siRNA) and microRNA (miRNA) has emerged as one of most powerful approaches for crop improvement. RNAi based on siRNA is one of the widely used tools of reverse genetics which aid in revealing gene functions in many species. This technology has been extensively applied to alter the gene expression in plants with an aim to achieve desirable traits. RNAi has been used for enhancing the crop yield and productivity by manipulating the gene involved in biomass, grain yield and enhanced shelf life of fruits and vegetables. It has also been applied for developing resistance against various biotic (bacteria, fungi, viruses, nematodes, insects) and abiotic stresses (drought, salinity, cold, etc.). Nutritional improvements of crops have also been achieved by enriching the crops with essential amino acids, fatty acids, antioxidants and other nutrients beneficial for human health or by reducing allergens or anti-nutrients. microRNAs are key regulators of important plant processes like growth, development, and response to various stresses. In spite of similarity in size (20–24 nt), miRNA differ from siRNA in precursor structures, pathway of biogenesis, and modes of action. This review also highlights the miRNA based genetic modification technology where various miRNAs/artificial miRNAs and their targets can be utilized for improving several desirable plant traits. microRNA based strategies are much efficient than siRNA-based RNAi strategies due to its specificity and less undesirable off target effects. As per the FDA guidelines, small RNA (sRNA) based transgenics are much safer for consumption than those over-expressing proteins. This review thereby summarizes the emerging advances and achievement in the field of sRNAs and its application for crop improvement.

INTRODUCTION

RNA interference (RNAi) is a gene silencing phenomenon that involves sequence-specific gene regulation induced by double stranded RNA (dsRNA) resulting in inhibition of translation or transcription. During the 1990s, post-transcriptional gene silencing (PTGS) was reported in many organisms including fungi, animals, ciliates, and plants (Baulcombe, 2000; Matzke et al., 2001). The phenomenon of gene silencing was discovered accidentally in petunia flowers where expression of chalcone synthase (pigment producing gene), resulted in variegated flowers instead of expected deep purple color. Since, the expression of both the transgene and the homologous endogenous gene was suppressed, the phenomenon was termed co-suppression (Napoli et al., 1990; Campbell, 2005). A similar phenomenon reported in the fungus *Neurospora crassa* was named quelling (Romano and Macino, 1992; Cogoni et al., 1996; Fire et al., 1998), and a related mechanism was identified in animals known as RNAi. The phenomenon of RNAi involves small non-coding RNAs called small interfering RNA (siRNA) and microRNA (miRNA) that are the cleavage product of dsRNA. A ribonuclease called DICER or a Dicer-like enzyme (DCL; Pare and Hobman, 2007) carry out this cleavage. The phenomenon of RNAi, in addition to small non-coding RNAs involves a RNA-induced silencing complex (RISC; Redfern et al., 2013; Wilson and Doudna, 2013) and Argonaute proteins (AGOs; Ender and Meister, 2010; Riley et al., 2012).

In plants, miRNAs were identified about 10 years later after discovery of animal miRNAs. In early, Gunnery and Datta (1987) for the first time reported small inhibitory RNA in plant system involved in translational control. They isolated a small molecular weight RNA from barley embryos that specifically inhibits translation initiation. In mid-2002, small RNAs (sRNAs) with miRNA characteristics were reported by different groups to be present in *Arabidopsis* (Llave et al., 2002b; Mette et al., 2002; Park et al., 2002; Reinhart et al., 2002).

In recent times, RNAi technology has evolved as an important tool of genetic engineering and functional genomics aimed for crop improvement.

miRNA vs. siRNA

microRNAs (miRNA) and short-interfering RNAs are known to be important regulators of gene expression (Bartel, 2004; Zamore and Haley, 2005; Vazquez, 2006). Though, they show similarity in size (20–24 nt), but differ in precursor structures, pathway of biogenesis, and modes of action (Axtell, 2013; Table 1). Both are processed from long RNA precursors by Dicer-like ribonucleases (Lee et al., 1993; Hamilton and Baulcombe, 1999; Hammond et al., 2000; Zamore et al., 2000; Bernstein et al., 2001; Hutvagner et al., 2001). Both regulate the target gene repression (Lee et al., 1993; Hammond et al., 2000; Zamore et al., 2000) through ribonucleoprotein silencing complexes.

Table 1: Comparison of miRNA and siRNA

Properties	miRNA	siRNA
Origin	Distinct genomic loci. Encoded by their own genes	Encoded by transposons, viruses, heterochromatin
Biogenesis (nature of precursor)	Single RNA molecules that include an imperfect stem–loop secondary structure	Long bimolecular RNA duplexes or extended hairpins
Evolutionary conservation	Nearly always conserved in related organisms	Rarely conserved in related organisms
Nature of regulatory target	Regulate different genes	Mediate the silencing of the same (or very similar) genes from which they originate

SHORT-INTERFERING RNAS (SIRNAS)

Long dsRNA or short-hairpin RNA (shRNA) precursors, which are homologous in sequence to the target gene to be silenced (Fire et al., 1998; Tuschl, 2001), initiate the process of RNAi. The entry of a virus, a genetic element like transposons or an introduced transgene provide long dsRNA can triggers the RNAi pathway. The dicer enzyme is recruited in the cell (Bernstein et al., 2001) leading to the cleavage of the dsRNA into siRNA (Hamilton and Baulcombe, 1999) that are short, 5′-phosphorylated dsRNAs (21–25 nt) with two nucleotide overhangs at the 3′ end (Bernstein et al., 2001; Elbashir et al., 2001). The recruitment of siRNA-induced silencing complex (siRISC) leads to degradation of sense strand of siRNA (having the same sequence as the target gene) The siRISC is then incorporated into the antisense strand of siRNA which in association with AGO and other effector proteins brings about cleavage of the target mRNA in sequence-specific manner. The activated RISC can repeatedly participate in mRNA degradation and protein synthesis inhibition resulting into PTGS.

On the basis of origin and mechanism of biogenesis, sRNAs in plants exhibit a range of size classes, predominantly from 20 to 24 nt. Four Dicer-like nucleases (DCLs) have been reported in *Arabidopsis*(Henderson et al., 2006). Depending on their origin and structure, dsRNA triggers of RNAi are cleaved by one or more DCLs (Zamore et al., 2000; Bernstein et al., 2001). DCL4 and DCL2 are involved in processing of dsRNA with perfectly complementary to produce 20- to 22-nt long short-interfering RNA (Hamilton and Baulcombe, 1999; Hamilton et al., 2002; Hannon, 2002; Tang et al., 2003; Bouche et al., 2006). MiRNAs that are 21-nt long are produced by cleaving of partially paired dsRNA precursors by DCL1 (Denli and Hannon, 2003; Bartel, 2004). DCL3 is involved in biogenesis of 24-nt long repeat-associated sRNAs known as heterochromatic siRNAs (hcsiRNAs; Xie et al., 2005). In a complex called RISC, siRNAs and miRNAs (20- to 21-nt long) in association with AGO proteins, direct sequence-specific cleavage of complementary target RNAs or interfere with their translation (Hammond et al., 2000; Hannon, 2002; Brodersen et al., 2008). Traditionally, sRNA species that are 24-nt long (derived from transposable elements and other repetitive sequences) are linked to transcriptional gene silencing via DNA methylation and repressive chromatin modification (Hamilton et al., 2002; Zilberman et al., 2003; Liu et al., 2004; Fagegaltier et al., 2009; Law and Jacobsen, 2010; Burkhart et al., 2011). SiRNA recruits several DNA- and histone-modifying proteins including the cytosine methyltransferase, chromomethylase3 (Ossowski et al., 2008) which in turn mediate the formation of a transcriptionally inactive silent chromatin state. However, 21 nt long siRNAs may also participate in siRNA-dependent methylation of genomic loci (Pontier et al., 2012). AGO is an important protein that complexes with sRNA to form the core of the RISC. In *Arabidopsis thaliana*, 10 different AGO proteins are known to mediate the effects of several distinct types of sRNAs (Vaucheret, 2008). Endonucleolytic cleavage (slicing) activity catalyzed by the AGO protein (Huntzinger and Izaurralde, 2011) is an important step in post-transcriptional silencing. Guide RNA confers sequence specificity to any RNA silencing reaction whereas the precise nature of silencing is determined by the properties of the associated AGO protein, including its ribonuclease activity, interacting proteins, and subcellular localization.

Movement of plant sRNAs falls into two main categories: cell-to-cell (short-range) and systemic (long-range) movement (Melnyk et al., 2011). Local movement is symplastic, i.e., from the site of initiation to neighboring cells through channels called plasmodesmata (Lough and Lucas, 2006). RNA silencing also spreads systemically over long distances through the vascular phloem tissue. This long distance or systemic movement of the silencing signal takes place over days (Voinnet et al., 1998) and it is generally from photosynthetic sources (i.e., leaves) to sucrose sinks (i.e., roots and growing points) through a bulk flow process. The systemic RNA silencing signal has been identified in plants by direct sampling of phloem sap (Yoo et al., 2004; Buhtz et al., 2008) and detection of RNAs in stocks and scions of grafted plants (Palauqui et al., 1997; Schwach et al., 2005; Brosnan et al., 2007; Dunoyer et al., 2010a; Molnar et al., 2010). Mobile silencing operates in a nucleotide-sequence-specific manner and its components include sRNA molecules (21–24 nt). Molnar et al. (2010) have reported mobile sRNAs of the 24-nt size class that are associated with DNA methylation of targeted loci. It was consistent with the analysis of viral suppressors of systemic silencing in *Nicotiana benthamiana* (Hamilton et al., 2002) and the presence of 24-nt sRNA in the phloem sap of oilseed rape (Buhtz et al., 2008) and pumpkin (Yoo et al., 2004). Dunoyer et al. (2010a,b) showed that the mechanically delivered, fluorescently labeled 21- and 24-nt siRNAs move from cell to cell and over long distances. Artificial miRNAs (amiRNAs) were shown to move short distances in leaves (Schwab et al., 2006) or between the pollen vegetative cell cytoplasm and the sperm cells (Slotkin et al., 2009). Endogenous 21-nt miRNAs (miR399) could also be mobile between shoots and roots (Pant et al., 2008) and within the roots (miR165a and miR166b; Carlsbecker et al., 2010).

RNA interference can be used to achieve desirable traits in crops by manipulating the gene expression (Table 2). RNAi mediated gene silencing technique mainly involves identification of the target gene(s) followed by generating RNAi construct with hairpin cassette (gene cloned in sense and antisense orientation flanking a spacer or intron), plant transformation and finally screening and evaluating the traits.

Table 2: Crop improvement by siRNA based RNA interference

Traits improvement	Targeted gene	Plant	Reference
Plant architecture			
Biomass	OsDWARF4	Rice	Feldmann (2006)
Grain yield	OsGA20ox2	Rice	Qiao et al. (2007)
Fruit improvement			
Carotenoid and flavonoid	DET1	Tomato	Davuluri et al. (2005)
β-Carotene and lycopene	SlNCED1	Tomato	Sun et al. (2012)
Carotenoid	Lycopene epsilon cyclase	Rapeseed	Yu et al. (2007)
Decreased Ethylene	ACC oxidase	Tomato	Xiong et al. (2005)
Decreased Ethylene	ACC synthase (ACS)	Tomato	Gupta et al. (2013)
Increased shelf life	α-Man/β- Hex	Tomato	Meli et al. (2010)
Seedless fruit	Chalcone synthase	Tomato	Schijlen et al. (2007)
Nutritional improvement			
Improved stearic- and oleic- fatty acids	Δ9-desaturase and oleoyl-phosphatidylcholine γ6-desaturase	Cotton	Liu et al. (2002)
Low glutenin content	GluB	Rice	Kusaba et al. (2003)
Amylose	SBE IIa and SBE IIb	Wheat	Regina et al. (2006)
Tearless onion	Lachrymatory factor synthase (LFS)	Onion	Eady et al. (2008)
Reduction of toxic terpenoid gossypol	Delta-cadinene synthase	Cotton	Sunilkumar et al. (2006)
Biotic stress			
Virus resistance			
Bean Golden Mosaic Virus (BGMV)	AC1	Bean	Bonfim et al. (2007)
Barley Yellow Dwarf Virus (BYDV)	BYDV-PAV	Barley	Wang et al. (2000)
Rice Dwarf Virus (RDV)	PNS12	Rice	Shimizu et al. (2009)
Turnip Yellow Mosaic Virus (TYMV)	P69	Tobacco	Niu et al. (2006)
Turnip Mosaic Virus (TuMV)	HC-Pro	Tobacco	Niu et al. (2006)
Insect resistance			
Helicoverpa armigera	CYPAE14	Cotton	Mao et al. (2007)
Corn rootworm	V-ATPase A	Maize	Baum et al. (2007)
Nematode resistance			
Meloidogyne incognita	Splicing factor and integrase	Tobacco	Yadav et al. (2006)
Meloidogyne incognita	16D10	Arabidopsis	Huang et al. (2006)
Bacterial resistance			
Xanthomonas citri subsp. citri (Xcc)	PDS and CalS1	Lemon	Enrique et al. (2011)

Agrobacterium tumefaciens	iaaM and ipt	Arabidopsis	Escobar et al. (2001), Dunoyer et al. (2006)
Fungal resistance			
Magnaporthe grisea Xanthomonas oryzae	OsSSI2	Rice	Jiang et al. (2009)
Magnaporthe grisea	OsFAD7 and OsFAD8	Rice	Yara et al. (2007)
Phytophthora infestans	SYR1	Potato	Eschen-Lippold et al. (2012)
Blumeria graminis f. sp. tritici	MLO	Wheat	Riechen (2007)
Enhanced drought tolerance			
	Farnesyl transferase	Canola	Wang et al. (2009)
	C-kinase 1 (RACK1)	Rice	Li et al. (2009)
	OsDSG1	Rice	Park et al. (2010)
	OsDIS1	Rice	Wang et al. (2011)

microRNA (miRNA)

Plant miRNAs are a class of small regulatory RNAs (20–22 nt) that are encoded by endogenous miRNA genes and transcribed by RNA polymerase II into primary miRNAs (pri-miRNA) having partially double- stranded stem–loop structures (Jones-Rhoades et al., 2006). Following transcription, pri-miRNA is cleaved by a DCL1 enzyme in a two step process resulting in production of a pre-miRNA and a mature miRNA duplex (miRNA/miRNA*). The mature miRNA is incorporated into RISC and mediates the degradation of mRNA target. miRNA can down-regulate the level of protein of their target genes through either translational repression (Aukerman and Sakai, 2003; Chen, 2004; Brodersen et al., 2008), through cleavage of transcript (Llave et al., 2002a; Xie et al., 2003), or transcriptional inhibition (Bao et al., 2004; Khraiwesh et al., 2010).

In animals, miRNA mostly act by translational inhibition, where they often bind motifs in the 3′ UTRs of their targets, which show several mismatches to the miRNA. However, in plants target motifs of miRNA have few mismatches which are most often found in the coding sequences and act mostly by transcript cleavage through a mechanism closely related to RNAi. In animals, miRNA usually consist of 20–22 nucleotides where as in plants it is 20–24 nucleotides (Reinhart et al., 2002; Bartel, 2004). Plant miRNAs are less conserved compared to animal miRNAs. In plants, mostly the mature miRNAs are conserved where as miRNA precursors are usually conserved in animals (Bartel, 2004). Similar to animlas, in *Arabidopsis*, a RNAse III-like enzymes carry out processing of miRNAs from primary precursors followed by incorporation into a protein complex named RISC. However, an additional step is included in the biogenesis of plant miRNAs, i.e., the methylation on the ribose of the last nucleotide of miRNA by the methyltransferase Hen1 (Park et al., 2002). Unlike animals, plant miRNAs show high degree of sequence complementarity to their target mRNAs (Rhoades et al., 2002). Because of this fact, bioinformatics prediction of plant miRNA targets is much easier and has facilitated prediction and subsequent validation of many plant miRNA targets.

A majority of plant miRNA targets includes genes encoding for transcription factors (Rhoades et al., 2002; Mitsuda and Ohme-Takagi, 2009), few code for hormone receptors (Navarro et al., 2006), some encode enzymes (Xie et al., 2003; Fujii et al., 2005). Many plant miRNAs are known to regulate post-transcriptional gene expression and play critical roles in many developmental processes, abiotic stresses and pathogen responses (Xin et al., 2010), in nutrient homeostasis etc. Functional characterization of several plant miRNA and their target need high-throughput sequencing at global genome-level (Wang et al., 2011). MiRNAs have been extensively studied in model plants such as rice, *A. thaliana* etc.

Schwab et al. (2006) have designed a series of amiRNAs targeting different endogenous mRNAs and compared their effects to those of natural miRNAs. amiRNAs can be generated by replacing the sequences of miRNA/miRNA*within the miRNA precursor without disrupting its structural features. Like natural miRNAs, amiRNAs had varying number of target mismatches and could silence both single and multiple target genes with high specificity as determined by genome-wide expression profiling. The direct targets of amiRNAs can be accurately predicted by parameters of target selection already determined for natural miRNAs (Schwab et al., 2005).

Thus, extensive base pairing with targets is required for plant miRNA function. amiRNAs can be efficiently used for gene silencing, especially when there is need to down regulate several related, but not identical, target gene. Transgenic plants harboring amiRNAs under constitutive and inducible promoters have shown specific and efficient down-regulation of target genes of interest with limited non-autonomous effect. Thus, amiRNAs have great potential for crop improvement (Table 3).

Table 3: Crop improvement by miRNA based RNA interference

Trait improved	microRNA	Target	Plant	Reference
Drought tolerance	miR169	NFYA5	Arabidopsis	Li et al. (2008b)
	miR169	GmNFYA3	Soybean and Arabidopsis	Ni et al. (2013)
Cold tolerance	miR319	PCF5/PCF8	Rice	Yang et al. (2013)
Heat stress tolerance	miR398	CSD1 CSD2 CCS	Arabidopsis	Guan et al. (2013)
Grain size, shape, and quality	miR156	SPL14/ SPL16	Rice	Jiao et al. (2010), Miura et al. (2010), Wang et al. (2012)
	miR397	OsLAC	Rice	Zhang et al. (2013)
Parthenocarpy	miR167	ARF8	Arabidopsis and tomato	Goetz et al. (2007)
Bacterial resistance	miR393	TIR1	Arabidopsis	Navarro et al. (2006)
Cauliflower mosaic virus (CMV) resistance	amiR171	2b of CMV	Tobacco	Qu et al. (2007).
Tomato leaf curl New Delhi virus (ToLCNDV) resistance	amiR-AV1-1	AV1/AV2	Tomato	Vu et al. (2013)
Turnip Yellow Mosaic Virus (TYMV) and Turnip Mosaic Virus (TuMV) resistance	amiR159	P69 of TYMV and HC-Pro of TuMV	Arabidopsis	Niu et al. (2006)

RNAI AND ITS ROLE IN CROP IMPROVEMENT

Plant Architecture

Biomass

RNA interference can be employed successfully to improve yield of crop and fruit plants by manipulating the basic agronomic traits of plant such as height, inflorescence, branching and size. RNAi mediated knockdown of a gene *OsDWARF4* in rice resulted in shorter plants with erect leaf architecture leading to increased photosynthesis in the lower leaves. Such plant has potential for improved yields under dense planting conditions (Feldmann, 2006).

Some plant materials show recalcitrance to a process of saccharification which causes major limitation for conversion of lignocellulosic biomass to ethanol. The problem of recalcitrance of plant cell wall to bioconversion can be overcome by genetic reduction of lignin content (Chen and Dixon, 2007). Independent down-regulation of each of six lignin biosynthesis enzymes in transgenic alfalfa lines, yielded nearly twice as much sugar from stem's cell walls as compared to wild-type plants (Reddy et al., 2005). Down-regulation of lignin genes like cinnamate 4-hydroxylase (C3H), shikimate hydroxycinnamoyl transferase (HCT) and 4-coumarate-coA ligase (4CL) in plants reduced total lignin content, improved accessibility of cellulases for cellulose degradation and increased dry matter degradability (Hisano et al., 2009). This lignin modification also facilitated bypassing the need for acid pretreatment (Chen and Dixon, 2007). Over-expression of the maize Corngrass1 (Cg1) miRNA that belongs to the MIR156 family caused prolonged vegetative phase and delayed flowering (Chuck et al., 2011) resulting in increased biomass. The transgenic plants showed up to 250% more starch and improved digestibility (Chuck et al., 2011). It was reported that the degrees of morphological changes and biomass yields were related to the expression level of an exogenous rice miRNA Osa-miR156b (Fu et al., 2012). Transgenic switch grass plants with relatively high levels of miR156 showed highly stunted growth, whereas those with moderate and low levels of miR156 expression had 58–101% more biomass production than wild-type controls as a result of increases in tiller numbers. It was also observed that over-expression of rice miR156 could improve the yield of solubilised sugar as well as forage digestibility. Over-expression of miR156 led to increased biomass in other plant species including *Arabidopsis* and rice (Schwab et al., 2005; Xie et al., 2006, 2012). Cg1 over-expressing, poplar transgenics plants showed significant increase in the growth of axillary meristem, shorterning of internode length, and reduction in stem lignin content (up to 30%) compared to that of the wild-type (Rubinelli et al., 2013).

Grain Yield

Plant architecture can be manipulated in order to achieve increased grain yield like in rice plants (Jiao et al., 2010; Miura et al., 2010; Springer, 2010; Wang et al., 2012). RNAi-mediated suppression of GA 20-oxidase (*OsGA20ox2*) gene resulted in semi-dwarf plants from a taller rice variety QX1. This RNAi transgenic exhibited significant increase in panicle length, increased number of seeds per panicle and higher test (1000 grain) weight (Qiao et al., 2007). OsSPL14 (Souamosa promoter binding protein-like 14) was reported to be the target of Osa-miR156 in rice (Jiao et al., 2010; Miura et al., 2010) and was shown to positively regulate the rice plant architecture leading to enhanced yield of rice grain (Jiao et al., 2010). Higher expression of OsSPL14 could modify the rice plant architecture resulting in decreased tiller number and increased grain yield (Jiao et al., 2010; Miura et al., 2010; Wang et al., 2012). Wang et al. (2012)showed OsSPL16 to be a positive regulator of cell proliferation with increase in grain width and yield in over-expressing rice plants. However, elevated expression of OsSPL16 resulted in decrease grain appearance quality. This was expected as grain quality and yield are usually negatively correlated. But, decreased expression of OsSPL16 resulted in slender grains with better quality. Thus, miR156 and its targets OsSPL14 or OsSPL16 can be used for modifying plant architecture to create superior rice cultivars with greater yield. Recently, Guo et al. (2013) showed interactions of OsmiR444a regulated OsMADS57, OsTB1 (TEO-SINTE BRANCHED1) and D14 (Dwarf14) in controlling rice tillering. Thus, manupulating OsmiR444a and its targets by genetic engineering can prove to be an important strategy to achieve high grain yield (Guo et al., 2013). Zhang et al. (2013) reported that over-expression of OsmiR397, which is expressed naturally in young panicles and grains of rice promotes panicle branching and enlargement of grain size, causing an increase in overall grain yield of up to 25% in a field trial. OsmiR397 increases grain yield by down-regulating its target, *OsLAC* coding for a laccase-like protein that is involved in the sensitivity of plants to brassinosteroids. Since, miR397 is highly conserved across different species; this strategy can be extended to other cereal crops for increasing grain yield.

Fruit Improvement

Enhanced Nutritional Value and Edibility

Tomato (*Lycopersicon esculentum*) is one of the most economically important fruit crop across the world rich in antioxidants, minerals, fibers, and vitamins (Rajam et al., 2007). RNAi has been utilized in development of tomato fruit with enhanced level of carotenoids and flavonoids which are highly beneficial for human health. RNAi in combination with fruit specific promoter has been used to suppress an endogenous *DET1* gene in tomato, a photomorphogenesis regulatory gene involved in repression of several light controlled signaling pathways (Davuluri et al., 2005). *DET1* was specifically degraded in transgenic tomatoes with suppressed *DET1*, accompanied by an increase in the level of flavonoid and carotenoid. Abscisic acid (ABA) plays important roles during tomato fruit ripening. The SlNCED1 gene of tomato encoding 9-*cis*-epoxycarotenoid dioxygenase), an important enzyme in the ABA biosynthesis, was suppressed by RNAi. The fruits from RNAi lines showed enhanced accumulation of upstream compounds in the pathway, chiefly lycopene and β carotene. Similarly, RNAi has been utilized to increase the carotenoid content of rapeseed (*Brassica napus*) by down-regulating the expression of lycopene epsilon cyclase (-*CYC*) gene. The seeds obtained from transgenic *Brassica* showed increased in zeaxanthin, violaxanthin, β-carotene, and lutein content (Yu et al., 2007). RNAi approach has also been used in apple to improve the fruit quality by enhanced self-life (Dandekar et al., 2004), reducing the amount of a major apple allergen (Gilisen et al., 2005) and silencing the leaf sorbitol synthesis which affects fruit quality such as starch accumulation and sugar-acid balance (Teo et al., 2006).

Enhanced Shelf Life

The post-harvest deterioration and spoilage of vegetables and fruits are major causes of economic loss. Therefore, increase in shelf life of vegetables and fruits by delayed ripening are another essential agronomic trait that is being addressed through RNAi/microRNA technology. Initiation of ripening in climacteric fruits like tomato is characterized by a climacteric burst of ethylene in, resulting in the regulation of the expression of ripening-specific genes (Osorio et al., 2011). Thus, delayed ripening can be achieved in climacteric fruits by manipulating the gene involved in ethylene production and perception by employing either sense or antisense technology (Hamilton et al., 1990; Oeller et al., 1991; Theologis et al., 1993; Ye et al., 1996; Wilkinson et al., 1997; Xiong et al., 2003, 2005). However, RNAi was shown to be more effective and specific in nature than strategies involving either sense or an antisense RNA (Fire et al., 1998). Xiong et al. (2005) introduced dsRNA targeting a single unit of 1-aminocyclopropane-1-carboxylate (ACC) oxidase, a gene of ethylene biosynthesis pathway in tomato and suppressed the expression of its gene. The rate of ethylene production was significantly inhibited in the ripened fruits of transgenic plants leading to prolonged shelf life. However, a RNAi strategy targeting suppression of more than one homolog would be certainly much effective than knockdown of a single homolog. Gupta et al. (2013) used RNAi technology to develop tomatoes with delayed ripening by targeting three homologs of 1-aminocyclopropane-1-carboxylate (ACC) synthase (ACS) gene during the course of ripening. Delayed ripening leading to extended shelf life for ≈45 days was achieved by effective repression of ethylene production in tomato fruits by using chimeric RNAi-ACS construct designed to target ACS homologs. Two ripening-specific ethylene induced *N*-glycoprotein modifying enzymes, α-mannosidase (α-Man) and β-D-*N*-acetylhexosaminidase (β-Hex) were identified and targeted by Meli et al. (2010) through RNAi technique. Enhanced shelf life of ≈30 days and ≈2–2.5 fold firmer fruits were observed after texture analysis of these transgenic tomatoes as rate of fruit softening was significantly reduced.

microRNAs are involved in tomato fruit development and ripening (Pilcher et al., 2007; Itaya et al., 2008; Moxon et al., 2008; Zhang et al., 2011b; Molesini et al., 2012; Karlova et al., 2013). MiR156 targets a important gene in fruit ripening, colorless never ripe (CNR; Moxon et al., 2008; Molesini et al., 2012). Transgenic tomato plants over-expressing sly-miR156 exhibited slightly lighter color than wild-type controls, but could completely ripen eventually (just later; Zhang et al., 2011b). Furthermore, over-expression of miR156 in tomato also resulted in smaller fruit, increased leaf numbers, but reduced height and decreased leaf size (Zhang et al., 2011b), suggesting that higher fruit yield and shorter ripening time in tomato could be expected through either reducing miR156 expression or releasing one or several miR156 target genes. This needs to be tested by transgenic studies. The most recent study shows that *CNR* is also negatively regulated by *APETALA2a* (*AP2a*), a target of miR172, which positively regulates fruit ripening and negatively regulates ethylene synthesis. At the same time, *CNR* positively regulates *AP2a*, suggesting a feedback loop regulation of fruit development and ripening (Karlova et al., 2013).

Seedless Fruit Development (Parthenocarpy)

Parthenocarpy or seedlessness is a process involving production of seedless fruits, developed from the ovary without pollination and fertilization. Parthenocarpy or seedlessness is a highly desirable agronomic trait for fruit which enhances its marketing values (Molesini et al., 2012) because high yield can be achieved even under environmental conditions unfavorable for pollination and fertilization. In tomato, seedless fruits have been achieved by down-regulating a chalcone synthase, a gene involved in first step of flavonoid biosynthesis (Schijlen et al., 2007). Phytohormones such as auxin and gibberellins are closely associated with the trait of parthenocarpy (Molesini et al., 2012) which in turn are regulated by many miRNAs. Thus, manipulating the level of phytohormones by controlling activities of miRNAs or their targets can prove to be an important approach to achieve parthenocarpy. For example, expression of an aberrant form of auxin response factor 8 (ARF8), a target of miR167 (Ru et al., 2006; Wu et al., 2006; Molesini et al., 2012), resulted in parthenocarpic fruit in both *Arabidopsis* and tomato (Goetz et al., 2007). Seedless fruits were observed in tomato plants when *ARF7* function was suppressed using RNAi (De Jong et al., 2009). Parthenocarpic fruits were also observed in tomato in which genes of the AUCSIA family coding for 53-amino-acid-long (protein or peptide) were functionally suppressed by RNAi (Molesini et al., 2009).

Biotic Stress Resistance

Phytopathogens are the cause many plant diseases leading to substantial damage to crop production and thereby heavy economic loss. Several RNAi strategies have been employed to improve the defense mechanism in crop plants against various biotic stresses including attack by virus, bacteria, fungi, nematodes, and insects.

Virus Resistance

Virus-induced gene silencing (VIGS) is a RNA-mediated PTGS mechanism that protects plants against foreign gene invasion (Beclin et al., 2002; Ding, 2010). VIGS has emerged as an extremely powerful functional genomics tool for knocking out gene expression of target plant genes in some plants. The concept of pathogen derived resistance (PDR) has promoted research aimed at achieving plants resistant to virus through genetic engineering (Simon-Mateo and García, 2011). PDR is either protein mediated involving protein encoded by the transgene or RNA-mediated, i.e., by the transcript produced from the transgene. In order to achieve PDR, hairpin dsRNA including small hairpin RNA (shRNA), self-complementary hpRNA, and intron-spliced hpRNA were formed *in vivo* using inverse repeat sequences from viral genomes. Among these, PTGS with the highest efficiency was elicited by the method involving self-complementary hairpin RNAs separated by an intron (Smith et al., 2000; Wesley et al., 2001). High resistance against viruses has been observed in plants even in the presence of inverted repeats of dsRNA-induced PTGS (IR-PTGS; Beclin et al., 2002; Pandolfini et al., 2003; Zrachya et al., 2007). Resistance to *Potato Spindle Tuber Viroid* (PSTVd) infection was achieved in transgenic tomato plants producing dsRNA against PSTVd sequences (Nora et al., 2009). Similar strategy was used to successfully engineer resistance in cassava plants against *African Cassava Mosaic Virus* (ACMV; Vanderschuren et al., 2009). Virus resistance has been engineered successfully in plants by targeting the coat protein (CP) gene through RNAi. Powell-Abel et al. (1986) showed that transgenic tobacco expressing the CP gene of *Tobacco Mosaic Virus* (TMV) was resistant to TMV and that the resistance was due to the expressed CP. Later, this strategy was extended to generate resistance against many different viruses as potato resistant to *Potato Virus Y*(PVY; Missiou et al., 2004), tobacco resistant to *Beet Necrotic Yellow Vein Virus* (BNYVV; Andika et al., 2005), *Cucumis* cv. *melo* resistant to *Papaya Ring Spot Virus* type W(PRSV-W; Krubphachaya et al., 2007), *N. benthamiana* resistant to *Cucumber Green Mottle Mosaic Virus* (CGMMV; Kamachi et al., 2007), *N. benthamiana* and *Prunus domestica* resistant to *Plum Pox virus* (PPV; Hily et al., 2007).

The use of RNA silencing strategy to engineer resistance is not limited to RNA viruses but can successfully be applied to DNA viruses. For example blackgram plants recovered efficiently from geminivirus *Vigna mungo yellow mosaic virus* (VMYMV) infection when inoculated with hpRNA construct containing the promoter sequence of VMYMV under the control of the 35S promoter (Pooggin et al., 2003). RNAi method has been used to generate common bean resistant to geminivirus *Beans Golden Mosaic Virus* (BGMV; Bonfim et al., 2007). A broad-spectrum resistance has been developed against tospoviruses in tomato plants by targeting multiple regions of a viral gene (Bucher et al., 2006). Zhou et al. (2012) have used sequences from disease-specific protein gene and CP gene from *Rice Stripe Virus* to develop resistance against Rice Stripe Disease.

Many viruses express viral silencing repressors (VSRs) proteins to counter host antiviral RNA silencing (Burgyan and Havelda, 2011). One of the strategies to improve virus resistance involves targeting the miRNA against these VSR. Niu et al. (2006) developed *Arabidopsis* plants with specific resistance against *Turnip Yellow Mosaic Virus* (TYMV) and *Turnip Mosaic Virus* (TuMV) by expressing amiRNAs based on miR159 precursor of *A. thaliana*. These amiRNAs target the sequence of two VSRs, P69 of TYMV and HC-Pro of TuMV. Transgenic tobacco resistant to *Cucumber Mosaic Virus* (CMV) was generated by targeting a VSR, 2b of CMV through expression of an amiRNA based on an *A. thaliana* miR171 precursor (Qu et al., 2007). TGBp1/p25 of *Potato Virus X* (PVX) was targeted in tobacco by expression of amiRNAs based on an *A. thaliana* miR159a, miR167b, and miR171a precursors (Ai et al., 2011). Singh et al. (2014) generated transgenic tobacco plants resistant to *Tomato leaf curl New Delhi virus* (ToLCNDV) by transformation with *trans*-acting siRNA generating constructs against RNAi suppressor proteins (AC2 and AC4) of the geminivirus. Another promising strategy to reduce the multiplication and spread of virus in the plant includes the use of amiRNAs targeting the viral genes involved in replication, transmission, and symptom development after viral infection. Vu et al. (2013) used two amiRNA targeting the middle region of the AV1 (coat protein) transcript (amiR-AV1-3) and the overlapping region of the AV1 and AV2 (pre-coat protein) transcripts (amiR-AV1-1) of a geminivirus, *Tomato leaf curl virus* (ToLCV). Transgenic tomato plants expressing amiR-AV1-1, were highly tolerant to ToLCNDV and could successfully propagate until the T2 generation.

Bacterial Resistance

Bacterial diseases are extremely difficult to control due to rapid rate of spreading. RNAi mediated supression of two genes of *Agrobacterium tumefaciens* involved in crown gall tumor formation (iaaM and ipt) could significantly reduce the production of tumors in *Arabidopsis* (Escobar et al., 2001; Dunoyer et al., 2006). This strategy could be further extended to other plants. Fatty acids and their derivatives are important signaling molecule reported to negatively regulate plant's resistance to bacterial disease (Li et al., 2008a; Jiang et al., 2009). *Arabidopsis* and soybean plants showing enhanced resistance to multiple pathogens were generated by RNAi mediated suppression of *SACPD* gene encoding a fatty acid desaturase (Jiang et al., 2009).

In *Arabidopsis*, miR393 was reported to repress auxin signaling by negatively regulating the F-box auxin receptors like *TIR1*, thereby restricting the infection by bacteria *Pseudomonas syringae* (Navarro et al., 2006). Transgenic *Arabidopsis* plants over-expressing miR393 had enhanced bacterial resistance but with some developmental alterations (Navarro et al., 2006). Two different miRNAs, miR398 (Jagadeeswaran et al., 2009) and miR825 (Fahlgren et al., 2007) were reported to be down-regulated by bacterial infections. In the case of miR398, the expression of miR398 targets coding for two Cu/Zn superoxide dismutases (CSD1 and CSD2) was analyzed, and CSD1 was up-regulated upon bacterial infection in accordance with the down-regulation of miR398 under biotic stress (Jagadeeswaran et al., 2009). Very recently, new insight into miRNA function was gained with the discovery, that several miRNA families target genes of plant innate immune receptors (NBS-

LRR) in Legumes (Zhai et al., 2011) and Solanaceae (Li et al., 2012). MiR482/2118 family of miRNAs were shown to target numerous NBS-LRR mRNAs encoding disease resistance proteins in tomato (*Solanum lycopersicum*) and other members of the Solanaceae (Shivaprasad et al., 2012). Viral and bacterial infection suppresses miR482-mediated silencing of *R* genes. Thus, these pathogen responsive miRNA are either up- or down-regulated in response to bacterial invasion and effect gene expression by suppressing negative regulators and inducing positive regulators of immune responses. Identification and characterization of the targets of these miRNAs would help decipher new players in the pathways of host defense. If these pathogen-regulated miRNAs serve as positive regulator of bacterial resistance, strategy to generate transgenics with enhanced bacterial resistance involves constitutive over-expression of miRNA or amiRNA. When miRNAs act as negative regulators, transgenic plants over-expressing these miRNAs become more sensitive to bacteria. In these cases, up-regulation of their target genes might be an effective strategy for improving plant stress tolerance, which can be achieved by over-expressing a miRNA-resistant form of its target or using amiRNA target mimic (Franco-Zorrilla et al., 2007).

Fungal Resistance

Targeting the genes of fatty acid metabolism through RNAi has proved to be a important strategy to generate diseases tolerance in various crop plants. RNAi-mediated suppression of a rice gene *OsSSI2* led to enhanced resistance to blast fungus *Magnaporthe grisea* and leaf blight bacterium *Xanthomonas oryzae*(Jiang et al., 2009). Besides, enhanced disease resistance against *M. grisea* in rice was achieved by suppression of two genes namely *OsFAD7* and *OsFAD8* which are Ω-3 fatty acid desaturase (Yara et al., 2007). RNAi mediated targeting of genes involved in lignin production, led to enhanced resistance of soybean to phytopathogen *Sclerotinia sclerotiorum* due to reduced lignin content (Peltier et al., 2009).

Very recently, 24 miRNAs were shown to be involved in responses to attack by the fungus *Blumeria graminis* f. sp. tritici (Bgt) in wheat which cause a devastating diseases of wheat-powdery mildew (Xin et al., 2010). A rice miRNA osa-miR7695 was found to negatively regulate a natural resistance-associated macrophage protein 6 (OsNramp6) in response to the blast fungus *Magnaporthe oryzae*. Improved resistance to rice blast infection was achieved by over-expression of Osa-miR7696 (Campo et al., 2013).

Insect and Nematode Resistance

RNA interference has also been applied to control insect pests which lead to substantial crop loss (Huvenne and Smagghe, 2010). The development of a new generation of insect-resistant crops involves feeding of dsRNA as a diet component to insect which was shown to efficiently down-regulate the targeted genes in insect (Price and Gatehouse, 2008). The strategy was used in corn plants by expression of dsRNAs for tubulin or vacuolar ATPase genes to develop western corn rootworm (WCR) resistant transgenic corn plants (Baum et al., 2007). Cotton bollworm (*Helicoverpa armigera*) larvae showed reduced growth when fed on plant material expressing dsRNA specific to cytochrome P450 gene (*CYP6AE14*; Mao et al., 2007). Silencing of genes involved in parasitism or housekeeping genes in the root-knot nematode by expression of dsRNA in host plant resulted in enhanced resistance to the nematode (Gheysen and Vanholme, 2007). Sindhu et al. (2009) has targeted four genes involved in parasitism of sugar beet cyst nematode (*Heterodera schachtii*), having host *A. thaliana* by RNAi. Though complete resistance was not achieved but the number of mature nematode females in different RNAi lines were reduced to 23–64%. Ibrahim et al. (2011) could successfully reduce the gall formation by *Meloidogyne incognita* in soybean roots by suppressing the genes encoding tyrosine phosphatase (TP) and mitochondrial stress-70 protein precursor (MSP).

microRNAs were also reported to be involved in plant nematode interactions. In response to infection by the nematode, *H. schachtii*, miR161, miR164, miR167a, miR172c, miR396c, miR396a,b, and miR398a were down-regulated in *Arabidopsis* (Hewezi et al., 2008; Khraiwesh et al., 2012). Comparative analysis of miRNA profiling in soybean indicated that 101 miRNAs belonging to 40 families were responsive to the infection of the soybean cyst nematode (SCN; *Heterodera glycines*), the most devastating pathogen in soybean. It was further revealed that 20 miRNAs were differentially expressed between SCN resistant and susceptible soybean cultivars (Li et al., 2012). Moreover, it has been suggested that nematode induced miRNAs and sRNAs likely to be involved in feeding site establishment and parasitism, respectively (Hewezi et al., 2008). Over-expression of these nematode induced candidate miRNAs, and/or silencing of their corresponding targets, may provide clear insight about plant-nematode parasitism, and also may provide nematode resistance to crop plants. Nematode resistance can also be achieved by expressing amiRNA, containing known miRNA genes with the replaced seed region of the vital gene (parasitism or housekeeping) of plant parasitic nematodes.

Abiotic Stress Tolerance

Drought and Salinity Tolerance

Water deficit or drought is one the major environmental stresses limiting crop productivity. RNAi has been applied successfully to develop drought tolerant crops. RNAi mediated down-regulation of farnesyl transferase in canola using the *AtHPR1* promoter showed more resistance to seed abortion during flowering induced by water deficiency without affecting yield during drought stress (Wang et al., 2009). Transgenic rice plants tolerant to drought stress were developed by silencing of receptor for activated C-kinase 1 (*RACK1*; Li et al., 2009). The ubiquitin ligase gene has been targeted for RNAi in rice to enhance drought tolerance. Rice knockdown of a RING finger E3 ligase gene-*OsDSG1* leads to enhanced drought tolerance (Park et al., 2010). Similarly, silencing of *OsDIS1* (for *Oryza sativa* drought-induced SINA protein 1), a C3HC4 RING finger E3 ligase by RNAi enhanced drought tolerance.

Wang et al. (2011) studied legume model plant, *Medicago truncatula* and identified several drought-responsive miRNAs. The already known and predicted targets of these miRNAs were found to be involved in wide variety of processes in plant cell including development, transcription, protein degradation, detoxification, etc. It was observed that miR393 was strongly up-regulated by dehydration, cold, high salinity, and ABA treatments (Sunkar and Zhu, 2004). On exposing *Arabidopsis* to varying degrees of different abiotic stress, up regulation was observed in miR402, miR319c, miR389a, and miR397b. Xin et al. (2010) identified and reported 12 miRNAs responsive to heat stress in wheat (*Triticum aestivum* L.). MiRNA responsive to various stresses (drought, cold, salinity, and ABA) were identified in rice seedlings (Jian et al., 2010). Among many stress responsive miRNAs discovered in rice, only two miR393 and miR169 were found to be responsive to abiotic stress like dehydration (Zhao et al., 2007). In *Arabidopsis*, miR169 was down-regulated by drought stress, and its target nuclear factor YA5 (NF-YA5) was significantly induced upon drought stress. Transgenic plants over-expressing miR169a were more sensitive to drought stress with increased leaf water loss as compared to wild-type controls, and over-expression of NF-YA5 led to enhanced drought tolerance in transgenic plants (Li et al., 2008a). Similarly, a newly characterized Soybean (*Glycine max* L.) gene *GmNFYA3*, a target of miR169, was demonstrated to positively modulate drought stress tolerance in transgenic *Arabidopsis* plants (Ni et al., 2013). However, over-expression of *GmNFYA3* in *Arabidopsis* resulted in increased sensitivity to salinity stress and exogenous ABA (Ni et al., 2013). These findings were contrary to the observation in tomato (*S. lycopersicum*), in which miR169 was induced by drought stress and its four targets NF-YA1/2/3 and multidrug resistance-associated protein gene 1 (*MRP1*) were all down-regulated by drought stress. Constitutive over-expression of tomato miR169c led to reduced stomatal openings, transpiration rate, and leaf water loss, thus enhanced drought tolerance in transgenic plants in comparison to wild-type controls (Zhang et al., 2011a). It is notable that transgenic tomato plants had no noticeable morphological and developmental alterations under field conditions, indicating that miR169 or its targets could be potential candidate genes for genetic engineering to achieve enhanced abiotic stress tolerance in transgenic plants (Zhang et al., 2011a). However, as mentioned above, miR169 positively regulates plant response to drought stress in tomato, but negatively in *Arabidopsis*. The opposite roles miR169 play in different plant species suggest that even conserved miRNAs might function in a species-specific manner and this should be taken into careful consideration when tailoring GM strategies for specific target species.

In a recent study, transgenic creeping bentgrass (*Agrostis stolonifera*) plants over-expressing a rice miR319 gene (Osa-miR319) exhibited enhanced tolerance to drought and salinity that was associated with increased leaf wax content and water retention but reduced sodium uptake. Gene expression analysis indicated that the enhanced abiotic stress tolerance can be attributed to a significant down-regulation of at least four putative turf miR319 target genes teosinte branched/

cycloidea/proliferating factors (TCP)—AsPCF5, AsPCF6, AsPCF8, and AsTCP14 and a homolog of a rice NAC domain gene AsNAC60 (Zhou et al., 2013). It is noticeable that over-expression of Osa-miR319 caused pleiotropic phenotypes including increased leaf size, enlarged stems and decreased tiller number, which are undesirable traits for creeping bent grass. To better utilize the miR319-based GM strategy, further functional characterization of miR319 targets or downstream genes in the related biological pathways needs to be undertaken. However, the mechanism of miR319-mediated plant abiotic stress response revealed from this study provides important information to develop novel strategies for plant genetic engineering and has the potential to be applied in other important crop species. MiR159 was reported to respond to hormone signaling and dehydration responses in *Arabidopsis* (Achard et al., 2004; Reyes and Chua, 2007).

Hu et al. (2009) showed that most of the rice histone deacetylases genes are responsive to drought or salt stresses with specific patterns of expression and divergent developmental functions compared to closely related homologs in *Arabidopsis*. Trindade et al. (2010) identified several conserved miRNAs in *M. truncatula* plants which are expressed differentially in water-deficit condition. MiR169 was shown to be down-regulated in roots, whereas miR398a/b and miR408 showed very high expression in shoots as well as roots. Hwang et al. (2011) reported a drought stress-responsive miR171 family in potato plants, *Solanum tuberosum* which miR171a, miR171b, and miR171c. The RNAi-mediated silencing of farnesyl transferase genes (*FTA* or *FTB*) in canola resulted in reduced transpiration rate due to decreased stomatal conductance, thereby promoting yield (Wang et al., 2005, 2009).

Cold and Heat Stress Tolerance

The expression of miR319 was reported to change in response to cold stress in *Arabidopsis* (Sunkar and Zhu, 2004; Liu et al., 2008), rice (Lv et al., 2010), and sugarcane (Thiebaut et al., 2011). Further transgenic studies using wild-type plants as controls indicated that over-expression of Osa-miR319 gene led to increased cold stress tolerance (4°C) after chilling acclimation(12°C) of plants (Yang et al., 2013). However, miR319 transgenic rice plants displayed severe developmental delay. To avoid the pleiotropic effect of miR319, two RNAi lines for the miR319 targets, OsPCF5 and OsTCP21, were generated. These RNAi lines, upon chilling acclimation, also exhibited better cold tolerance than wild-type controls, but were phenotypically normal. It should be noted that improvement of plant cold tolerance in the miR319 over-expression lines was more significant than that in the OsPCF5 and OsTCP21 RNAi lines (Yang et al., 2013), most likely due to target function redundancy in the latter case. This is also a challenge when manipulating miRNA targets instead of miRNAs themselves for plant trait modification, because in some cases multiple targets have to be simultaneously down-regulated to achieve the same level of effect as over-expression of an individual miRNA gene. Guan et al. (2013) discovered a novel plant thermotolerance mechanism, especially for the protection of reproductive organs. It involves induction of miR398 to downregulate its targets CSD (copper/zinc superoxide dismutase) genes, *CSD1* and *CSD2* as well as *CCS* (a gene encoding copper chaperone for both *CSD1* and *CSD2*; Guan et al., 2013). They found that csd1, csd2, and ccs mutants displayed higher heat stress tolerance than wild-type plants associated with increased accumulation of heat stress transcription factors and heat shock proteins and reduced damage to flowers (Guan et al., 2013). These results strongly suggest that manipulating miR398 or its targets can be an applicable strategy to increase heat tolerance in crop species, especially in corn, which suffers damage to its reproductive tissues by prolonged periods of high summer temperatures.

Oxidative Stress Tolerance

Plants tend to accumulate reactive oxygen species (ROS) in response to environmental stimuli such as high intensity light, extreme temperatures, UV radiation, heavy metals, salinity, drought stresses, and mechanical stresses (Khraiwesh et al., 2012). Superoxide dismutases (SODs) in plants can detoxify superoxide radicals by converting them into molecular oxygen and hydrogen peroxide. Several studies have been conducted to improve plant stress tolerance by over-expression of superoxide dismutase (Cu/Zn-SODs) in transgenic plants which detoxify superoxide radicals (Tepperman and Dunsmuir, 1990; Pitcher et al., 1991; Gupta et al., 1993; Perl et al., 1993; Sunkar et al., 2006). However, in some of the studies, the transgenic plants exhibited minimal or no increase in stress tolerance (Tepperman and Dunsmuir, 1990; Pitcher et al., 1991). Sunkar et al. (2006) offered an improved strategy to solve this problem by over-expressing miR398-resistant form of *CSD2*, which led to increased tolerance to high intensity light, heavy metals and other oxidative stresses (Table 1; Sunkar et al., 2006). The exploration of interactions between miR398 and its targets (*CSD1* and *CSD2*) also provided a possible explanation to the previously failed attempts. The expression of the introduced SOD transgenes containing the miR398 target sites were negatively impacted by miR398-mediated gene regulation.

Nutritional Improvement

Biofortification

Essential fatty acids have important role in maintaining heart health in human. RNAi technology has also been used successfully or modification of fatty acid composition of oil. Two key fatty acid desaturase genes encoding stearoyl-acyl-carrier protein Δ9-desaturase and oleoyl-phosphatidylcholine γ6-desaturase (Liu et al., 2002) targeted by RNAi for nutritional enhancement of cotton seed oil with high-oleic (HO) and high-stearic (HS) contents. Soybean oil flavor and stability can be improved and process of hydrogenation can be avoided by reducing the content of alpha-linolenic acid (18:3). Flores et al. (2008) have used hairpin RNA mediated RNAi strategy (glycinin promoterwas used for seed-specific silencing) to down-regulate omega-3 fatty acid desaturase [an enzyme converting linoleic acid (18:2) to alpha-linolenic acid (18:3)]. Transgenic soybean seed were reported to significant reduction in alpha-linoleic acid content (1–3%) compared to non-transgenic soybean seed (7–10%).

Seed-specific RNAi approaches have also successfully been used to generate dominant high lysine corn by suppressing the expression of 22-kDa maize zein storage proteins, a group of abundant proteins in maize seed but poor in lysine content (Segal et al., 2003). RNAi was also used to down-regulate the starch-branching enzyme resulting in high-amylose wheat, which has a great potential to improve human health (Regina et al., 2006). RNAi technology can be applied to increase the starch content in the leaves. The process of phosphorylation and dephosphorylation of starch are crucial steps of leaf starch degradation. Based on this fact, starch content has been increased by manipulation of phosphate metabolism genes in *Arabidopsis* and maize (Weise et al., 2012). RNAi-mediated gene silencing has been utilized to reduce glutenin content in rice which is difficult to digest by kidney patients. To achieve low glutenin content GluB hairpin RNA was employed by Kusaba et al. (2003) to produce a rice variety called *LGC-1* (low glutenin content 1).

The two main enzymatic steps involved in sucrose biosynthesis involve synthesis of sucrose-6-phosphate (Suc6P) by sucrose-phosphate synthase (SPS) which is hydrolysed to sucrose and inorganic phosphate (Pi) by an another enzyme sucrose phosphatase (SPP). At low temperature (4°C), potato tubers undergo an unfavorable phenomenon of, i.e., 'cold sweetening' which involves conversion of starch to reducing sugars glucose and fructose. To address this problem of cold induced sweetening, Chen et al. (2008) have targeted *NtSPP2* gene of tobacco for RNAi mediated suppression of SPP in transgenic potato tubers. The transgenic potato tubers showed accumulation of Suc6P when exposed to cold storage. In SPP-silenced tubers, the process of sucrose-to-hexose conversion was reduced as cold induced expression of vacuolar invertase (VI) was found to be blocked.

RNA interference technology was used by Gil-Humanes et al. (2008) to significantly reduce the level of γ-gliadins in different wheat cultivars by silencing the expression of specific γ-gliadins. Gil-Humanes et al. (2012) further, reported slight increase in the protein content of transgenic lines due to compensatory effect produced by down-regulation of γ-gliadins on the rest of gluten proteins. The glutenin content was found to increase whereas no significant changes were observed in the total gliadins content.

Allergen and Anti-Nutrient Elimination

Certain food like peanuts, apple, etc. consists of certain allergens that lead to allergic response when consumed. Besides many commonly consumed plants consist of natural toxin which is harmful to human health so, reduction or elimination of these allergens or toxins from food by RNAi will enhance its edibility and food quality. Expression of one of the major apple allergen Mald1 was reduced significantly by Gilisen et al. (2005) through RNAi-based gene silencing. RNAi was used by Le et al. (2006) to suppress the expression of a tomato allergen-Lyce3 which is a non-specific lipid transfer protein from tomato peel. Dodo et al. (2008) reduced the content of Arah2, a major peanut allergen by 25% in crude peanut extract. In onion, Eady et al. (2008) using hp RNAi suppressed the tear inducing lachrymatory factor synthase (LFS) gene which is involved in conversion of 1-propenyl sulfenic acid to propanthial *S*-oxide [tear-inducing, lachrymatory factor (LF)] resulting in the production of tearless onion. The RNAi technology has also been used to reduce the level of, a neurotoxin β-*N*-oxalyl-amino-alanine-L-alanine (BOAA) in the Grass pea. (*Lathyrus sativus*) or chickling pea which can cause paralytic disease called lathyrism (Angaji et al., 2010). Though cotton seeds are rich source of dietary protein but consist of a toxic terpenoid-gossypol which plays an important role in plant defense mechanism. RNAi has been used to lower gossypol content in cotton seeds by reducing seed specific expression of delta-cadinene synthase, an enzyme in gossypol biosynthesis pathway (Sunilkumar et al., 2006). Recently, Rathore et al. (2012)reported that RNAi-mediated silencing of delta-cadinene synthase gene resulted in cotton plants that produced ultra-low gossypol cottonseed (ULGCS). Lewis et al. (2008) used RNAi for silencing the nicotine demethylase gene to suppress the conversion of nicotine to nor-nicotine by six fold which is precursor of a carcinogenic compound in tobacco (*N. tabacum*).

CONCLUSION AND FUTURE PROSPECTS

RNA interference technology involving siRNA and miRNA have emerged as an attractive tool used by plant biologists not only to decipher the plant function but also to develop plants with improved and novel traits by manipulation of both desirable and undesirable genes. It is a powerful tool to understand the functions of individual genes and also proved useful to molecular breeders in producing improved crop varieties. sRNAs have been utilized either for silencing a gene of interest by RNAi or altering the expression of a gene by over-expressing a miRNA or introduction of artificially synthesized miRNA targeting it. The technology has well been utilized in crops not only to improve the food productivity (biomass and grain yield) but its nutritional value (cereals, fruits, and vegetables enriched with essential minerals, vitamins, fatty acids, and amino acids). RNAi technology has also been exploited to develop plants with improved resistance against various environmental stresses (especially drought) and biotic stresses like pathogen attack (virus, bacteria, insects, and nematodes).

Thus applications of research related with small non-coding RNAs has the potential to contribute to food safety by generating crop cultivars with improved agronomic traits leading to increasing yield and nutritional value. Research focussing on interaction of miRNAs and their targets can provides valuable insight into mechanisms of post-transcriptional gene regulation and multiple molecular pathways controlling plant stress responses. After completion of several recent genome sequencing projects, the study of RNAi has become increasingly more expanding and challenging. Researchers can manipulate several newly explored RNAi pathways in order to amend gene expression in crops, which can result in new and better quality traits.

Although RNAi technology can serve as potential tools for crop improvement but certain limitations are also associated with it. Altering the expression of a target gene or its miRNA might lead to undesirable pleiotropic changes in plant morphology and development. Therefore transgenic strategies should be designed only after completely understanding the mechanism of miRNA regulation. This will minimize undesirable trade-off in transgenic products. MiRNA strategy useful for one plant species may not work for another species, since regulations of evolutionarily conserved miRNAs may vary in different species (Li et al., 2008a; Zhang et al., 2011a). A family of related genes can be silenced with only one RNAi construct due to the sequence heterogeneity of siRNAs, but can cause off-target effects leading to unintentional silencing of genes with regions of homology to the intended target. Besides, post-transcriptional silencing in plants is mobile that can be induced locally and then spreads throughout the plant. Thus, siRNA-based RNAi strategies might not be suitable for some applications requiring tissue-specific silencing of genes. Atificial miRNAs can be used to overcome these limitations as it brings about much more specific silencing of genes. There is a reduced chance of off-target effects with miRNA-based RNAi, since miRNAs use only a single 21- or 22-nt sequence to identify the target, whereas siRNAs comprise a population of sequences. Thus, with miRNA-based RNAi, it is easier to target individual genes, even in a closely related gene family. amiRNAs are also suitable for tissue-specific silencing because miRNA-directed silencing does not spread throughout the plant. However, silencing might not be very durable with amiRNA because only a single 21- or 22-nt specificity determinant is involved.

Mutation of the target can lead to easier escape from miRNA-directed silencing compared to siRNA-directed silencing which targets much larger sequence. This problem can be overcome by using two (or more) different amiRNAs against the target.

Crops engineered with constructs encoding heterologous proteins or over-expressing protein are fundamentally different from those consisting of RNA-mediated gene suppression cassettes because these constructs do not encode any protein(s) and are intended to express only non-coding RNAs. Transgenic proteins are recommended to undergo digestibility studies, to evaluate their potential for digestion, as allergenic and toxic proteins may be refractory to digestion (Astwood et al., 1996). RNA is not known to produce oral toxicity in humans but proteins in some rare cases can produce oral toxicity. According to the US FDA, "Introduced nucleic acids [in biotech crops], in and of themselves, do not raise safety concerns (FDA, 1992)." Furthermore, with regards to RNA-mediated gene regulation in genetically engineered crops, the US FDA states, "thus, for example, the introduction of a gene encoding an anti-sense ribonucleic acid (RNA) would not raise concerns about either the gene or the anti-sense RNA. Any safety considerations would focus on the intended effects of the anti-sense RNA." Thus, transgenic crops with introduced RNA based traits are much safe for human consumptions than crops over-expressing proteins and need not require acute oral toxicity studies and evaluation of digestibility of the introduced RNA component. Certain bio safety concerns do arise on use of RNAi transgenic as transcriptional gene silencing by chromatin modification might lead to hereditary changes associated with adverse effects. However, crop plants developed with thoughtfully designed RNAi strategy and proper assessment of risk related to food safety may help overcome the biosafety concerns. In spite of few limitations, crop improvement strategies based on small non-coding RNAs have enormous potential to increase productivity as well as nutritional value.

ACKNOWLEDGMENTS

We would like to thank the reviewers for their comments and suggestions which helped to improve the manuscript. We sincerely apologize to those authors whose work we have inadvertently omitted, or could not review at length due to space limitations.

REFERENCES

1. Achard, P., Herr, A., Baulcombe, D. C., and Harberd, N. P. (2004). Modulation of floral development by gibberellin-regulated microRNA. *Development* 131, 3357–3365. doi: 10.1242/dev.01206
2. Ai, T., Zhang, L., Gao, Z., Zhu, C. X., and Guo, X. (2011). Highly efficient virus resistance mediated by artificial microRNAs that target the suppressor of PVX and PVY in plants. *Plant Biol.* 13, 304–316. doi: 10.1111/j.1438-8677.2010.00374.x
3. Andika, I. B., Kondo, H., and Tamada, T. (2005). Evidence that RNA silencing-mediated resistance to beet necrotic yellow vein virus is less effective in roots than in leaves. *Mol. Plant Microbe Interact.* 18, 194–204. doi: 10.1094/MPMI-18-0194
4. Angaji, S. A., Hedayati, S. S., Hosein poor, R., Samad poor, S., Shiravi, S., and Madani, S. (2010). Application of RNA interference in plants. *Plant Omics J.* 3, 77–84.
5. Astwood, J. D., Leach, J. N., and Fuchs, R. L. (1996). Stability of food allergens to digestion in vitro. *Nat. Biotechnol.* 14, 1269–1273. doi: 10.1038/nbt1096-1269
6. Aukerman, M. J., and Sakai, H. (2003). Regulation of flowering time and floral organ identity by a microRNA and its APETALA2-like target genes. *Plant Cell* 15, 2730–2741. doi: 10.1105/tpc.016238
7. Axtell, M. J. (2013). Classification and comparison of small RNAs from plants. *Annu. Rev. Plant Biol.* 64, 137–159. doi: 10.1146/annurev-arplant-050312-120043
8. Bao, N., Lye, K. W., and Barton, M. K. (2004). MicroRNA binding sites in *Arabidopsis* class III HD-ZIP mRNAs are required for methylation of the template chromosome. *Dev. Cell* 7, 653–662. doi: 10.1016/j.devcel.2004.10.003
9. Bartel, D. P. (2004). MicroRNAs: genomics, biogenesis, mechanism, and function. *Cell* 116, 281–297. doi: 10.1016/S0092-8674(04)00045-5
10. Baulcombe, D. (2000). Unwinding RNA silencing. *Science* 290, 1108–1109. doi: 10.1126/science.290.5494.1108
11. Baum, J. A., Bogaert, T., Clinton, W., Heck, G. R., Feldmann, P., Ilagan, O., et al. (2007). Control of coleopteran insect pests through

RNA interference. *Nat. Biotechnol.* 25, 1322–1326. doi: 10.1038/nbt1359

12. Beclin, C., Boutet, S., Waterhouse, P., and Vaucheret, H. (2002). A branched pathway for transgene-induced RNA silencing in plants. *Curr. Biol.* 12, 684–688. doi: 10.1016/S0960-9822(02)00792-3
13. Bernstein, E., Caudy, A. A., Hammond, S. M., and Hannon, G. J. (2001). Role for a bidentate ribonuclease in the initiation step of RNA interference. *Nature* 409, 363–366. doi: 10.1038/35053110
14. Bonfim, K., Faria, J. C., Nogueira, E. O., Mendes, E. A., and Aragao, F. J. (2007). RNAi-mediated resistance to bean golden mosaic virus in genetically engineered common bean (*Phaseolus vulgaris*). *Mol. Plant Microbe Interact.* 20, 717–726. doi: 10.1094/MPMI-20-6-0717
15. Bouche, N., Lauressergues, D., Gasciolli, V., and Vaucheret, H. (2006). An antagonistic function for *Arabidopsis* DCL2 in development and a new function for DCL4 in generating viral siRNAs. *EMBO J.* 25, 3347–3356. doi: 10.1038/sj.emboj.7601217
16. Brodersen, P., Sakvarelidze-Achard, L., Bruun-Rasmussen, M., Dunoyer, P., Yamamoto, Y. Y., Sieburth, L., et al. (2008). Widespread translational inhibition by plant miRNAs and siRNAs. *Science* 320, 1185–1190. doi: 10.1126/science.1159151
17. Brosnan, C. A., Mitter, N., Christie, M., Smith, N. A., Waterhouse, P. M., and Carroll, B. J. (2007). Nuclear gene silencing directs reception of long-distance mRNA silencing in *Arabidopsis*. *Proc. Natl. Acad. Sci. U.S.A.* 104, 14741–14746. doi: 10.1073/pnas.0706701104
18. Bucher, E., Lohius, D., van Poppel, P. M., Geerts-Dimitriadou, C., Gold-bach, R., and Prins, M. (2006). Multiple virus resistance at a high frequency using a single transgene construct. *J. Gen. Virol.* 87, 697–701. doi: 10.1099/vir.0.82276-0
19. Buhtz, A., Springer, F., Chappell, L., Baulcombe, D. C., and Kehr, J. (2008). Identification and characterization of small RNAs from the phloem of *Brassica napus*. *Plant J.* 53, 739–749. doi: 10.1111/j.1365-313X.2007.03368.x
20. Burgyan, J., and Havelda, Z. (2011). Viral suppressors of RNA silencing. *Trends Plant Sci.* 16, 265–272. doi: 10.1016/j.

tplants.2011.02.010

21. Burkhart, K. B., Guang, S., Buckley, B. A., Wong, L., Bochner, A. F., and Kennedy, S. (2011). A pre-mRNA-associating factor links endogenous siRNAs to chromatin regulation. *PLoS Genet.* 7:e1002249. doi: 10.1371/journal.pgen.1002249
22. Campbell, T. N. (2005). Choy FYM, RNA interference: past, present and future. *Curr. Issues Mol. Biol.* 7, 1–6.
23. Campo, S., Peris-Peris, C., Sire, C., Moreno, A. B., Donaire, L., Zytnicki, M., et al. (2013). Identification of a novel microRNA (miRNA) from rice that targets an alternatively spliced transcript of the Nramp6 (Natural resistance-associated macrophage protein 6) gene involved in pathogen resistance. *New Phytol.* 199, 212–227. doi: 10.1111/nph.12292
24. Carlsbecker, A., Lee, J. Y., Roberts, C. J., Dettmer, J., Lehesranta, S., Zhou, J., et al. (2010). Cell signalling by microRNA165/6 directs gene dose-dependent root cell fate. *Nature* 465, 316–321. doi: 10.1038/nature08977
25. Chen, F., and Dixon, R. A. (2007). Lignin modification improves fermentable sugar yields for biofuel production. *Nat. Biotechnol.* 25, 759–761. doi: 10.1038/nbt1316
26. Chen, S., Hajirezaei, M. R., Zanor, M. I., Hornyik, C., Debast, S., Lacomme, C., et al. (2008). RNA interference-mediated repression of sucrose-phosphatase in transgenic potato tubers (*Solanum tuberosum*) strongly affects the hexose-to-sucrose ratio upon cold storage with only minor effects on total soluble carbohydrate accumulation. *Plant Cell Environ.* 31, 165–176.
27. Chen, X. (2004). A microRNA as a translational repressor of APETALA2 in *Arabidopsis* flower development. *Science* 303, 2022–2025. doi: 10.1126/science.1088060
28. Chuck, G. S., Tobias, C., Sun, L., Kraemer, F., Li, C., Dibble, D., et al. (2011). Overexpression of the maize corngrass1 microRNA prevents flowering, improves digestibility, and increases starch content of switchgrass. *Proc. Natl. Acad. Sci. U.S.A.* 108, 17550–17555. doi: 10.1073/pnas.1113971108
29. Cogoni, C., Irelan, J. T., Schumacher, M., Schmidhauser, T., Selker, E. U., and Macino, G. (1996). Transgene silencing of the al-1 gene

in vegetative cells of *Neurospora* is mediated by a cytoplasmic effector and does not depend on DNA-DNA interactions or DNA methylation. *EMBO J.* 15, 3153–3163.

30. Dandekar, A. M., Teo, G., Defilippi, B. G., Uratsu, S. L., Passey, A. J., Kader, A. A., et al. (2004). Effect of down-regulation of ethylene biosynthesis on fruit flavor complex in apple fruit. *Transgenic Res.* 13, 373–384. doi: 10.1023/B:TRAG.0000040037.90435.45
31. Davuluri, G. R., Tuinen, A., Fraser, P. D., Manfredonia, A., Newman, R., Burgess, D., et al. (2005). Fruit-specific RNAi-mediated suppression of DET1 enhances carotenoid and flavonoid content in tomatoes. *Nat. Biotechnol.* 23, 890–895. doi: 10.1038/nbt1108
32. De Jong, M., Wolters-Arts, M., Feron, R., Mariani, C., and Vriezen, W. H. (2009) The *Solanum lycopersicum* auxin response factor 7 (SlARF7) regulates auxin signaling during tomato fruit set and development. *Plant J.* 5, 160–170. doi: 10.1111/j.1365-313X.2008.03671.x
33. Denli, A., and Hannon, G. (2003) RNAi: an ever-growing puzzle. *Trends Biochem. Sci.* 28, 196–201. doi: 10.1016/S0968-0004(03)00058-6
34. Ding, S. W. (2010) RNA-based antiviral immunity. *Nat. Rev. Immunol.* 10, 632–644. doi: 10.1038/nri2824
35. Dodo, H. W., Konan, K. N., Chen, F. C., Egnin, M., and Viquez, O. M. (2008) Alle viating peanut allergy using genetic engineering: the silencing of the immunodominant allergen Ara h 2 leads to its significant reduction and a decrease in peanut allergenicity. *Plant Biotechnol. J.* 6, 135–145. doi: 10.1111/j.1467-7652.2007.00292.x
36. Dunoyer, P., Himber, C., and Voinnet, O. (2006) Induction, suppression and requirement of RNA silencing pathways in virulent *Agrobacterium tumefaciens* infections. *Nat. Genet.* 38, 258–263. doi: 10.1038/ng1722
37. Dunoyer, P., Schott, G., Himber, C., Meyer, D., Takeda, A., Carrington, J. C., et al. (2010a) Small RNA duplexes function as mobile silencing signals between plant cells. *Science* 328, 912–916. doi: 10.1126/science.1185880
38. Dunoyer, P., Brosnan, C. A., Schott, G., Wang, Y., Jay, F., Alioua, A., et al. (2010b) An endogenous, systemic RNAi pathway in

plants. *EMBO J.* 29, 1699–1712. doi: 10.1038/emboj.2010.65

39. Eady, C. C., Kamoi, T., Kato, M., Porter, N. G., Davis, S., Shaw, M., et al. (2008) Silencing onion lachrymatory factor synthase causes a significant change in the sulfur secondary metabolite profile. *Plant Physiol.* 147, 2096–2106. doi: 10.1104/pp.108.123273
40. Elbashir, S. M., Lendeckel, W., and Tuschl, T. (2001) RNA interference is mediated by 21-and 22-nucleotide RNAs. *Genes Dev.* 15, 188–200. doi: 10.1101/gad.862301
41. Ender, C., and Meister, G. (2010) Argonaute proteins at a glance. *J. Cell Sci.* 123, 1819–1823. doi: 10.1242/jcs.055210
42. Enrique, R., Siciliano, F., Favaro, M. A., Gerhardt, N., Roeschlin, R., Rigano, L., et al. (2011) Novel demonstration of RNAi in citrus reveals importance of citrus callose synthase in defence against *Xanthomonas citrisubsp.*Citri. *Plant Biotechnol. J.* 9, 394–407. doi: 10.1111/j.1467-7652.2010.00555.x
43. Eschen-Lippold, L., Landgraf, R., Smolka, U., Schulze, S., Heilmann, M., Heilmann, I., et al. (2012) Activation of defense against *Phytophthora infestans* in potato by down-regulation of syntaxin gene expression. *New Phytol.* 193, 985–996. doi: 10.1111/j.1469-8137.2011.04024.x
44. Escobar, M. A., Civerolo, E. L., Summerfelt, K. R., and Dandekar, A. M. (2001) RNAi-mediated oncogene silencing confers resistance to crown gall tumorigenesis. *Proc. Natl. Acad. Sci. U.S.A.* 98, 13437–13442. doi: 10.1073/pnas.241276898
45. Fagegaltier, D., Bouge, A. L., Berry, B., Poisot, E., Sismeiro, O., Coppee, J. Y., et al. (2009) The endogenous siRNA pathway is involved in heterochromatin formation in *Drosophila*. *Proc. Natl. Acad. Sci. U.S.A.* 106, 21258–21263. doi: 10.1073/pnas.0809208105
46. Fahlgren, N., Howell, M. D., Kasschau, K. D., Chapman, E. J., Sullivan, C. M., Cumbie, J. S., et al. (2007) High-throughput sequencing of *Arabidopsis* microRNAs: evidence for frequent birth and death of MIRNA Genes. *PLoS ONE* 2:e219. doi: 10.1371/journal.pone.0000219
47. FDA. (1992) Statement of policy: foods derived from new plant varieties. *Fed. Regist.* 57, 22984–23005.
48. Feldmann, K. A. (2006) Steroid regulation improves crop yield. *Nat.*

Biotechnol. 24, 46–47. doi: 10.1038/nbt0106-46

49. Fire, A., Xu, S., Montgomery, M. K., Kostas, S. A., Driver, S. E., and Mello, C. C. (1998) Potent and specific genetic inteference by double-stranded RNA in *Caenorhabditis elegans*. *Nature* 391, 806–811. doi: 10.1038/35888
50. Flores, T., Karpova, O., Su, X., Zeng, P., Bilyeu, K., Sleper, D. A., et al. (2008) Silencing of GmFAD3 gene by siRNA leads to low alpha-linolenic acids (18:3) of fad3-mutant phenotype in soybean [*Glycine max* (Merr.)]. *Transgenic Res.* 17, 839–850. doi: 10.1007/s11248-008-9167-6
51. Franco-Zorrilla, J. M., Valli, A., Todesco, M., Mateos, I., Puga, M. I., Rubio-Somoza, I., et al. (2007) Target mimicry provides a new mechanism for regulation of microRNA activity. *Nat. Genet.* 39, 1033–1037. doi: 10.1038/ng2079
52. Fu, C., Sunkar, R., Zhou, C., Shen, H., Zhang, J. Y., Matts, J., et al. (2012) Overexpression of miR156 in switchgrass (*Panicum virgatum*L.) results in various morphological alterations and leads to improved biomass production. *Plant Biotechnol. J.* 10, 443–452. doi: 10.1111/j.1467-7652.2011.00677.x
53. Fujii,H.,Chiou,T.J.,Lin,S.I.,Aung,K.,andZhu,J.K.(2005)AmiRNA involved in phosphate-starvation response in *Arabidopsis*. *Curr. Biol.* 15, 2038–2043. doi: 10.1016/j.cub.2005.10.016
54. Gheysen, G., and Vanholme, B. (2007) RNAi from plants to nematodes. *Trends Biotechnol.* 25, 89–92. doi: 10.1016/j.tibtech.2007.01.007
55. Gil-Humanes, J., Piston, F., Hernando, A., Alvarez, J. B., Shewry, P. R., and Barro, F. (2008) Silencing of γ-gliadins by RNA interference (RNAi) in bread wheat. *J. Cereal Sci.* 48, 565–568. doi: 10.1016/j.jcs.2008.03.005
56. Gil-Humanes, J., Pistón, F., Rosell, C. M., and Barro, F. (2012) Significant down-regulation of γ-gliadins has minor effect on gluten and starch properties of bread wheat. *J. Cereal Sci.* 56, 161–170. doi: 10.1016/j.jcs.2012.02.009
57. Gilisen, L. J., Bolhaar, S. T., Matos, C. I., Rouwendal, G. J., Boone, M. J., Krens, F. A., et al. (2005) Silencing of major apple allergen Mal d 1 by using the RNA interference approach. *J. Allergy Clin.*

Immunol. 115, 364–369. doi: 10.1016/j.jaci.2004.10.014

58. Goetz, M., Hooper, L. C., Johnson, S. D., Rodrigues, J. C., Vivian-Smith, A., and Koltunow, A. M. (2007) Expression of aberrant forms of auxin response factor8 stimulates parthenocarpy in *Arabidopsis* and tomato. *Plant Physiol.* 145, 351–366. doi: 10.1104/pp.107.104174
59. Guan, Q., Lu, X., Zeng, H., Zhang, Y., and Zhu, J. (2013) Heat stress induction of miR398 triggers a regulatory loop that is critical for thermo tolerance in *Arabidopsis*. *Plant J.* 74, 840–851. doi: 10.1111/tpj.12169
60. Gunnery, S., and Datta, A. (1987) An inhibitor RNA of translation from *Barley* embryo. *Biochem. Biophys. Res. Comm.* 142, 383–388. doi: 10.1016/0006-291X(87)90285-3
61. Guo, S., Xu, Y., Liu, H., Mao, Z., Zhang, C., Ma, Y., et al. (2013) The interaction between OsMADS57 and OsTB1 modulates rice tillering via DWARF14. *Nat. Comm.* 4:1566. doi: 10.1038/ncomms2542
62. Gupta, A. S., Heinen, J. L., Holaday, A. S., Burke, J. J., and Allen, R. D. (1993) Increased resistance to oxidative stress in transgenic plants that overexpress chloroplastic Cu/Zn superoxide dismutase. *Proc. Natl. Acad. Sci. U.S.A.* 90, 1629–1633. doi: 10.1073/pnas.90.4.1629
63. Gupta, A., Pal, R. K., and Rajama, M. V. (2013) Delayed ripening and improved fruit processing quality in tomato by RNAi-mediated silencing of three homologs of 1-aminopropane-1-carboxylate synthase gene. *J. Plant Physiol.* 170, 987–995. doi: 10.1016/j.jplph.2013.02.003
64. Hamilton, A. J., and Baulcombe, D. C. (1999) A novel species of small antisense RNA in posttranscriptional gene silencing. *Science* 286, 950–952. doi: 10.1126/science.286.5441.950
65. Hamilton, A. J., Lycett, G. W., and Grierson, D. (1990) Antisense gene that inhibits synthesis of the hormone ethylene in transgenic plants. *Nature* 346, 284–287. doi: 10.1038/346284a0
66. Hamilton, A., Voinnet, O., Chappell, L., and Baulcombe, D. C. (2002) Two classes of short interfering RNA in RNA silencing. *EMBO J.*21, 4671–4679. doi: 10.1093/emboj/cdf464
67. Hammond, S. M., Bernstein, E., Beach, D., and Hannon, G. J. (2000) An RNA-directed nuclease mediates post-transcriptional

gene silencing in *Drosophila* cells. *Nature* 404, 293–296. doi: 10.1038/35005107

68. Hannon, G. J. (2002) RNA interference. *Nature* 418, 244–251. doi: 10.1038/418244a
69. Henderson, I. R., Zhang, X., Lu, C., Johnson, L., Meyers, B. C., Green, P. J., et al. (2006) Dissecting *Arabidopsis thaliana* DICER function in small RNA processing, gene silencing and DNA methylation patterning. *Nat. Genet.* 38, 721–725. doi: 10.1038/ng1804
70. Hewezi, T., Howe, P., Maier, T. R., and Baum, T. J. (2008) *Arabidopsis* small RNAs and their targets during cyst nematode parasitism. *Mol. Plant Microbe Interact.* 21, 1622–1634. doi: 10.1094/MPMI-21-12-1622
71. Hily, J. M., Ravelonandro, M., Damsteegt, V., Basset, C., Petri, C., Liu, Z., et al. (2007) *Plum pox virus* coat protein gene intron-hair pin-RNA (ihpRNA) constructs provide resistance to *Plum pox virus* in *Nicotiana bethamiana* and *Prunus domestica*. *J. Am. Soc. Hort. Sci.*132, 850–858.
72. Hisano, H., Nandakumar, R., and Wang, Z. Y. (2009) Genetic modification of lignin biosynthesis for improved biofuel production. *Vitro Cell. Dev. Biol. Plant* 45, 306–313. doi: 10.1007/s11627-009-9219-5
73. Hu, Y., Qin, F., Huang, L., Sun, Q., Li, C., Zhao, Y., et al. (2009) Rice histone deacetylase genes display specific expression patterns and developmental functions. *Biochem. Biophys. Res. Commun.* 388, 266–271. doi: 10.1016/j.bbrc.2009.07.162
74. Huang, G., Allen, R., Davis, E. L., Baum, T. J., and Hussey, R. S. (2006) Engineering broad root-knot resistance in transgenic plants by RNAi silencing of a conserved and essential root-knot nematode parasitism gene. *Proc. Natl. Acad. Sci. U.S.A.* 103, 14302–14306. doi: 10.1073/pnas.0604698103
75. Huntzinger, E., and Izaurralde, E. (2011) Gene silencing by micro-RNAs: contributions of translational repression and mRNA decay. *Nat. Rev. Genet.* 12, 99–110. doi: 10.1038/nrg2936
76. Hutvagner, G., McLachlan, J., Pasquinelli, A. E., Balint, E., Tuschl, T., and Zamore, P. D. (2001) A cellular function for the RNA-

interference enzyme Dicer in the maturation of the let-7 small temporal RNA. *Science* 293, 834–838. doi: 10.1126/science.1062961

77. Huvenne, H., and Smagghe, G. (2010) Mechanisms of dsRNA uptake in insects and potential of RNAi for pest control: a review. *J. Insect. Physiol.* 56, 227–235. doi: 10.1016/j.jinsphys.2009.10.004
78. Hwang, E. W., Shin, S. J., Yu, B. K., Byun, M. O., and Kwon, H. B. (2011) miR171 family members are involved in drought response in *Solanum tuberosum*. *J. Plant Biol.* 54, 43–48. doi: 10.1007/s12374-010-9141-8
79. Ibrahim, H. M., Alkharouf, N. W., Meyer, S. L., Aly, M. A., El-Din, A. K., Hussein, E. H., et al. (2011) Post-transcriptional gene silencing of root-knot nematode in transformed soybean roots. *Exp. Parasitol.* 127, 90–99. doi: 10.1016/j.exppara.2010.06.037
80. Itaya, A., Bundschuh, R., Archual, A. J., Joung, J. G., Fei, Z., Dai, X., et al. (2008) Small RNAs in tomato fruit and leaf development. *Biochim. Biophys. Acta* 1779, 99–107. doi: 10.1016/j.bbagrm.2007.09.003
81. Jagadeeswaran, G., Saini, A., and Sunkar, R. (2009) Biotic and abiotic stress down-regulate miR398 expression in *Arabidopsis*. *Planta* 229, 1009–1014. doi: 10.1007/s00425-009-0889-3
82. Jian, X., Zhang, L., Li, G., Zhang, L., Wang, X., Cao, X., et al. (2010) Identification of novel stress-regulated microRNAs from *Oryza sativa* L. *Genomics* 95, 47–50. doi: 10.1016/j.ygeno.2009.08.017
83. Jiang, C. J., Shimono, M., Maeda, S., Inoue, H., Mori, M., Hasegawa, M., et al. (2009) Suppression of the rice fatty-acid desaturase gene OsSSI2 enhances resistance to blast and leaf blight diseases in rice. *Mol. Plant Microbe Int.* 22, 820–829. doi: 10.1094/MPMI-22-7-0820
84. Jiao, Y., Wang, Y., Xue, D., Wang, J., Yan, M., Liu, G., et al. (2010) Regulation of OsSPL14 by OsmiR156 defines ideal plant architecture in rice. *Nat. Genet.* 42, 541–544. doi: 10.1038/ng.591
85. Jones-Rhoades, M. W., Bartel, D. P., and Bartel, B. (2006) MicroRNAs and their regulatory roles in plants. *Annu. Rev. Plant Biol.* 57, 19–53. doi: 10.1146/annurev.arplant.57.032905.105218
86. Kamachi, S., Mochizuki, A., Nishiguchi, M., and Tabei, Y. (2007) Transgenic *Nicotiana benthamiana* plants resistant to cucumber

green mottle mosaic virus based on RNA silencing. *Plant Cell Rep.* 26, 1283–1288. doi: 10.1007/s00299-007-0358-z

87. Karlova, R., van Haarst, J. C., Maliepaard, C., van de Geest, H., Bovy, A. G., Lammers, M., et al. (2013) Identification of microRNA targets in tomato fruit development using high-throughput sequencing and degradome analysis. *J. Exp. Bot.* 64, 1863–1878. doi: 10.1093/jxb/ert049
88. Khraiwesh, B., Arif, M. A., Seumel, G. I., Ossowski, S., Weigel, D., Reski, R., et al. (2010) Transcriptional control of gene expression by microRNAs. *Cell* 140, 111–122. doi: 10.1016/j.cell.2009.12.023
89. Khraiwesh, B., Zhu, J. K., and Zhu, J. (2012) Role of miRNAs and siRNAs in biotic and abiotic stress responses of plants. *Biochim. Biophys. Acta* 1819, 137–148. doi: 10.1016/j.bbagrm.2011.05.001
90. Krubphachaya, P., Jurícek, M., and Kertbundit, S. (2007) Induction of RNA mediated resistance to papaya ring spot virus type W. *J. Biochem. Mol. Biol.* 40, 401–411. doi: 10.5483/BMBRep.2007.40.3.404
91. Kusaba, M., Miyahara, K., Iida, S., Fukuoka, H., Takano, T., Sassa, H., et al. (2003) Low glutelin content 1: a dominant mutation that suppresses the glutelin multigene family via RNA silencing in rice. *Plant Cell* 15, 1455–1467. doi: 10.1105/tpc.011452
92. Law, J. A., and Jacobsen, S. E. (2010) Establishing, maintaining and modifying DNA methylation patterns in plants and animals. *Nat. Rev. Genet.* 11, 204–220. doi: 10.1038/nrg2719
93. Le, L., Lorenz, Y., Scheurer, S., Fotisch, K., Enrique, E., Bartra, J., et al. (2006) Design of tomato fruits with reduced allergenicity by dsRNAi-mediated inhibition of ns-LTP (Lyc e 3) expression. *Plant Biotechnol. J.* 4, 231–242. doi: 10.1111/j.1467-7652.2005.00175.x
94. Lee, R. C., Feinbaum, R. L., and Ambros, V. (1993) The *C. elegans* heterochronic gene lin-4 encodes small RNAs with antisense complementarity to lin-14. *Cell* 75, 843–854. doi: 10.1016/0092-8674(93)90529-Y
95. Lewis, R. S., Jack, A. M., Morris, J. W., Robert, V. J., Gavilano, L. B., Siminszky, B., et al. (2008) RNA interference (RNAi)-induced suppression of nicotine demethylase activity reduces levels of a key carcinogen in cured tobacco leaves. *Plant Biotechnol. J.* 6, 346–

354. doi: 10.1111/j.1467-7652.2008.00324.x

96. Li, D. H., Liu, H., Yang, Y. l., Zhen, P. P., and Liang, J. S. (2009) Down-regulated expression of RACK1 gene by RNA interference enhances drought tolerance in rice. *Rice Sci.* 16, 14–20. doi: 10.1016/S1672-6308(08)60051-7

97. Li, K., Yuh-Shuh, W., Srinivasa Rao, U., Keri, W., Yuhong, T., Vatsala, V., et al. (2008a) Overexpression of a fatty acid amide hydrolase compromises innate immunity in *Arabidopsis*. *Plant J.* 56, 336–349. doi: 10.1111/j.1365-313X.2008.03603.x

98. Li, W. X., Oono, Y., Zhu, J., He, X. J., Wu, J. M., Iida, K., et al. (2008b) The *Arabidopsis* NFYA5 transcription factor is regulated transcriptionally and post transcriptionally to promote drought resistance. *Plant Cell* 20, 2238–2251. doi: 10.1105/tpc.108.059444

99. Li, X., Wang, X., Zhang, S., Liu, D., Duan, Y., and Dong, W. (2012) Identification of soybean microRNAs involved in soybean cyst nematode infection by deep sequencing. *PLoS ONE* 7:e39650. doi: 10.1371/journal.pone.0039650

100. Liu, H. H., Tian, X., Li, Y. J., Wu, C. A., and Zheng, C. C. (2008) Microarray-based analysis of stress-regulated microRNAs in *Arabidopsis thaliana*. *RNA* 14, 836–843. doi: 10.1261/rna.895308

101. Liu, J., He, Y., Amasino, R., and Chen, X. (2004) siRNA s targeting an intronic transposon in the regulation of natural flowering behavior in *Arabidopsis*. *Genes Dev.* 18, 2873–2878. doi: 10.1101/gad.1217304

102. Liu, Q., Singh, P. S., and Green, A. G. (2002) High-stearic and high-oleic cotton seed oils produced by hairpin RNA-mediated post-transcriptional gene silencing. *Plant Physiol.* 129, 1732–1743. doi: 10.1104/pp.001933

103. Llave, C., Kasschau, K. D., Rector, M. A., and Carrington, J. C. (2002a) Endogenous and silencing associated small RNAs in plants. *Plant Cell* 14, 1605–1619. doi: 10.1105/tpc.003210

104. Llave, C., Xie, Z., Kasschau, K. D., and Carrington, J. C. (2002b) Cleavage of Scarecrow-like mRNA targets directed by a class of *Arabidopsis* miRNA. *Science* 297, 2053–2056. doi: 10.1126/science.1076311

105. Lough, T. J., and Lucas, W. J. (2006) Integrative plant biology: role of

phloem long-distance macromolecular trafficking. *Annu. Rev. Plant Biol.* 57, 203–232. doi: 10.1146/annurev.arplant.56.032604.144145

106. Lv, D. K., Bai, X., Li, Y., Ding, X. D., Ge, Y., Cai, H., et al. (2010) Profiling of cold-stress-responsive miRNAs in rice by microarrays. *Gene*459, 39–47. doi: 10.1016/j.gene.2010.03.011
107. Mao, Y. B., Cai, W. J., Wang, J. W., Hong, G. J., Tao, X. Y., Wang, L. J., et al. (2007) Silencing a cotton bollworm P450 monooxygenase gene by plant-mediated RNAi impairs larval tolerance of gossypol. *Nat. Biotechnol.* 25, 1307–1313. doi: 10.1038/nbt1352
108. Matzke, M., Matzke, A., Pruss, G., and Vance, V. (2001) RNA-based silencing strategies in plants. *Curr. Opin. Genet. Dev.* 11, 221–227. doi: 10.1016/S0959-437X(00)00183-0
109. Meli, V. S., Ghosh, S., Prabha, T. N., Chakraborty, N., Chakraborty, S., and Datta, A. (2010) Enhancement of fruit shelf life by suppressing *N*-glycan processing enzymes. *Proc. Natl. Acad. Sci. U.S.A.* 107, 2413–2418. doi: 10.1073/pnas.0909329107
110. Melnyk, C. W., Molnar, A., and Baulcombe, D. C. (2011) Intercellular and systemic movement of RNA silencing. *EMBO J.* 30, 3553–3563 doi: 10.1038/emboj.2011.274
111. Mette, M. F., van der Winden, J., Matzke, M., and Matzke, A. J. (2002) Short RNAs can identify new candidate transposable element families in *Arabidopsis*. *Plant Physiol.* 130, 6–9. doi: 10.1104/pp.007047
112. Missiou, A., Kalantidis, K., Boutla, A., Tzortzakaki, S., Tabler, M., and Tsagris, M. (2004) Generation of transgenic potato plants highly resistant to potato virus Y (PVY) through RNA silencing. *Mol. Breed.* 14, 185–197. doi: 10.1023/B:MOLB.0000038006.32812.52
113. Mitsuda, N., and Ohme-Takagi, M. (2009) Functional analysis of transcription factors in *Arabidopsis*. *Plant Cell Physiol.* 50, 1232–1248. doi: 10.1093/pcp/pcp075
114. Miura, K., Ikeda, M., Matsubara, A., Song, X. J., Ito, M., Asano, K., et al. (2010) OsSPL14 promotes panicle branching and higher grain productivity in rice. *Nat. Genet.* 42, 545–549. doi: 10.1038/ng.592
115. Molesini, B., Pandolfini, T., Rotino, G. L., Dani, V., and Spena, A. (2009) Aucsia gene silencing causes parthenocarpic fruit development in tomato. *Plant Physiol.* 149, 534–548. doi: 10.1104/

pp.108.131367

116. Molesini, B., Pii, Y., and Pandolfini, T. (2012) Fruit improvement using intragenesis and artificial microRNA. *Trends Biotechnol.* 30, 80–88. doi: 10.1016/j.tibtech.2011.07.005

117. Molnar, A., Melnyk, C. W., Bassett, A., Hardcastle, T. J., Dunn, R., and Baulcombe, D. C. (2010) Small silencing RNAs in plants are mobile and direct epigenetic modification in recipient cells. *Science* 328, 872–875. doi: 10.1126/science.1187959

118. Moxon, S., Jing, R., Szittya, G., Schwach, F., Pilcher, R. L. R., Moulton, V., et al. (2008) Deep sequencing of tomato short RNAs identifies microRNAs targeting genes involved in fruit ripening. *Genome Res.* 18, 1602–1609. doi: 10.1101/gr.080127.108

119. Napoli, C., Lemieux, C., and Jorgensen, R. (1990) Introduction of chimeric chalcone synthase gene into *Petunia* results in reversible cosuppression of homologous genes in trans. *Plant Cell* 2, 279–289. doi: 10.1105/tpc.2.4.279

120. Navarro, L., Dunoyer, P., Jay, F., Arnold, B., Dharmasiri, N., Estelle, M., et al. (2006) A plant miRNA contributes to antibacterial resistance by repressing auxin signaling. *Science* 312, 436–439. doi: 10.1126/science.1126088

121. Ni, Z., Hu, Z., Jiang, Q., and Zhang, H. (2013) GmNFYA3, a target gene of miR169, is a positive regulator of plant tolerance to drought stress. *Plant Mol. Biol.* 82, 113–129. doi: 10.1007/s11103-013-0040-5

122. Niu, Q. W., Lin, S. S., Reyes, J. L., Chen, K. C., Wu, H. W., Yeh, S. D., et al. (2006) Expression of artificial microRNAs in transgenic *Arabidopsis thaliana* confers virus resistance. *Nat. Biotechnol.* 24, 1420–1428. doi: 10.1038/nbt1255

123. Nora, S., MichÈLe, Z., Asuka, I., Biao, D., Ming-Bo, W., Gabi, K., et al. (2009) RNAi-mediated resistance to *Potato spindle* tuber viroid in transgenic tomato expressing a viroid hairpin RNA construct. *Mol. Plant Pathol.* 10, 459–469. doi: 10.1111/j.1364-3703.2009.00546.x

124. Oeller, P. W., Lu, M. W., Taylor, L. P., Pike, D. A., and Theologis, A. (1991) Reversible inhibition of tomato fruit senescence by antisense RNA. *Science* 254, 437–439. doi: 10.1126/science.1925603

125. Osorio, S., Alba, R., Damasceno, C. M., Lopez-Casado, G., Lohse,

M., Zanor, M. I., et al. (2011) Systems biology of tomato fruit development: combined transcript, protein, and metabolite analysis of tomato transcription factor (nor, rin) and ethylene receptor (Nr) mutants reveals novel regulatory interactions. *Plant Physiol.* 157, 405–425. doi: 10.1104/pp.111.175463

126. Ossowski, S., Schwab, R., and Weigel, D. (2008) Gene silencing in plants using artificial microRNAs and other small RNAs. *Plant J.* 53, 674–690. doi: 10.1111/j.1365-313X.2007.03328.x
127. Palauqui, J.-C., Elmayan, T., Pollien, J.-M., and Vaucheret, H. (1997) Systemic acquired silencing: transgene-specific post-transcriptional silencing is transmitted by grafting from silenced stocks to non-silenced scions. *EMBO J.* 16, 4738–4745. doi: 10.1093/emboj/16.15.4738
128. Pandolfini, T., Molesini, B., Avesani, L., Spena, A., and Polverari, A. (2003) Expression of self-complementary hairpin RNA under the control of the rolC promoter confers systemic disease resistance to plum pox virus without preventing local infection. *BMC Biotechnol.*3:7. doi: 10.1186/1472-6750-3-7
129. Pant, B. D., Buhtz, A., Kehr, J., and Scheible, W. R. (2008) MicroRNA399 is a long-distance signal for the regulation of plant phosphate homeostasis. *Plant J.* 53, 731–738. doi: 10.1111/j.1365-313X.2007.03363.x
130. Pare, J. M., and Hobman, T. C. (2007) "Dicer: structure, function and role in RNA-dependent gene-silencing pathways," in *Industrial Enzymes*, eds J. Polaina and A. P. MacCabe (Berlin: Springer), 421–438.
131. Park, G. G., Park, J. J., Yoon, J., Yu, S. N., and An, G. (2010) A Ring finger E3 ligase gene, *Oryza sativa* delayed seed germination 1 (OsDSG1), controls seed germination and stress responses in rice. *Plant Mol. Biol.* 74, 467–478. doi: 10.1007/s11103-010-9687-3
132. Park, W., Li, J., Song, R., Messing, J., and Chen, X. (2002) Carpel factory, a dicer homolog, and HEN1, a novel protein, act in microRNA metabolism in *Arabidopsis thaliana*. *Curr. Biol.* 12, 1484–1495. doi: 10.1016/S0960-9822(02)01017-5
133. Peltier, A. J., Hatfield, R. D., and Grau, C. R. (2009) Soybean

stem lignin concentration relates to resistance to *Sclerotinia sclerotiorum*. *Plant Dis.* 93, 149–154. doi: 10.1094/PDIS-93-2-0149

134. Perl, A., Perl-Treves, R., Galili, S., Aviv, D., Shalgi, E., Malkin, S., et al. (1993) Enhanced oxidative-stress defense in transgenic potato expressing tomato Cu, Zn superoxide dismutases. *Theor. Appl. Genet.* 85, 568–576. doi: 10.1007/BF00220915
135. Pilcher, R. L., Moxon, S., Pakseresht, N., Moulton, V., Manning, K., Seymour, G., et al. (2007) Identification of novel small RNAs in tomato (*Solanum lycopersicum*). *Planta* 226, 709–717. doi: 10.1007/s00425-007-0518-y
136. Pitcher, L. H., Brennan, E., Hurley, A., Dunsmuir, P., Tepperman, J. M., and Zilinskas, B. A. (1991) Overproduction of petunia chloroplastic copper/zinc superoxide dismutase does not confer ozone tolerance in transgenic tobacco. *Plant Physiol.* 97, 452–455. doi: 10.1104/pp.97.1.452
137. Pontier, D., Picart, C., Roudier, F., Garcia, D., Lahmy, S., Azevedo, J., et al. (2012) NERD, a plant-specific GW protein, defines an additional RNAi-dependent chromatin based pathway in *Arabidopsis*. *Mol. Cell* 48, 121–132. doi: 10.1016/j.molcel.2012.07.027
138. Pooggin, M., Shivaprasad, P. V., Veluthambi, K., and Hohn, T. (2003) RNAi targeting of DNA virus in plants. *Nat. Biotechnol.* 21, 131–132. doi: 10.1038/nbt0203-131b
139. Powell-Abel, P., Nelson, R. S., De, B., Hoffmann, N., Rogers, S. G., Fraley, R. T., et al. (1986) Delay of disease development in transgenic plants that express the tobacco mosaic virus coat protein gene. *Science* 232, 738–743. doi: 10.1126/science.3457472
140. Price, D. R. G., and Gatehouse, J. A. (2008) RNAi-mediated crop protection against insects. *Trends Biotechnol.* 26, 393–400. doi: 10.1016/j.tibtech.2008.04.004
141. Qiao, F., Yang, Q., Wang, C. L., Fan, Y. L., Wu, X. F., and Zhao, K. J. (2007) Modification of plant height via RNAi suppression of OsGA20ox2 gene in rice. *Euphytica* 158, 35–45. doi: 10.1007/s10681-007-9422-6
142. Qu, J., Ye, J., and Fang, R. (2007) Artificial microRNA-mediated virus resistance in plants. *J. Virol.* 81, 6690–6699. doi: 10.1128/

JVI.02457-06

143. Rajam, M. V., Madhulatha, P., Pandey, R., Hazarika, P. J., and Razdan, M. K. (2007) "Applications of genetic engineering in tomato," in *Genetic Improvement of Solanaceae Crops: Tomato*, Vol. 2, eds M. K. Razdan and A. K. Mattoo (Enfield, NH: Science Publishers), 285–311.

144. Rathore, K. S., Sundaram, S., Sunilkumar, G., Campbell, L. M., Puckhaber, L., Marcel, S., et al. (2012) Ultra-low gossypol cottonseed: generational stability of the seed-specific, RNAi-mediated phenotype and resumption of terpenoid profile following seed germination. *Plant Biotechnol. J.* 10, 174–183. doi: 10.1111/j.1467-7652.2011.00652.x

145. Reddy, M. S. S., Chen, F., Shadle, G., Jackson, L., Aljoe, H., and Dixon, R. A. (2005) Targeted down-regulation of cytochrome P450 enzymes for forage quality improvement in alfalfa (*Medicago sativa* L.). *Proc. Natl. Acad. Sci. U.S.A.* 102, 16573–16578. doi: 10.1073/pnas.0505749102

146. Redfern, A. D., Colley, S. M., Beveridge, D. J., Ikeda, N., Epis, M. R., Li, X., et al. (2013) RNA-induced silencing complex (RISC) proteins PACT, TRBP, and Dicer are SRA binding nuclear receptor coregulators. *Proc. Natl. Acad. Sci. U.S.A.* 110, 6536–6541. doi: 10.1073/pnas.1301620110

147. Regina, A., Bird, A., Topping, D., Bowden, S., Freeman, J., Barsby, T., et al. (2006) High amylose wheat generated by RNA interference improves indices of large-bowel health in rats. *Proc. Natl. Acad. Sci. U.S.A.* 103, 3546–3551. doi: 10.1073/pnas.0510737103

148. Reinhart, B. J., Weinstein, E. G., Rhoades, M. W., Bartel, B., and Bartel, D. P. (2002) MicroRNAs in plants. *Genes Dev.* 16, 1616–1626. doi: 10.1101/gad.1004402

149. Reyes, J. L., and Chua, N. H. (2007) ABA induction of miR159 controls transcript levels of two MYB factors during *Arabidopsis* seed germination. *Plant J.* 49, 592–606. doi: 10.1111/j.1365-313X.2006.02980.x

150. Rhoades, M. W., Reinhart, B. J., Lim, L. P., Burge, C. B., Bartel, B., and Bartel, D. P. (2002) Prediction of plant microRNA targets. *Cell*110, 513–520. doi: 10.1016/S0092-8674(02)00863-2

151. Riechen, J. (2007) Establishment of broad-spectrum resistance against *Blumeria graminis* f. sp. tritici in *Triticum aestivum* by RNAi-mediated knock-down of MLO. *J. Verbrauch. Lebensm.* 2:120. doi: 10.1007/s00003-007-0282-8

152. Riley, K. J., Yario, T. A., and Steitz, J. A. (2012) Association of argonaute proteins and microRNAs can occur after cell lysis. *RNA* 18, 1581–1585. doi: 10.1261/rna.034934.112

153. Romano, N., and Macino, G. (1992) Quelling: transient inactivation of gene expression in *Neurospora crassa* by transformation with homologous sequences. *Mol. Microbiol.* 6, 3343–3353. doi: 10.1111/j.1365-2958.1992.tb02202.x

154. Ru, P., Xu, L., Ma, H., and Huang, H. (2006) Plant fertility defects induced by the enhanced expression of microRNA167. *Cell Res.* 16, 457–465. doi: 10.1038/sj.cr.7310057

155. Rubinelli, P. M., Chuck, G., Li, X., and Meilan, R. (2013) Constitutive expression of the Corn-grass1 microRNA in poplar affects plant architecture and stem lignin content and composition. *Biomass Bioenerg.* 54, 312–321. doi: 10.1016/j.biombioe.2012.03.001

156. Schijlen, E. G. W. M., de Vos, R. C. H., Martens, S., Jonker, H. H., Rosin, F. M., Molthoff, J. W., et al. (2007) RNA Interference silencing of chalcone synthase, the first step in the flavonoid biosynthesis pathway, leads to parthenocarpic tomato fruits. *Plant Physiol.* 144, 1520–1530. doi: 10.1104/pp.107.100305

157. Schwab, R., Ossowski, S., Riester, M., Warthmann, M., and Weigel, D. (2006) Highly specific gene silencing by artificial microRNAs in *Arabidopsis*. *Plant Cell* 18, 1121–1133. doi: 10.1105/tpc.105.039834

158. Schwab, R., Palatnik, J. F., Riester, M., Schommer, C., Schmid, M., and Weigel, D. (2005) Specific effects of microRNAs on the plant transcriptome. *Dev. Cell* 8, 517–527. doi: 10.1016/j.devcel.2005.01.018

159. Schwach, F., Vaistij, F. E., Jones, L., and Baulcombe, D. C. (2005) An RNA dependent RNA-polymerase prevents meristem invasion by *Potato virus* X and is required for the activity but not the production of a systemic silencing signal. *Plant Physiol.* 138, 1842–1852. doi: 10.1104/pp.105.063537

160. Segal, G., Song, R., and Messing, J. (2003) A new opaque variant of maize by a single dominant RNA-interference inducing transgene. *Genetics* 165, 387–397.

161. Shimizu, T., Yoshii, M., Wei, T., Hirochika, H., and Omura, T. (2009) Silencing by RNAi of the gene for Pns12, a viroplasm matrix protein of Rice dwarf virus, results in strong resistance of transgenic rice plants to the virus. *Plant Biotechnol. J.* 7, 24–32. doi: 10.1111/j.1467-7652.2008.00366.x

162. Shivaprasad, P. V., Chen, H. M., Patel, K., Bond, D. M., Santos, B. A., and Baulcombe, D. C. (2012) A microRNA superfamily regulates nucleotide binding site leucine rich repeats and other mRNAs. *Plant Cell* 24, 859–874. doi: 10.1105/tpc.111.095380

163. Simon-Mateo, C., and García, J. A. (2011) Antiviral strategies in plants based on RNA silencing. *Biochim. Biophys. Acta* 1809, 722–731. doi: 10.1016/j.bbagrm.2011.05.011

164. Sindhu, A. S., Maier, T. R., Mitchum, M. G., Hussey, R. S., Davis, E. L., and Baum, T. J. (2009) Effective and specific in planta RNAi in cyst nematodes: expression interference of four parasitism genes reduces parasitic success. *J. Exp. Bot.* 60, 315–324. doi: 10.1093/jxb/ern289

165. Singh, A., Taneja, J., Dasgupta, I., and Mukherjee, S. K. (2014) Development of plants resistant to tomato geminiviruses using artificial trans-acting small interfering RNA. *Mol. Plant Patholol.* doi: 10.1111/mpp.12229 [Epub ahead of print].

166. Slotkin, R. K., Vaughn, M., Borges, F., Tanurdzic, M., Becker, J. D., Feijo, J. A., et al. (2009) Epigenetic reprogramming and small RNA silencing of transposable elements in pollen. *Cell* 136, 461–472. doi: 10.1016/j.cell.2008.12.038

167. Smith, N. A., Singh, S. P., Wang, M. B., Stoutjesdijk, P. A., Green, A. G., and Waterhouse, P. M. (2000) Total silencing by intron-spliced hairpin RNAs. *Nature* 407, 319–320. doi: 10.1038/35036500

168. Springer, N. (2010) Shaping a better rice plant. *Nat. Genet.* 42:475. doi: 10.1038/ng0610-475

169. Sun, L., Yuan, B., Zhang, M., Wang, L., Cui, M., Wang, Q., et al. (2012) Fruit-specific RNAi mediated suppression of SlNCED1 increases both lycopene and E-carotene contents in tomato fruit. *J.*

Exp. Bot. 63, 3097–3108. doi: 10.1093/jxb/ers026

170. Sunilkumar, G., Campbell, L. M., Puckhaber, L., Stipanovic, R. D., and Rathore, K. S. (2006) Engineering cottonseed for use in human nutrition by tissue-specific reduction of toxic gossypol. *Proc. Natl. Acad. Sci. U.S.A.* 103, 18054–18059. doi: 10.1073/pnas.0605389103
171. Sunkar, R., Kapoor, A., and Zhu, J. K. (2006) Posttranscriptional induction of two Cu/Zn superoxide dismutase genes in *Arabidopsis* is mediated by downregulation of miR398 and important for oxidative stress tolerance. *Plant Cell* 18, 2051–2065. doi: 10.1105/tpc.106.041673
172. Sunkar, R., and Zhu, J. K. (2004) Novel and stress-regulated micro RNAs and other small RNAs from *Arabidopsis*. *Plant Cell* 16, 2001–2019. doi: 10.1105/tpc.104.022830
173. Tang, G., Reinhart, B. J., Bartel, D. P., and Zamore, P. D. (2003) A biochemical framework for RNA silencing in plants. *Genes Dev.* 17, 49–63. doi: 10.1101/gad.1048103
174. Teo, G., Suziki, Y., Uratsu, S. L., Lampien, B., Ormonde, N., Hu, W. K., et al. (2006) Silencing leaf sorbitol synthesis alters long-distance portioning and apple fruit quality. *Proc. Natl. Acad. Sci. U.S.A.* 103, 18842–18847. doi: 10.1073/pnas.0605873103
175. Tepperman, J. M., and Dunsmuir, P. (1990) Transformed plants with elevated levels of chloroplastic SOD are not more resistant to superoxide toxicity. *Plant Mol. Biol.* 14, 501–511. doi: 10.1007/BF00027496
176. Theologis, A., Oeller, P. W., Wong, L. M., Rottmann, W. H., and Gantz, D. M. (1993) Use of a tomato mutant constructed with reverse genetics to study fruit ripening, a complex developmental process. *Dev. Gen.* 14, 282–295. doi: 10.1002/dvg.1020140406
177. Thiebaut, F., Rojas, C. A., Almeida, K. L., Grativol, C., Domiciano, G. C., Lamb, C. R., et al. (2011) Regulation of miR319 during cold stress in sugarcane. *Plant Cell Environ.* 35, 502–512. doi: 10.1111/j.1365-3040.2011.02430.x
178. Trindade, I., CapitCapitão, C., Dalmay, T., Fevereiro, M. P., and Santos, D. M. (2010) miR398 and miR408 are up-regulated in response to water deficit in *Medicago truncatula*. *Planta* 231, 705–716. doi: 10.1007/s00425-009-1078-0

179. Tuschl, T. (2001) RNA interference and small interfering RNAs. *Chembiochem* 2, 239–145. doi: 10.1002/1439-7633(20010401)2:4<239::AID-CBIC239>3.0.CO;2-R
180. Vanderschuren, H., Alder, A., Zhang, P., and Gruissem, W. (2009) Dose-dependent RNAi-mediated geminivirus resistance in the tropical root crop cassava. *Plant Mol. Biol.* 70, 265–272. doi: 10.1007/s11103-009-9472-3
181. Vaucheret, H. (2008) Plant ARGONAUTES. *Trends Plant Sci.* 13, 350–358. doi: 10.1016/j.tplants.2008.04.007
182. Vazquez, F. (2006) *Arabidopsis* endogenous small RNAs: highways and byways. *Trends Plant Sci.* 11, 460–468. doi: 10.1016/j.tplants.2006.07.006
183. Voinnet, O., Vain, P., Angell, S., and Baulcombe, D. C. (1998) Systemic spread of sequence-specific transgene RNA degradation is initiated by localised introduction of ectopic promoterless DNA. *Cell* 95, 177–187. doi: 10.1016/S0092-8674(00)81749-3
184. Vu, T. V., Choudhury, N. R., and Mukherjee, S. K. (2013) Transgenic tomato plants expressing artificial microRNAs for silencing the pre-coat and coat proteins of a begomovirus, tomato leaf curl New Delhi virus, show tolerance to virus infection. *Virus Res.* 172, 35–45. doi: 10.1016/j.virusres.2012.12.008
185. Wang, M., Abbott, D. C., and Waterhouse, P. M. (2000) A single copy of a virus-derived transgene encoding hairpin RNA gives immunity to barley yellow dwarf virus. *Mol. Plant Pathol.* 1, 347–356. doi: 10.1046/j.1364-3703.2000.00038.x
186. Wang, S., Wu, K., Yuan, Q., Liu, X., Liu, Z., Lin, X., et al. (2012) Control of grain size, shape and quality by OsSPL16 in rice. *Nat. Genet.*44, 950–954. doi: 10.1038/ng.2327
187. Wang, T., Chen, L., Zhao, M., Tian, Q., and Zhang, W. H. (2011) Identification of drought-responsive microRNAs in *Medicago truncatula*by genome-wide high-throughput sequencing. *BMC Genomics* 12:367–378. doi: 10.1186/1471-2164-12-367
188. Wang, Y., Beaith, M., Chalifoux, M., Ying, J., Uchacz, T., Sarvas, C., et al. (2009) Shoot-specific down-regulation of protein farnesyltransferase (a-subunit) for yield protection against drought

in canola. *Mol. Plant* 2, 191–200. doi: 10.1093/mp/ssn088

189. Wang, Y., Ying, J., Kuzma, M., Chalifoux, M., Sample, A., McArthur, C., et al. (2005) Molecular tailoring of farnesylation for plant drought tolerance and yield protection. *Plant J.* 43, 413–424. doi: 10.1111/j.1365-313X.2005.02463.x
190. Weise, S. E., Aung, K., Jarou, Z. J., Mehrshahi, P., Li, Z., Hardy, A. C., et al. (2012) Engineering starch accumulation by manipulation of phosphate metabolism of starch. *Plant Biotechnol. J.* 10, 545–554. doi: 10.1111/j.1467-7652.2012.00684.x
191. Wesley, S. V., Helliwell, C. A., Smith, N. A., Wang, M. B., Rouse, D. T., Liu, Q., et al. (2001) Construct design for efficient, effective and high throughput gene silencing in plants. *Plant J.* 27, 581–590. doi: 10.1046/j.1365-313X.2001.01105.x
192. Wilkinson, J. Q., Lanahan, M. B., Clark, D. G., Bleecker, A. B., Chang, C., Meyerowitz, E. M., et al. (1997) A dominant mutant receptor from *Arabidopsis* confers ethylene insensitivity in heterologous plants. *Nat. Biotechnol.* 15, 444–447. doi: 10.1038/nbt0597-444
193. Wilson, R. C., and Doudna, J. A. (2013) Molecular mechanisms of RNA interference. *Annu. Rev. Biophys.* 42, 217–239. doi: 10.1146/annurev-biophys-083012-130404
194. Wu, M. F., Tian, Q., and Reed, J. W. (2006) *Arabidopsis* microRNA167 controls patterns of ARF6 and ARF8 expression, and regulates both female and male reproduction. *Development* 133, 4211–4218. doi: 10.1242/dev.02602
195. Xie, K., Shen, J., Hou, X., Yao, J., Li, X., Xiao, J., et al. (2012) Gradual increase of miR156 regulates temporal expression changes of numerous genes during leaf development in rice. *Plant Physiol.* 158, 1382–1394. doi: 10.1104/pp.111.190488
196. Xie, K., Wu, C., and Xiong, L. (2006) Genomic organization, differential expression, and interaction of Squamosa promoter-binding- like transcription factors and microRNA156 in rice. *Plant Physiol.* 142, 280–293. doi: 10.1104/pp.106.084475
197. Xie, Z., Allen, E., Wilken, A., and Carrington, J. C. (2005) DICER-LIKE 4 functions in trans-acting small interfering RNA biogenesis and vegetative phase change in *Arabidopsis thaliana*. *Proc. Natl.*

Acad. Sci. U.S.A. 102, 12984–12989. doi: 10.1073/pnas.0506426102

198. Xie, Z., Kasschau, K. D., and Carrington, J. C. (2003) Negative feedback regulation of Dicer-Like1 in *Arabidopsis* by microRNA-guided mRNA degradation. *Curr. Biol.* 13, 784–789. doi: 10.1016/S0960-9822(03)00281-1
199. Xin, M., Wang, Y., Yao, Y., Xie, C., Peng, H., Ni, Z., et al. (2010) Diverse set of microRNAs are responsive to powdery mildew infection and heat stress in wheat (*Triticum aestivum* L.). *BMC Plant Biol.* 10:123–134. doi: 10.1186/1471-2229-10-123
200. Xiong, A. S., Yao, Q. H., Li, X., Fan, H. Q., and Peng, R. H. (2003) Double antisense ACC oxidase and ACC synthase fusion gene introduced into tomato by *Agrobacterium*-mediated transformation and analysis the ethylene production of transgenic plants. *Acta Biol. Exp. Sin.* 36, 35–41.
201. Xiong, A. S., Yao, Q. H., Peng, R. H., Li, X., Han, P. L., and Fan, H. Q. (2005) Different effects on ACC oxidase genesilencing triggered by RNA interference in transgenic tomato. *Plant Cell Rep.* 23, 639–646. doi: 10.1007/s00299-004-0887-7
202. Yadav, B. C., Veluthambi, K., and Subramaniam, K. (2006) Host-generated double stranded RNA induces RNAi in plant-parasitic nematodes and protects the host from infection. *Mol. Biochem. Parasitol.* 148, 219–222. doi: 10.1016/j.molbiopara.2006.03.013
203. Yang, C., Li, D., Mao, D., Liu, X., Li, C., Li, X., et al. (2013) Overexpression of microRNA319 impacts leaf morphogenesis and leads to enhanced cold tolerance in rice (*Oryza sativa* L.). *Plant Cell Environ.* 36, 2207–2218. doi: 10.1111/pce.12130
204. Yara, A., Yaeno, T., Hasegawa, M., Seto, H., Montillet, J. L., Kusumi, K., et al. (2007) Disease resistance against *Magnaporthe grisea* is enhanced in transgenic rice with suppression of o-3 fatty acid desaturases. *Plant Cell Physiol.* 48, 1263–1274. doi: 10.1093/pcp/pcm107
205. Ye, Z. B., Li, H. X., Zheng, Y. L., and Liu, H. L. (1996) Inhibition of introducing antisense ACC oxidase gene into tomato genome on expression of its endogeous gene. *J. Huazhong Agric. Univ.* 15, 305–309.
206. Yoo, B.-C., Kragler, F., Varkonyi-Gasic, E., Haywood, V.,

Archer-Evans, S., Lee, Y. M., et al. (2004) A systemic small RNA signaling system in plants. *Plant Cell* 16, 1979–2000. doi: 10.1105/tpc.104.023614

207. Yu, B., Lydiate, D. J., Young, L. W., Schafer, U. A., and Hannoufa, A. (2007) Enhancing the carotenoid content of *Brassica napus* seeds by downregulating lycopene epsilon cyclase. *Transgenic Res.* 17, 573–585. doi: 10.1007/s11248-007-9131-x
208. Zamore, P. D., and Haley, B. (2005) Ribo-genome: the big world of small RNAs. *Science* 309, 1519–1524. doi: 10.1126/science.1111444
209. Zamore, P. D., Tuschl, T., Sharp, P. A., and Bartel, D. P. (2000) RNAi: double-stranded RNA directs the ATP-dependent cleavage of mRNA at 21 to 23 nucleotide intervals. *Cell* 101, 25–33. doi: 10.1016/S0092-8674(00)80620-0
210. Zhai, J., Jeong, D. H., De Paoli, E., Park, S., Rosen, B. D., Li, Y., et al. (2011) MicroRNAs as master regulators of the plant NBLRR defense gene family via the production of phased, trans-acting siRNAs. *Genes Dev.* 25, 2540–2553. doi: 10.1101/gad.177527.111
211. Zhang, X., Zou, Z., Gong, P., Zhang, J., Ziaf, K., Li, H., et al. (2011a) Over-expression of microRNA169 confers enhanced drought tolerance to tomato. *Biotechnol. Lett.* 33, 403–409. doi: 10.1007/s10529-010-0436-0
212. Zhang, X., Zou, Z., Zhang, J., Zhang, Y., Han, Q., Hu, T., et al. (2011b) Over-expression of sly-miR156a in tomato results in multiple vegetative and reproductive trait alterations and partial phenocopy of the sft mutant. *FEBS Lett.* 585, 435–439. doi: 10.1016/j.febslet.2010.12.036
213. Zhang, Y. C., Yu, Y., Wang, C. Y., Li, Z. Y., Liu, Q., Xu, J., et al. (2013) Overexpression of microRNA OsmiR397 improves rice yield by increasing grain size and promoting panicle branching. *Nat. Biotechnol.* 31, 848–852. doi: 10.1038/nbt.2646
214. Zhao, B. T., Liang, R. Q., Ge, L. F., Li, W., Xiao, H. S., Lin, H. X., et al. (2007) Identification of drought-induced microRNAs in rice. *Biochem. Biophy. Res. Comm.* 354, 585–590. doi: 10.1016/j.bbrc.2007.01.022
215. Zhou, M., Li, D., Li, Z., Hu, Q., Yang, C., Zhu, L., et al. (2013) Constitutive expression of a miR319 gene alters plant development

and enhances salt and drought tolerance in transgenic creeping bentgrass. *Plant Physiol.* 161, 1375–1391. doi: 10.1104/pp.112.208702

216. Zhou, Y., Yuan, Y., Yuan, F., Wang, M., Zhong, H., Gu, M., et al. (2012) RNAi-directed down-regulation of RSV results in increased resistance in rice (*Oryza sativa* L.). *Biotechnol. Lett.* 34, 965–978. doi: 10.1007/s10529-012-0848-0
217. Zilberman, D., Cao, X., and Jacobsen, S. E. (2003) ARGONAUTE4 control of locus specific siRNA accumulation and DNA and histone methylation. *Science* 299, 716–719. doi: 10.1126/science.1079695
218. Zrachya, A., Glick, E., Levy, Y., Arazi, T., Citovsky, V., and Gafni, Y. (2007) Suppressor of RNA silencing encoded by tomato yellow curl virus-Israel. *Virology* 358, 159–165. doi: 10.1016/j.virol.2006.08.016

CHAPTER 9

AN RNAi-BASED CONTROL OF FUSARIUM GRAMINEARUM INFECTIONS THROUGH SPRAYING OF LONG DSRNAS INVOLVES A PLANT PASSAGE AND IS CONTROLLED BY THE FUNGAL SILENCING MACHINERY

Aline Koch[1] , Dagmar Biedenkopf[1] , Alexandra Furch[2] , Lennart Weber[3] , Oliver Rossbach[4] , Eltayb Abdellatef[1] , Lukas Linicus[1] , Jan Johannsmeier[1] , Lukas Jelonek[5] , Alexander Goesmann[5] , Vinitha Cardoza[6] , John McMillan[6] , Tobias Mentzel[7] , KarlHeinz Kogel[1]

[1] Institute for Phytopathology, Centre for BioSystems, Land Use and Nutrition, Justus Liebig University, Giessen, Germany

[2] Institute of General Botany and Plant Physiology, Friedrich-Schiller-University, Jena, Germany

[3] Institute for Microbiology and Molecular Biology, Centre for BioSystems, Land Use and Nutrition, Justus Liebig University, Giessen, Germany

[4] Institute of Biochemistry, Centre for BioSystems, Land Use and Nutrition, Justus Liebig University, Giessen, Germany

[5] Institute for Bioinformatics and Systems Biology, Centre for BioSystems, Land Use and Nutrition, Justus Liebig University, Giessen, Germany

[6] BASF Plant Science LP, Research Triangle Park, Durham, North Carolina, United States of America

[7] BASF SE, Limburgerhof, Germany

ABSTRACT

Meeting the increasing food and energy demands of a growing population will require the development of ground-breaking strategies that promote sustainable plant production. Host-induced gene silencing has shown great potential for controlling pest and diseases in crop plants. However, while delivery of inhibitory noncoding double-stranded (ds)RNA by transgenic expression is a promising concept, it requires the generation of transgenic crop plants which may cause substantial delay for application strategies depending on the transformability and genetic stability of the crop plant species. Using the agronomically important barley—*Fusarium graminearum* pathosystem, we alternatively demonstrate that a spray application of a long noncoding dsRNA (791 nt *CYP3*-dsRNA), which targets the three fungal cytochrome P450 *lanosterol C-14 -demethylases*, required for biosynthesis of fungal ergosterol, inhibits fungal growth in the directly sprayed (local) as well as the non-sprayed (distal) parts of detached leaves. Unexpectedly, efficient spray-induced control of fungal infections in the distal tissue involved passage of *CYP3*-dsRNA via the plant vascular system and processing into small interfering (si)RNAs by fungal DICER-LIKE 1 (*FgDCL-1*) after uptake by the pathogen. We discuss important consequences of this new finding on future RNA-based disease control strategies. Given the ease of design, high specificity, and applicability to diverse pathogens, the use of target-specific dsRNA as an anti-fungal agent offers unprecedented potential as a new plant protection strategy.

AUTHOR SUMMARY

RNA interference has emerged as a powerful genetic tool for scientific research. The demonstration that agricultural pests, such as insects and nematodes, are killed by exogenously supplied RNA targeting their essential genes has raised the possibility that plant predation can be controlled by lethal RNA signals. We show that spraying barley with a 791 nt long dsRNA (*CYP3*-dsRNA) targeting the three fungal ergosterol biosynthesis genes (*CYP51A*, *CYP51B*, *CYP51C*), whose respective proteins also are known as azole fungicide targets, efficiently inhibited the necrotrophic fungus *Fusarium graminearum* in directly sprayed and systemic leaf tissue. Strong inhibition of fungal growth required an operational fungal RNA interference mechanism as demonstrated by the fact that a Fusarium DICER-LIKE-1 mutant was insensitive to *CYP3*-dsRNA in systemic, non-sprayed leaf areas. Our findings will help in the efficient design of RNAi-based plant disease control. We provide essential information on a fundamentally new plant protection strategy, thereby opening novel avenues for improving crop yields in an environmentally friendly and sustainable manner.

INTRODUCTION

According to the FAO [1], more than half of the world's harvested area is allotted to cereals such as rice, maize and wheat (ca. 2.3 billion tons in 2010). Diseases of cereal crops such as Fusarium head blight (FHB) and Fusarium seedling blight (FSB), caused by necrotrophic fungi of the genus Fusarium, exert a particularly great economic and agronomic impact on global grain production and the grain industry [2,3]. Food safety can be compromised by contamination of agricultural products with mycotoxins, which are produced during FHB and FSB development [4] and represent a serious threat to human and animal health. Currently, the major strategies to control Fusarium diseases include resistance breeding, crop rotation, and biological control along with the application of DMI (demethylation inhibitors) fungicides [5]. DMI fungicides, such as tebuconazole, triadimefon, and prochloraz inhibit ergosterol biosynthesis by binding to cytochrome P450 lanosterol C-14 α-demethylase (CYP51), thereby disrupting fungal membrane integrity [6]. However, heavy reliance on DMI fungicides since their discovery in the mid-1970s holds a risk of the emergence of DMI-tolerant strains of plant pathogens. Conventional plant breeding strategies have been only partly successful, as the quantitative nature of FHB and FSB resistance does not allow straightforward breeding programs. Since the discovery in 1998 that exogenous double-stranded (ds)RNA triggers suppression of gene activity in a homology-dependent manner [7], along with the identification of small RNAs (sRNAs) as a new class of regulatory molecules [8] that functions via RNA interference (RNAi), our understanding of the essential cellular function of gene silencing has increased considerably [9–10]. Mobile RNA silencing signals are capable of translocating from the host to its interacting organism, and vice versa [11–14]. Recent evidence supports the significant contribution of sRNAs and RNAi to the communication between plant hosts and a pathogenic fungus [15]. Exploiting the RNAi mechanism in plants also has a strong potential for agriculture. Indeed, expression of inhibitory dsRNAs in the corresponding host plant conferred protection from predation or infection by targeted gene silencing [16–18], a phenomenon that has been termed host-induced gene silencing (HIGS).

Recently, we demonstrated that in Arabidopsis (*Arabidopsis thaliana*) and barley (*Hordeum vulgare*), transgenic expression of *CYP3*-dsRNA, a 791 nt long dsRNA targeting the three fungal *CYP51* genes involved in ergosterol biosynthesis, confers resistance to infection with *Fusarium graminearum* [19]. While these results provided proof-of-concept that RNAi-based plant protection is an effective strategy for controlling diseases caused by devastating necrotrophic pathogens, the broad applicability of this transgenic method remains questionable due to the persisting weak acceptance of GMO strategies for food and feed production in many countries. More important, a broad application of this transgenic approach is hampered by the lack of transformability of various crop plants and the missing genetic stability of the silencing trait. Here we investigate the potential and the mechanism of an RNAi-based crop protection strategy using direct spray applications of *CYP3*-dsRNA to target *F. graminearum*. We show that the 791 nt long dsRNA is taken up by the plant and transferred in an unmodified form via the vascular system to fungal infection sites where it is processed by the fungal RNAi machinery as a prerequisite for its antifungal activity. We show a strong correlation between accumulation of *CYP3*-dsRNA at infection sites, silencing of *CYP51* expression, and fungal inhibition.

RESULTS

Spray-induced Gene Silencing (SIGS) of Fusarium Genes

To provide a proof of concept, we conducted an experiment targeting the expression of the jellyfish *green fluorescent protein* (*GFP*) in the *GFP*-expressing *F. graminearum* strain Fg-IFA65$_{GFP}$ [20] by using a *GFP*-specific 720 nt long dsRNA (*GFP*-dsRNA). Detached barley leaves were locally sprayed with 20 ng μL^{-1} *GFP*-dsRNA or Tris-EDTA buffer (TE, control) and drop-inoculated 48 h later with Fg-IFA65$_{GFP}$ in the distal (non-sprayed) leaf segment. Confocal microcopy showed strong GFP fluorescence associated with fungal mycelia on TE-treated control leaves at six days post inoculation (dpi) (Fig 1A). In contrast, fluorescence (Fig 1B) and *GFP* transcripts (Fig 1C) were largely absent in mycelia grown on leaves that were locally sprayed with *GFP*-dsRNA, although mycelial growth was unrestricted as evidenced by light microscopy. This observation clearly demonstrates the possibility of targeting a gene of an attacking microbe via SIGS.

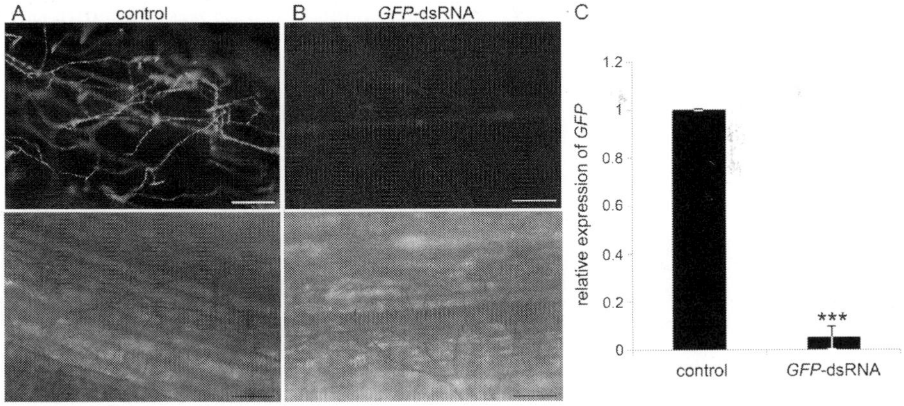

Figure 1: (A-C) Spray-induced gene silencing (SIGS) of *GFP* **expression in** *Fusarium graminearum* **strain Fg-IFA65$_{GFP}$.**

Detached second leaves of three-week-old barley plants were locally sprayed with Tris-EDTA (TE, A, control) or *GFP*-dsRNA (B). Forty-eight hours after spraying, distal, non-sprayed leaf segments were drop-inoculated with Fg-IFA65$_{GFP}$ (20 μL of a solution containing 2 x 10^4 conidia mL^{-1}). *GFP* silencing efficiency was visualized 6 dpi using confocal microscopy. (C) *GFP* transcripts were quantified by qPCR at 6 dpi. The reduction in fungal *GFP* expression on leaves sprayed with *GFP*-dsRNA and infected with Fg-IFA65$_{GFP}$ compared with TE-sprayed controls was statistically significant (***$P < 0.001$; Student′s *t* test). Bars represent mean values ± SDs of three independent experiments. Scale bars represent 100 µm.

https://doi.org/10.1371/journal.ppat.1005901.g001

To further explore the potential of SIGS, we assessed the silencing efficiency of *CYP3*-dsRNA, which targets the three Fusarium genes *CYP51A*, *CYP51B*, and *CYP51C*. The 791 nt long *CYP3*-dsRNA contains complementary fragments of these genes starting with N-terminal *CYP51B*, followed by *CYP51A* and *CYP51*C [19]. Leaves were sprayed with *CYP3*-dsRNA and 48 h later drop-inoculated directly onto the sprayed area with Fg-IFA65. At six dpi, *CYP3*-dsRNA-treated leaves developed brownish lesions that were substantially smaller than those on TE- or *GFP*-dsRNA-sprayed leaves that served as control in this experiment (Fig 2A). Quantitative real-time PCR (qPCR) analysis of fungal DNA levels, based on the ratio between fungal tubulin and plant ubiquitin, confirmed reduced fungal growth on *CYP3*-dsRNA-treated leaves (Fig 2B). To confirm that inhibition of Fusarium growth by *CYP3*-dsRNA was provoked by sequence-specific gene silencing, expression of all the three fungal *CYP51* genes was assessed. At six dpi, total RNA was isolated from infected leaves and the levels of *CYP51A*, *CYP51B* and *CYP51C* transcripts were measured by qPCR and normalized to the expression of the fungal *ß-tubulin* gene. Consistent with the concept of spray-induced gene silencing, we found that the relative amounts of *CYP51* transcripts were reduced on average by 58% (*CYP51A*), 50% (*CYP51B*), and 48% (*CYP51C*) in leaves sprayed with *CYP3*-dsRNA vs. the *GFP*-dsRNA control (Fig 2C).

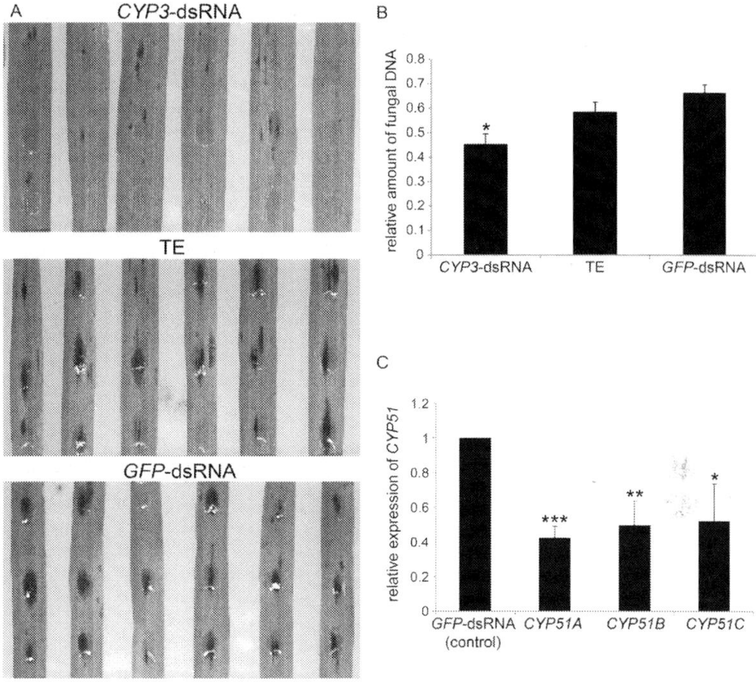

Figure 2: (A-C) SIGS-mediated control of *F. graminearum* **on leaves sprayed with** *CYP3*-dsRNA.

(A) Detached second leaves of three-week-old barley were sprayed evenly with *CYP3*-dsRNA, TE (mock control), and *GFP*-dsRNA (negative control), respectively. After 48 hours, leaves were drop-inoculated with 2×10^4 conidia mL^{-1} of Fg-IFA65 onto the sprayed area and evaluated for necrotic lesions at 6 dpi. (B) The relative amount of fungal DNA at 6 dpi as measured by qPCR was reduced in *CYP3*-dsRNA-treated leaves compared to control leaves. Bars represent mean values ± SDs of three independent experiments. The reduction of fungal growth on *CYP3*-dsRNA vs. TE- or *GFP*-dsRNA-sprayed leaves was statistically significant (*$P < 0.05$; Student´s t test). (C) Gene-specific qPCR analysis of fungal *CYP51A*, *CYP51B*, and *CYP51C* transcripts at 6 dpi (corresponding to 8 d after spraying). The reduction in fungal *CYP51* gene expression on *CYP3*-dsRNA-sprayed leaves as compared with *GFP*-dsRNA-sprayed controls was statistically significant (*$P < 0.05$, **$P < 0.01$, ***$P < 0.001$; Student´s t test).

https://doi.org/10.1371/journal.ppat.1005901.g002

SIGS Confers Strong Resistance against Fusarium in Distal Leaf Parts

Mobile cell non-autonomous inhibitory RNAs that spread gene silencing into adjacent cells and tissues have been observed in various plants [21–23]. Encouraged by the observed reduction in GFP fluorescence in Fg-IFA65$_{GFP}$ upon infection of leaf segments that did not receive a direct *GFP*-dsRNA spray (see Fig 1), we tested whether locally sprayed *CYP3*-dsRNA confers gene silencing in Fusarium infecting distal, non-sprayed segments of barley leaves. To this end, the upper part of detached leaves (local tissue) was sprayed with 20 ng μL^{-1} *CYP3*-dsRNA, *GFP*-dsRNA, or TE, while the lower part (distal tissue) was covered by a plastic tray to prevent direct dsRNA contamination. After 48 h, the distal, non-sprayed part of the leaves was drop-inoculated with Fg-IFA65$_{GFP}$; six days later, resistance to fungal infection was assessed. Distal leaf areas of *CYP3*-dsRNA-treated leaves developed substantially smaller lesions as compared to leaves sprayed with *GFP*-dsRNA or TE indicating that the silencing signal was basipetally transported. Consistent with this finding, the amount of fungal DNA as determined by qPCR was greatly reduced in the distal leaf area as compared to the control treatments (Fig 3A). The relative amounts of fungal *CYP51A*, *CYP51B* and *CYP51C* transcripts were strongly reduced on average by 72% (*CYP51A*), 90% (*CYP51B*), and 71% (*CYP51C*) as compared with control (*GFP*-dsRNA) treatment (Fig 3B). Confocal microscopy of fungal inoculation sites in distal leaf areas confirmed that, on TE-treated leaves, Fg-IFA65$_{GFP}$ conidia had germinated and colonized tissue next to the inoculation site (Fig 3C). In contrast, fungal mycelia on *CYP3*-dsRNA-treated leaves were only visible at the inoculation sites, and the surrounding leaf tissue was free of infection hyphae (Fig 3D). Consistent with this, the large number of fungal conidia with very short germ tubes at the inoculation sites of *CYP3*-dsRNA-treated leaves indicated that fungal germination was strongly impaired. We also tested whether the silencing signal was transported in the acropetal direction. Segments were sprayed with 20 ng μL^{-1} *CYP3*-dsRNA and subsequently drop-inoculated in the distal leaf area. Fusarium infections were also reduced in the acropetal experimental set up as shown by macroscopic inspection and qPCR quantification of fungal DNA.

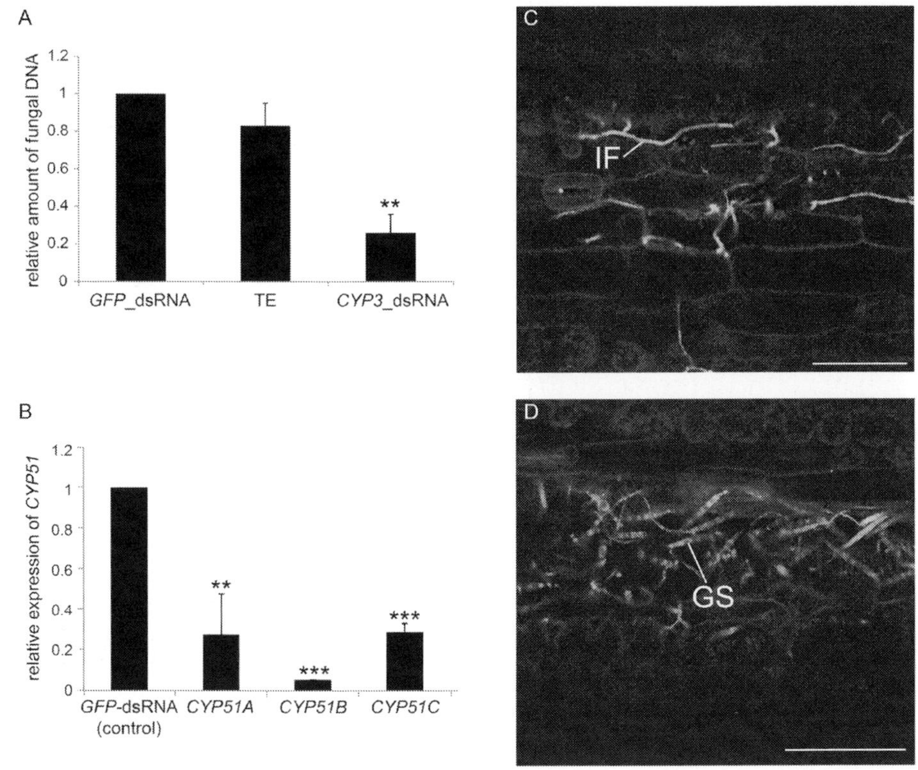

Figure 3: (A-D) SIGS-mediated semi-systemic control of *Fusarium graminearum*.

(A) Upper parts of detached second leaves of three-week-old barley were sprayed evenly with *CYP3*-dsRNA, TE, and *GFP*-dsRNA, respectively. After 48 h, the non-inoculated, semi-systemic (distal) tissue was drop-inoculated with 2×10^4 conidia $mL^{-1}$of Fg-IFA65$_{GFP}$; the relative amount of fungal DNA in distal tissue, as measured by qPCR at 6 dpi, was reduced in *CYP3*-dsRNA-treated leaves. Bars represent mean values ± SDs of three independent experiments. The reduction of fungal growth on *CYP3*-dsRNA-sprayed leaves was statistically significant (**$P < 0.01$; Student´s t test). (B) Gene-specific qPCR analysis of *CYP51A*, *CYP51B*, and *CYP51C* transcripts at 6 dpi in distal leaf areas. Bars represent mean values ±SDs of three independent sample collections. The reduction in *CYP51* expression in leaves sprayed with *CYP3*-dsRNA compared with *GFP*-dsRNA-sprayed controls was statistically significant (**$P < 0.01$, ***$P < 0.001$; Student´s t test). (C,D) Microscopy of fungal growth at semi-systemic sites of drop-inoculation with Fg-IFA65$_{GFP}$. (C) Successful fungal colonization (green) on TE-sprayed leaves. Profuse hyphal growth is seen inside the cells (plasma membrane stained with RH414 is highlighted in red) (D) Hyphal formation is strongly reduced and confined to the inoculated leaf area on *CYP3*-dsRNA-sprayed leaves. Impaired spore germination was observed in the area around the inoculation site while the surrounding cells remained free of colonization. (IF, infection hyphae; GS, germinating spore). Photographs for C and D were taken at 6 dpi.

https://doi.org/10.1371/journal.ppat.1005901.g003

Uptake and Processing of Sprayed *CYP3*-dsRNA

To further explore the SIGS mechanism, we investigated whether the spray-applied long *CYP3*-dsRNA is translocated in the plant tissue and/ or processed by the plant's silencing machinery independent of fungal infections. Following *CYP3*-dsRNA treatment, local (sprayed) and distal (non-sprayed) leaf segments were harvested separately at 24, 48, 72, or 168 h after spraying. Northern blot analysis detected unprocessed 791 nt *CYP3*-dsRNA in both local and distal tissue (Fig 4A), showing that the long dsRNA is systemically translocated within the plant. In the local (sprayed) segment, *CYP3*-dsRNA was detected over the full time range, while it accumulated only transiently at early time points (24 h) after spraying in the distal (non-sprayed segments). This accumulation profile is consistent with the idea that the vast bulk of the *CYP3*-dsRNA fraction was absorbed via the cut surface of the detached leaf. Moreover, *CYP3*-dsRNA-derived 21 nt and 22 nt small interfering (si)RNAs also accumulated over the whole time range after spraying in the local leaf segments, demonstrating that *CYP3*-dsRNA was partly processed by the plant (Fig 4B). In this experiment, Northern analysis could not detect siRNAs in distal leaf parts, probably because the technique was not sensitive enough.

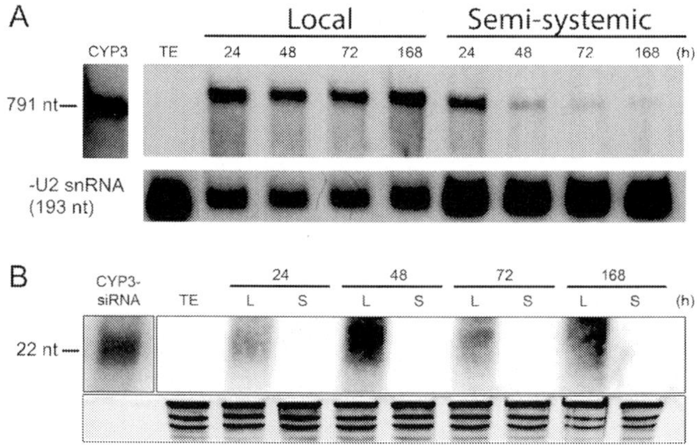

Figure 4: (A,B) Northern gel blot analysis of *CYP3*-dsRNA and *CYP3*-dsRNA-derived siRNA accumulation in local and distal (semi-systemic) barley leaf areas.

(A) Detection of 791 nt long *CYP3*-dsRNA precursor in pooled leaf tissue from non-infected leaves using [α-32P]-dCTP labeled *CYP3*-dsRNA as probe. Local (L) and distal (semi-systemic [S]) leaf segments were sampled separately at the indicated times after spraying with *CYP3*-dsRNA. No signal was detected in samples from TE-sprayed plants. (B) Recording *CYP3*-dsRNA-derived small RNAs in local and distal (semi-systemic) leaf areas using [α-32P]-dCTP labeled *CYP3*-dsRNA as probe. In this experiment, small RNAs could not be detected in distal (non-sprayed) tissues. siRNA generated *in vitro* by a commercial Dicer preparation from *CYP3*-dsRNA was used as positive control. No signal was detected in samples from TE-sprayed plants. Ethidium bromide-stained rRNA served as the loading control. Signals originate from the same membrane but different exposure times.

https://doi.org/10.1371/journal.ppat.1005901.g004

To further investigate uptake and transport of sprayed *CYP3*-dsRNA, it was labeled with the green fluorescent dye ATTO 488 (*CYP3*-dsRNA$_{A488}$) and sprayed onto barley leaves. The biological activity of *CYP3*-dsRNA$_{A488}$ was indistinguishable from non-labeled *CYP3*-dsRNA as evidenced by reduced fungal infection and strong silencing of fungal *CYP51* genes upon spray application. Moreover, using confocal laser scanning microscopy, a green fluorescent signal was detected in the vascular tissue at 24 hours after spraying leaves with 20 ng μl^{-1} *CYP3*-dsRNA$_{A488}$. In leaf cross-sections, fluorescence was seen in the xylem (Fig 5A–5C). Inspection of longitudinal leaf sections revealed that the fluorescence was not confined to the apoplast but also was present in the symplast of phloem parenchyma cells, companion cells, and mesophyll cells, as well as in trichomes and stomata (Fig 5D–5I). When *CYP3*-dsRNA$_{A488}$-sprayed leaves were inoculated with Fg-IFA65, the fluorescent signal also was detectable inside fungal conidia and germ tubes (Fig 5H) and fungal mycelium (Fig 5J). Together these data show that *CYP3*-dsRNA is taken up by the plant and is transferred via the plant vascular system. Systemic translocation within the plant and accumulation by the fungus also raised the possibility that *CYP3*-dsRNA is processed by the fungus into inhibitory siRNAs to eventually target fungal *CYP51* genes.

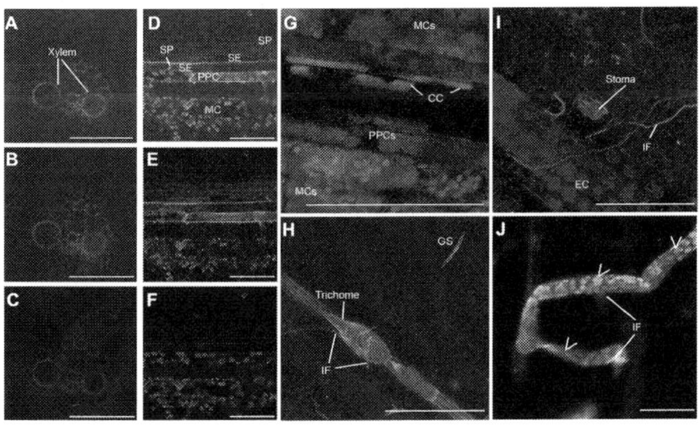

Figure 5: (A-J) Confocal laser scanning microscopy of ATTO 488-labeled *CYP3*-dsRNA$_{A488}$ **in locally sprayed barley leaves.**

(A-C) Detection of *CYP3*-dsRNA$_{A488}$ (green) in xylem vessels of vascular bundles 24 h after spraying. (D-G) Longitudinal sections reveal uptake of *CYP3*-dsRNA$_{A488}$ by cells of the phloem tissue at 24 h after spraying. SE, sieve element; CC, companion cell; SP, sieve plate; PPC, phloem parenchyma cell; MC, mesophyll cell. The red cells result from the autofluorescence of chloroplasts (F,G). (H-J) Leaf hair cells (trichome), stomata, germinating spores (GS) and fungal hyphae strongly accumulated *CYP3*-dsRNA$_{A488}$. Fungal hyphae (IF) are stained with chitin-specific dye WGA-Alexa Fluor 594 (red) 24 h after inoculation. EC, epidermal cells. RNA signals in germinated conidia are marked by arrow heads.

Scale bars 100 µm (A-H), 20 µm (F), and 10 µm (J).

https://doi.org/10.1371/journal.ppat.1005901.g005

To test this possibility, we first profiled *CYP3*-dsRNA-derived siRNAs in infected and non-infected leaves. Small RNA sequencing (RNAseq) analysis revealed distinctly different *CYP3*-dsRNA-derived siRNA profiles in mock- vs. Fg-IFA65-infected local and distal (non-sprayed) leaf segments (Fig 6A and 6B) with higher numbers of reads of *CYP3*-dsRNA-derived siRNAs in infected leaves, and highest numbers of reads in locally-inoculated vs. distally-inoculated leaves. These data suggest that *CYP3*-dsRNA also is processed by the fungus and that the fungal silencing machinery is involved in SIGS and reduced fungal infections. Detection by RNAseq of *CYP3*-dsRNA-derived siRNA in the distal (non-sprayed) part of leaves also supported our interpretation that northern analysis failed to detect low amounts of these siRNAs due to sensitivity problems.

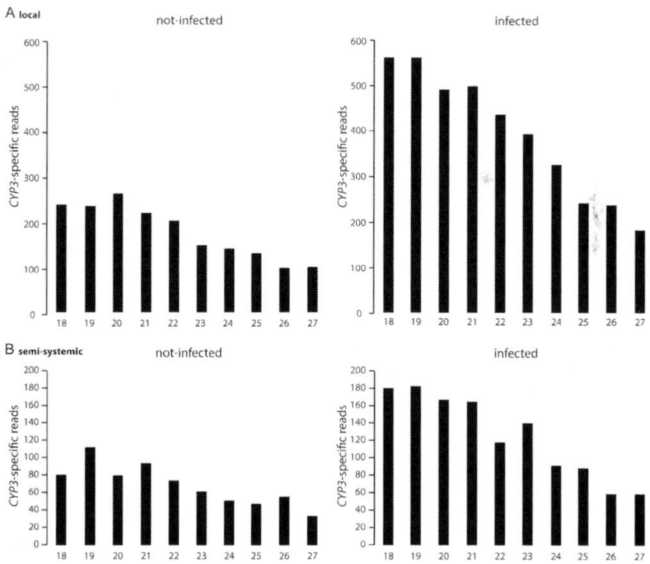

Figure 6: (A,B) RNA profiling (RNAseq analysis) of *CYP3*-dsRNA-derived RNAs in local and distal tissue of *CYP3*-dsRNA-treated barley leaves on Illumina HiSeq.

Higher numbers of reads of *CYP3*-dsRNA-derived sRNAs were found in infected vs. non-infected leaves both in the sprayed (local [A]) and non-sprayed (semi-systemic [B]) leaf area at 6 dpi with strain Fg-IFA65.

https://doi.org/10.1371/journal.ppat.1005901.g006

Fungal DICER-LIKE-1 is required for Efficient SIGS in Systemic Leaf Areas

To further substantiate involvement of the fungal silencing machinery, we generated a fungal *dcl-1* mutant (Fg-IFA65$_{\Delta dcl\text{-}1}$) that is deficient for DICER-LIKE 1, a critical component of the fungal silencing machinery that produces siRNA from long dsRNA stretches. Fg-IFA65$_{\Delta dcl\text{-}1}$ and the wild type Fg-IFA65 were indistinguishably virulent on TE-sprayed barley leaves (Fig 7A), showing that fungal DCL-1 is not required for successful leaf infections. However, in contrast to Fg-IFA65, the mutant Fg-IFA65$_{\Delta dcl\text{-}1}$ also heavily infected distal areas of *CYP3*-dsRNA-treated barley leaves (Fig 7B), suggesting that the mutant strain is not amenable to SIGS. We concluded that the fungal silencing machinery appears to be indispensable for *CYP3*-dsRNA-mediated SIGS at systemic areas in the barley-*Fusarium graminearum* pathosystem. To further confirm that FgDCL-1 is required for *CYP51* target gene silencing, levels of *CYP51A*, *CYP51B* and *CYP51C* transcripts were compared by qPCR in the wild type vs. the *dcl-1* mutant on infection of *CYP3*-dsRNA sprayed leaves. The relative amounts of transcripts were reduced in Fg-IFA65 on average by 50% (*CYP51A*), 70% (*CYP51B*), and 40% (*CYP51C*) as compared with TE (control) treatment. In contrast, expression of *CYP51* targets was not reduced in the Fg-IFA65$_{\Delta dcl\text{-}1}$ mutant (Fig 7C).

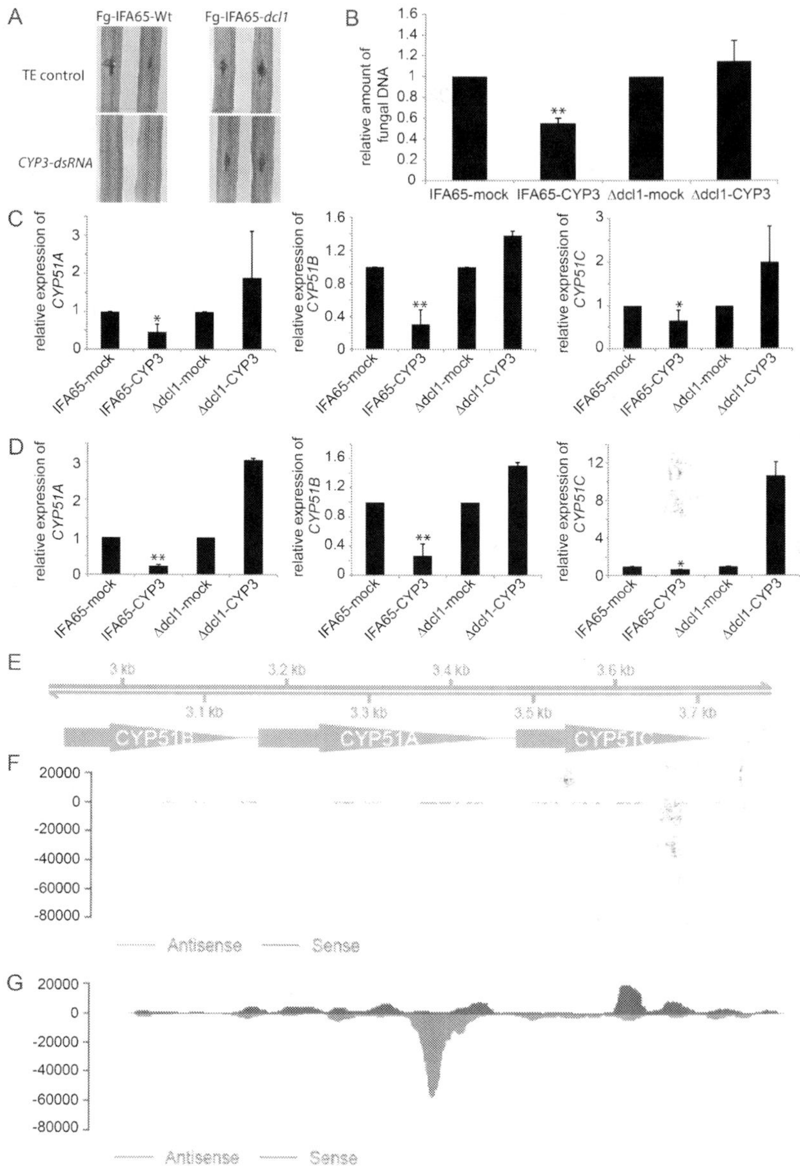

Figure 7: (A-E) The fungal silencing machinery is required for efficient SIGS in distal leaf parts.

(A,B) The fungal *dicer-like-1* mutant Fg-IFA65$_{\Delta dcl\text{-}1}$ heavily infected barley leaves despite a prior spray-treatment with *CYP3*-dsRNA. Photographs were taken at 6 dpi. (C)Gene-specific qPCR analysis of *CYP51A*, *CYP51B*, and *CYP51C* transcripts in the wild type Fg-IFA65 and the mutant Fg-IFA65$_{\Delta dcl-1}$ at 6 dpi in the distal, semi-systemic leaf areas. (D) Inhibition of *CYP51* gene expression upon *CYP3*-dsRNA treatment of axenically grown Fg-IFA65. Bars represent mean values ±SDs of three independent sample collections. The reduction in *CYP51* expression in samples treated with *CYP3*-dsRNA compared with mock-treated controls was statistically significant (*P < 0.05, **P < 0.01; Student´s t test). (E-G) Profiling of *CYP3*-dsRNA-derived sRNAs in axenically grown Fg-IFA65. (E) Scaffold of the 791 nt long *CYP3*-dsRNA. The fragments of *CYP51*genes are indicated. (F,G) Total sRNAs were isolated from axenically-cultured Fg-IFA65. sRNA reads of fungal sRNAs from untreated (F) and *CYP3*-dsRNA-treated (G) fungal cultures are mapped to the sequence of *CYP3*-dsRNA.

https://doi.org/10.1371/journal.ppat.1005901.g007

We additionally conducted an *in vitro* experiment to further demonstrate the requirement of FgDCL-1 for *CYP3*-dsRNA-mediated silencing of fungal *CYP51* genes. Mycelia of axenic cultures of Fg-IFA65 and Fg-IFA65$_{\Delta dcl-1}$ were treated with *CYP3*-dsRNA. Subsequently, expression of *CYP51* genes was recorded. Consistent with the leaf assay, the relative amounts of fungal *CYP51A*, *CYP51B* and *CYP51C* transcripts were reduced in the wild type Fg-IFA65 but not in the Fg-IFA65$_{\Delta dcl-1}$ mutant (Fig 7D). Confirmatory total sRNAs profiling by RNAseq in axenically-grown Fg-IFA65 revealed a range of sRNAs originating from *CYP3*-dsRNA (Fig 7E–7G), further proving that the fungus can process *CYP3*-dsRNA. Suspiciously, the majority of siRNA species mapped to sites in the *CYP51A* gene fragment of the *CYP3*-dsRNA. Further work must show if this profile is a result of the physical structure of the dsRNA.

CYP3-dsRNA-derived siRNAs also Confer SIGS

The failure to detect *CYP3*-dsRNA-derived siRNA in the distal area of *CYP3*-dsRNA-sprayed leaves by northern analysis along with the compromised SIGS phenotype of the mutant Fg-IFA65Δdcl-1 suggested that the concentration of siRNA in the distal leaf parts was too low to mediate silencing of *CYP51* genes in the fungus. Alternatively, Fusarium is generally unable to absorb siRNA from barley leaves. To address these possibilities, we sprayed barley leaves with high concentration of *CYP3*-dsRNA-derived siRNAs (20 ng μl^{-1}) and subsequently inoculated local (sprayed) and distal (non-sprayed) leaf segments with Fg-IFA65. We found that the fungus was strongly inhibited by these siRNAs both in the local and distal leaf segments as compared with a control (*GFP*-dsRNA-derived siRNAs). Consistent with this, *CYP3*-dsRNA-derived siRNA also reduced the expression of *CYP51*genes of the fungus growing on local and distal leaves segments, which shows that *F. graminearum* also can ingest inhibitory siRNAs from plant tissue. In clear support of this notion and consistent with the finding that *CYP3*-dsRNA-derived siRNA accumulated to higher concentration in leaf areas directly sprayed with *CYP3*-dsRNA, mutant Fg-IFA65$_{\Delta dcl\text{-}1}$was not compromised in SIGS when drop-inoculated directly to the sprayed leaf area.

SIGS-Mediated Fungal Inhibition is Independent of Innate Immune Responses

In mammalian cells, perception of certain dsRNAs via toll-like receptors triggers an inflammation response [24,25]. Therefore, we assessed whether *CYP3*-dsRNA elicits an innate immune response called pattern-triggered immunity (PTI) [26], when sprayed onto barley leaf segments. To this end, expression of barley genes that are indicative of the canonical defense-related salicylate- and jasmonate-dependent pathways [27] was evaluated. Expression of salicylate-responsive *pathogenesis-related 1* (*HvPR1*) and jasmonate-responsive S-*adenosyl-l-methionine:jasmonic acid carboxyl methyltransferase* (*HvJMT*) in TE-treated leaves was strongly induced by Fg-IFA65, but not by *CYP3*-dsRNA treatment (Fig 8). Furthermore, Fg-IFA65-induced expression of either gene was much lower in *CYP3*-dsRNA-treated leaves as compared with TE-treated leaves. This result strongly suggests that *CYP3*-dsRNA does not induce PTI in barley, and that the SIGS mechanism does not rely on activation of canonical defense pathways. The finding also is relevant when considering the fitness cost, and thus yield, of the SIGS strategy.

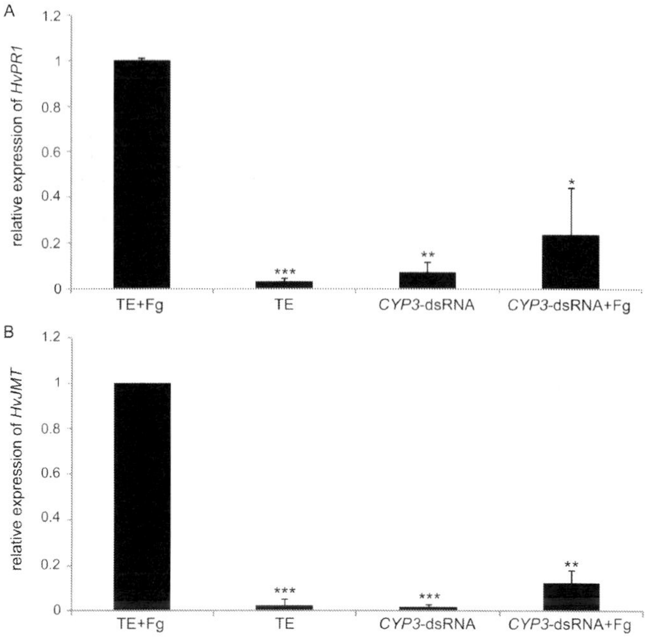

Figure 8: (A,B) Defense-related salicylate- and jasmonate-responsive genes are not induced by *CYP3*-dsRNA.

Detached second leaves of three-week-old barley were sprayed with 20 ng µL^{-1} *CYP3*-dsRNA or TE (control), respectively, and 48 h later drop-inoculated with Fg-IFA65. Leaves were harvested 6 dpi and analyzed for gene expression by qPCR: (A)*Pathogenesis-related* 1 (*HvPR1*) and (B) *S-adenosyl-l-methionine:jasmonic acid carboxyl methyltransferase* (*HvJMT*). Both genes are highly responsive to Fg-IFA65 but not to *CYP3*-dsRNA or TE treatment. Please note that a combined treatment of *CYP3*-dsRNA followed by Fg-IFA65 48 h later also did not induce these marker genes, which shows independently that fungal development on *CYP3*-dsRNA-treated leaves is strongly inhibited.

https://doi.org/10.1371/journal.ppat.1005901.g008

DISCUSSION

In this study, we show that delivery of long noncoding double-stranded RNA targeting the three *CYP51* genes of the necrotrophic ascomycete fungus *Fusarium graminearum* via spray application effectively reduces the development of the pathogen on barley leaves. Thus, our work further supports the idea that RNA could be used as a chemical treatment to control plant diseases.

Next to the economic and ecologic consideration about the deployment of antimicrobial RNAs as a new plant protection measure, elucidating the molecular mechanisms of RNA-based disease control is a key for successful future implementation. While plant-derived transgene-mediated silencing of target genes in plant pathogens and pests (a mechanism known as host-induced gene silencing [HIGS]) has been frequently reported [12,14,19], few examples have demonstrated the efficiency of exogenous RNA delivery to kill plant attackers. HIGS virtually is based on the plant's silencing machinery, whereas the mechanism of gene silencing by exogenously delivered dsRNA constitutes a more complex situation. For instance, diverging questions are possible involvement of the silencing machineries of host and/or the pathogen/pest, the requirement of local and remote transport of channeled dsRNA molecules, and the problem of dsRNA transport at the apoplast–symplast interface.

Using the *F. graminearum*—barley pathosystem as a model to study the mechanism of exogenously applied inhibitory dsRNA was motivated by the fact that Fusarium Head Blight and Crown Rot cause serious problems worldwide including food and feed safety issues due to mycotoxin contamination of cereals and maize. Focusing on fungal *CYP51* genes as targets for silencing was reasonable because our previous work provided proof-of-concept that transgene-mediated silencing of these genes effectively reduced fungal development in Arabidopsis and barley. More than that, direct treatment of *F. graminearum* with inhibitory dsRNA matching *CYP51* gene sequences had been demonstrated to inhibit fungal development in axenic cultures [19].

The previous finding of impaired fungal growth in axenic cultures, when treated with a 791 nt dsRNA (*CYP3*-dsRNA) targeting the three fungal genes *CYP51A*, *CYP51B*, and *CYP51C*, let us speculate that the fungus can process *CYP3*-dsRNA into siRNA that interfere with the expression of *CYP51* genes. Gene annotation of *F. graminearum*'s genome (http://www.broadinstitute.org) predicted genes coding for two ARGONAUTE-like proteins, two DICER-like proteins and five RNA-dependent RNA Polymerases (RDR), suggesting that the RNAi pathway is functional [28]. Consistent with these findings, RNAseq analysis of axenically grown *F. graminearum*, treated with *CYP3*-dsRNA, showed high numbers of reads of *CYP3*-dsRNA-derived siRNA (Fig 7). Together these data showed that *F. graminearum* possesses a functional gene silencing system, which is a prerequisite for disease control by SIGS.

To test the antifungal activity of *CYP3*-dsRNA and their siRNA derivatives, we used a detached leaf assay that enabled us to assess fungal growth in local (directly sprayed) and distal (semi-systemic, non-sprayed) leaf segments. Using this approach, we could demonstrate that inhibitory dsRNA translocated via the plant vascular system and eventually was absorbed by the pathogen from leaf tissue (Fig 5). The profile of inhibitory dsRNA accumulation, as demonstrated by northern blot analysis (Fig 4) and RNAseq (Fig 6), showed that both long *CYP3*-dsRNA and plant-processed *CYP3*-dsRNA-derived siRNA accumulate in the plant vascular system, though translocation of siRNA seems to be less efficient and thus siRNA concentration at the remote infection sites probably was not high enough to induced SIGS. Consistent with this notion, *in planta* produced *CYP3*-dsRNA-derived siRNAs was detected in the distal leaf parts only by the more sensitive RNAseq technique but not by northern analysis. Nevertheless, spraying high concentrations of *CYP3*-dsRNA-derived siRNA (20 ng μL^{-1}) induced the SIGS process, demonstrating that the fungus is able to absorb siRNAs from barley leaves.

Because of the less efficient movement of siRNAs, transport and translocation of the unprocessed 791 nt dsRNA has a critical role in the SIGS process, demonstrated by a compromised SIGS phenotype of the Fusarium mutant Fg-IFA65$_{\Delta dcl\text{-}1}$ at distal leaf segments (Fig 7) but not at local, directly sprayed leaf segments. Compromised DICER activity resulted in the fungus inability to cleave *CYP3*-dsRNA into siRNA, thus interrupting the RNA interference mechanism in case the concentration of *CYP3*-dsRNA-derived siRNA is not sufficiently high. Our finding that the unprocessed long dsRNA could be absorbed from leaf tissue has further implications for future disease control strategies using dsRNA. There are good arguments that application of longer dsRNAs might be more efficient than application of siRNAs given there more efficient translocation. Long dsRNA would be processed into many different inhibitory siRNA by the fungus, which not least also could be an issue when considering the risk of compound resistance emerging in a pathogen under field test conditions. Thus, based on our findings, further research is required to establish rules for optimal dsRNA structures, considering e.g. molecule lengths, combinatorial order of gene fragments, target sites in a given gene target, and the number of genes targeted by one dsRNA. Supporting the requirement for more information as to the design of dsRNA probes, RNAseq analysis revealed that most of the *CYP3*-dsRNA-derived siRNAs, accumulating in the axenic fungal mycelium, were not equally distributed at the *CYP3*-dsRNA scaffold, but accumulated at the *CYP51A*gene fragment (Fig 7). Further analysis is required to explain this bias in the production of siRNAs from *CYP3*-dsRNA.

Our results are also consistent with the view that inhibitory dsRNA is more effectively absorbed by the fungus through infection hyphae that have intimate contact to plant tissue (compare *CYP51* gene expression in Figs 2 and 3). How these hyphae differ from the fungal germ tubes and extracellular hyphae is however yet unclear although distinct biochemical modifications of fungal hyphae that penetrate host plants and are involved in nutrient uptake have been demonstrated [29,30]. Thus it is likely that these specialized, leaf tissue colonizing hyphae show dsRNA uptake that is superior over germ tubes.

Our data are consistent with reports showing that silencing signals in plants are mobile [31,32]. The design of our experiments based on the previous finding that sRNAs, just as viroids [33], preferably move via the vascular system in the source-to-sink direction although some reports discussed transport in the opposite route (for review see [21]). Source-to-sink movement is one reason why the phloem rather than the xylem is generally considered as the conduit for movement of the silencing signal. This hypothesis is supported by the finding that the xylem sap, which transports water and ions, commonly is free of RNA [34]. However, spray application of dsRNA onto detached leaves cannot be compared with the situation in an intact leaf. Exogenously applied dsRNA first reached the apoplast, including the xylem (Fig 5), and subsequently translocated into the symplast by a yet unknown mechanism. Consistent with this, we could also demonstrate acropetal movement of the silencing signal that resulted in the inhibition of the fungus in distal not sprayed leaf areas. Apoplastic movement of RNA has been proposed, e.g. to explain how maternally expressed siRNAs could be transferred from the endosperm of developing seeds into the symplastically isolated embryo [35]. Regardless of how target-specific inhibitory RNAs are applied–by transgene delivery (HIGS) or spray (SIGS)–the use of target-specific inhibitory RNAs to confer plant protection potentially is an alternative to conventional chemicals because they are i) highly specific and solely depending on their nucleotide sequence and ii) can be developed against an unlimited range of pathogens provided that the RNAi machinery is in place. Given the accumulation of dsRNA in the plant phloem (Fig 5), sucking insects also are realistic SIGS targets as their efficient control by HIGS has been largely demonstrated [36,37]. Certainly, many questions have to be addressed in the future to eventually judge the agronomical potential of SIGS, including the costs of RNA applications and their stability under field conditions.

Hence, more research is required to develop application strategies, including improved uptake by compound design and use of chemical formulations. Another yet unassessed issue is the risk that microbial strains become insensitive to a commercial dsRNA product. Such scenario could probably be resolved by application of dsRNA mixtures that target different regions in one gene or even different genes. Moreover, a commercial dsRNA product should be designed not to have off-target effects in other organisms that might be relevant in the respective agroecosystem, including beneficial fungi and bacteria. In this respect, it is important to understand that both plants [12] and fungi [38] support the production of secondary siRNAs, meaning there is a potential for transitivity and amplification. It is therefore possible that low abundance inhibitory dsRNA sprayed onto the plant triggered a large silencing effect in Fusarium via these secondary RNAs. Importantly, when considering the regulatory issue of RNA-based plant protection it is crucial to emphasize that the principles of SIGS and HIGS rely on the mechanisms found for trans-kingdom communication in mutualistic and parasitic interactions, and thus is a natural phenomenon [12,13,14].

Apart from the dsRNAs prospects in future plant protection strategies, there is an additional technological potential in developing new pesticides. The simple phenotyping adopted by the SIGS screens renders them a powerful tool for genetic studies to assess compound targets with high efficiency and low costs. Thus, the present data provide essential scientific information on a fundamentally new plant protection strategy, thereby opening novel avenues for improving crop yields in an environmentally friendly and sustainable manner.

MATERIALS AND METHODS

Plant and Fungal Materials

The spring barley (*Hordeum vulgare*) cultivar (cv.) Golden Promise was grown in a climate chamber under 16 h light photoperiod (240 μmol m^{-2} s^{-1} photon flux density) at 18°C/14°C (day/night) and 65% relative humidity. The wild type *Fusarium graminearum* strain Fg-IFA65 was described earlier [20]. Plates were incubated at room temperature under constant illumination from one near-UV tube (Phillips TLD 36 W/08) and one white-light tube (Phillips TLD 36 W/830HF). Fungal strains were cultured on synthetic nutrient-poor agar (SNA)-medium [39].

Generation of transgenic *F. graminearum* (Fg-IFA65$_{GFP}$), expressing a jellyfish *green fluorescence protein* (*GFP*) gene under the *Neurospora crassa isocitrate lyase* gene promoter (PCII) [40], was performed as described [41].

For generation of the *DICER LIKE-1* Fg-IFA65$_{\Delta dcl\text{-}1}$ knock-out mutant, a homologous recombination strategy was used. The two homologous recombination segments USS and DSS (~1 kb each), representing promoter and termination regions of the *FgDCL-1* gene, were selected based on the sequence information available at the *Fusarium graminearum* genome database (http://www.broadinstitute.org), and were PCR amplified. Primers used for USS (DCL_1_USS_KpnI_F and DCL_1_USS_KpnI_R), and DSS (DCL_1_DSS_HindIII_F and DCL_1_DSS_HindIII_R). The USS and DSS were cloned into flanking sites of the *hph* cassette of the pPK2 binary vector [38] using USER enzyme mix (New England Biolab, Inc., Ipswich, MA, USA) in *Escherichia coli*. The resultant plasmid was confirmed for proper orientation of cloned inserts in the vector by PCR conducted using USS/DSS and vector-specific primers and then by sequencing the PCR products.

The pPK2::Δ*Fgdcl-1* binary plasmid containing two *Fgdcl-1* USS and DSS was transformed into *Agrobacterium tumefaciens* strain LBA4404 by electroporation, and transformants were analyzed by conducting restriction analysis. The ATMT of *F. graminearum* was based on a modified protocol [42]. Briefly, *A. tumefaciens* LBA4404 containing the pPK2::Δ*Fgdcl-1* plasmid was grown overnight in LB medium at 28°C (25 μg mL^{-1} kanamycin, 25 μg mL^{-1} rifampicin, and 5 μg mL^{-1} tetracycline). The next day, 10 ml of LB medium supplemented with above mentioned antibiotics and 200 μM acetosyringone was inoculated with 100 μl of the *A. tumefaciens* culture. This *A. tumefaciens* cell suspension with an OD_{600} of 0.5 to 0.7 was mixed with *F. graminearum* conidial suspensions (10^5–10^6 mL^{-1}) in liquid SNA medium in equal proportions [1:1(v/v)]. Aliquots of 200 μl of the mixture were spread on black filter paper circles (Grade 551; Whatman Inc., Piscataway, NJ, USA), which were overlaid on SNA plates containing 200 μM acetosyringone and incubated for 2 days in the dark at RT until mycelial growth was observed on the filter paper. Transformants were selected on SNA medium supplemented with 50 μg mL^{-1} of hygromycin B (Sigma, St. Louis, MO, USA) and 300 μg mL^{-1}ticarcillin (Fisher Scientific, Pittsburgh, PA, USA).

dsRNA Synthesis

The stacked clone (*CYP51 B-A-C*) [19] covering sequences of the three cytochrome P450 lanosterol C-14α-demethylase genes *CYP51A* (FGSG_04092), *CYP51B* (FGSG_01000), and *CYP51C* (FGSG_11024) from *F. graminearum* was used as template for the synthesis of a 791 nt long *CYP3*-dsRNA [19]. The pLH6000-Ubi::sGFP plasmid [43] was used as template for the synthesis of a 720 nt long *GFP*-dsRNA. dsRNA was generated using MEGAscript RNAi Kit (Invitrogen) following MEGAscript protocols. Primer pairs T7_F and T7_R with T7 promoter sequence at the 5`end of both forward and reverse primers were designed for amplification of dsRNA.

sRNAs Synthesis

sRNAs were generated using PowerCut Dicer (Thermo Scientific). Following the PowerCut Dicer protocol, *CYP3*-dsRNA or *GFP*-dsRNA was used as template for Dicer cleavage.

Spray Application

Detached leaf assay: Detached leaves of three-week-old barley plants were transferred into square Petri dishes (120 x 120 x 17mm) containing

1% agar. For spray application, dsRNA was diluted in 500 μL water to a final concentration of 20 ng μL^{-1}, corresponding to 10 μg dsRNA per plate. For sRNA application, the reaction mixture of DICER-cleaved dsRNA was used at a final concentration of 10 μg siRNA diluted in 500 μL^{-1} water per plate. Leaves were sprayed using a spray flask (10 mL capacity). Each dish containing six detached leaves was evenly sprayed. For the semi-systemic setup, detached leaves were covered before spraying with a plastic tray leaving only the upper part (approximately 1 cm) uncovered. After spraying, dishes were kept open until the surface of each leaf was dried (approximately 2 h). After an indicated lag time, leaves were drop-inoculated with 20 μL of 2×10^4 fungal conidia mL^{-1}. Closed dishes were incubated for 6 d at approximately 20°C on the lab bench. Alternatively, one-week-old barley seedlings were sprayed in the first leaf stage with 20 ng μL^{-1} dsRNA, and spray-inoculated three weeks later with 2×10^4 conidia mL^{-1} of Fg-IFA65. Inoculated plants were grown for three weeks in a growth chamber before evaluating the infection symptoms.

In vitro Axenic Cultures

Fg-IFA65 and Fg-IFA65$_{\Delta dcl\text{-}1}$ were cultured on synthetic nutrient SNA-medium. Plates were incubated at room temperature under constant illumination from one near-UV tube (Phillips TLD 36 W/08) and one white-light tube (Phillips TLD 36 W/830HF). Conidia were harvested from one-weak-old cultures with a sterile glass rod and sterile water [19]. *CYP3*-dsRNA was added to the fungal samples. Plates were incubated at room temperature. Gene expression studies were performed 24 h post *CYP3*-dsRNA treatment.

Quantification of Fungal Infection and Transcript Analysis

The relative amount of fungal DNA was measured using qPCR to quantify fungal infection. DNA was extracted using the CTAB method [44]. Expression analysis of the three fungal *CYP51*genes as well as plant defense marker genes *PR1* and *JMT*, respectively, was performed using qPCR. RNA extraction from infected leaves was performed with TRIzol (Invitrogen) following the manufacturer's instructions. Freshly extracted mRNA was used for cDNA synthesis using QuantiTect

Reverse-Transcription kit (Qiagen). For qPCR 10 ng of cDNA was used as template in the Applied Biosystems 7500 FAST realtime PCR system. Amplifications were performed in 7.5 μL of SYBER green JumpStart Taq ReadyMix (Sigma-Aldrich) with 0.5 pmol oligonucleotides. Each sample had three repetitions.

To quantify the amount of fungal DNA, primers were used for assessing expression of the fungal β-tubulin gene (FGSG_09530) with reference to barley ubiquitin gene. Primers were used for assessing expression of target *CYP51* genes with reference to β-tubulin gene. After an initial activation step at 95°C for 5 min, 40 cycles (95°C for 30 sec, 57°C for 30 sec, 72°C for 30 sec) were performed. Ct values were determined with the 7500 Fast software supplied with the instrument. Transcript levels of β -tubulin gene were determined via the 2-Δ Δ Ct method [45] by normalizing the amount of target transcript to the amount of reference transcript.

Small RNA Library Production and Sequence Analysis

RNA enriched for the sRNA fraction was purified from plant and fungal samples using the mirVana miRNA Isolation Kit (Life Technologies). Indexed sRNA libraries were constructed from these enriched sRNA fractions with the NEBNext Multiplex Small RNA Library Prep Set for Illumina (New England Biolabs) according to the manufacturer's instructions. Indexed sRNA libraries were pooled and sequenced on the Illumina HiSeq and NextSeq 500 platforms and the sequences sorted into individual datasets based on the unique indices of each sRNA library. The adapters and indices were trimmed using Trimmomatic [46] version 0.2.2 and the reads were mapped to the *CYP3*-dsRNA vector sequence using bowtie2 [47] version 2.1.0. to identify sRNAs with a perfect match. Each library contained at least 5 million total reads.

Northern Blot Analysis

For Northern blot analysis, 8 ng of total RNA from local region and 80 ng of systemic region or negative control (TE-mock) and 10 pg of *in vitro* transcribed *CYP3*-dsRNA was loaded onto a 6% denaturing polyacrylamide gel with DNA Molecular Weight Marker VIII (Roche), transferred to a nylon membrane. *CYP3*-dsRNA and U2 snRNA were detected using the DIG Labeling and Detection System (Roche) following the manufacturer's instructions. Chemiluminescence was detected using X-ray films.

The *CYP3*-dsRNA probe was created by PCR with *CYP3*-dsRNA forward and reverse primer on the stacked clone (*CYP51 B-A-C*) [19] using PCR DIG Labeling Mix (Roche). U2 snRNA loading control was amplified from cDNA created from total RNA using qScript Flex cDNA Kit (Quanta BioSciences) and primers U2 forward (TACCTTTCTCGGCCTTTTGG and U2 reverse (CAGCAGCAAGCTACTGTGGT). Gel purified probes were hybridized in NorthernMax Prehybridization/Hybridization Buffer (Ambion) at 45°C over night.

Northern blots for the detection of *CYP3*-dsRNA-derived siRNA were performed as described [19] using a 791 nt [α-32P]-dCTP labeled *CYP3*-dsRNA as probe.

Confocal Microscopy of Fluorophore Distribution

Twenty-four h after spraying fluorescing dsRNA were imaged using a Leica TCS SP2 (Leica Microsystems, Wetzlar, Germany) equipped with a 75-mW argon/krypton laser (Omnichrome, Chino, CA) and a water immersion objective (HCX APO L40x0.80 W U-V-l objective). Fluorescing dsRNA were imaged using a LSM 880 (Zeiss Microscopy GmbH, Jena, Germany) with the 488 nm laser line of an argon multiline laser (11.5 mW) and a HeNe 594 nm (1.3 mW) laser. Images were taken with a 20x objective (Plan-Apochromat 20x/0.8). Lambda stacks were created using the 32 channel GaAsP detector. Reference spectra with each pure fluorescence dye were recorded. The sample was inspected in Online Fingerprinting mode. Specific areas of the sample were imaged in lambda mode followed by Linear Unmixing with ZEN software (Zeiss, Jena, Germany). Fluorescent labeling of the dsRNA was performed using the Atto 488 RNA Labeling Kit (Jena Bioscience, Jena, Germany) following the manufacturer's instructions. Leaves were sprayed with the labeled dsRNA and 24 h later drop-inoculated with 2×10^4 Fg-IFA65 conidia mL^{-1}. To assess whether dsRNA has an effect on fungal morphology, leaves were inoculated with Fg-$IFA65_{GFP}$ and infected leaves were analyzed at 6 dpi.

For observation of phloem tissue, cortical cell layers were removed down to the phloem from the lower side of the main vein of a mature leaf. The leaf surface and longitudinal- as well as cross sections were stained with 4.3 μM of the membrane dye RH-414 (-N-(3-triethylammoniumpropyl)-4-(4-(4-(diethylamino)phenyl)butadienyl) pyridiniumdibromid) and/or with 5 μg mL^{-1} of the fungal hyphae dye wheat germ agglutinin (WGA) Alexa Fluor® 594 conjugate (Invitrogen) for at least 10 min.

RH-414, WGA Alexa Fluor 594, and the autofluorescence of cell walls and chloroplasts were excited by the 564-nm line of the argon/krypton laser, while GFP and ATTO 488 were excited with the 488-nm line. For observation at the 590 nm and 510 nm wave lengths, respectively, a long pass filter was used. Digital images were processed with Adobe Photoshop to optimize brightness, contrast, and color and to enable an overlay of the photomicrographs.

Statistical Analysis

Analyses were performed in SigmaPlot 12 (Systat Software) using Student´s t-tests after data were tested for normality distribution (Shapiro-Wilk test).

Genes Mentioned in the Text

FgCYP51A (FGSG_04092); *FgCYP51B* (FGSG_01000); *FgCYP51C* (FGSG_11024); *FgDCL1*(FGSG_09025); *β-tubulin* (FGSG_09530); *HvPR1* (X74940); *HvJMT* (BAD33074.1); *HvUBQ*(M60175)

ACKNOWLEDGMENTS

We thank P. Ganz for excellent technical assistance. C. Birkenstock, U. Schnepp and V. Weisel for excellent plant cultivation. We also thank Dr. Jens Steinbrenner for assisting in microscopic analyses.

REFERENCES

1. FAO. FAO Statistical Yearbook. World Food and Agriculture. Part 3: Feeding the World (ISBN 978-92-5-107396-4). Accessed from Food and Agriculture Organization of the United Nations website: http://www. fao. org (2013)
2. Bai G, Shaner G. Management and resistance in wheat and barley

to Fusarium head blight. Annu Rev Phytopathol. 2004; 42:135–61. pmid:15283663

3. Broekaert N, Devreese M, De Baere S, De Backer P, Croubels S. Modified *Fusarium*mycotoxins unmasked: From occurrence in cereals to animal and human excretion. Food Chem Toxicol. 2015; 80:17–31. pmid:25725190
4. Merhej J, Richard-Forget F, Barreau C. Regulation of trichothecene biosynthesis in Fusarium: Recent advances and new insights. Appl Microbiol Biotechnol. 2011; (3):519–28. pmid:21691790
5. Kazan K, Gardiner DM, Manners JM. On the trail of a cereal killer: recent advances in *Fusarium graminearum* pathogenomics and host resistance. Mol Plant Pathol. 2012; 13(4):399–413. pmid:22098555
6. Kuck KH, Stenzel K, Vors JP (2012) Sterol biosynthesis inhibitors modern crop protection compounds (pp. 761–805): Wiley-VCH Verlag.
7. Fire A, Xu S, Montgomery MK, Kostas SA, Driver SE, Mello CC. Potent and specific genetic interference by double-stranded RNA in *Caenorhabditis elegans*. Nature. 1998; 391(6669):806–11. pmid:9486653
8. Hamilton AJ, Baulcombe DC. A species of small antisense RNA in posttranscriptional gene silencing in plants. Science. 1999; 286(5441):950–2. pmid:10542148
9. Vaucheret H, Fagard M. Transcriptional gene silencing in plants: targets, inducers and regulators. Trends Genet. 2001; 17(1):29–35. pmid:11163919
10. Castel SE, Martienssen RA. RNA interference in the nucleus: roles for small RNAs in transcription, epigenetics and beyond. Nat Rev Genet. 2013; 14(2):100–12. pmid:23329111
11. Tomilov AA, Tomilova NB, Wroblewski T, Michelmore R, Yoder JI. Trans-specific gene silencing between host and parasitic plants. Plant J. 2008; 56:389–397 pmid:18643992
12. Baulcombe DC. VIGS, HIGS and FIGS: small RNA silencing in the interactions of viruses or filamentous organisms with their plant hosts. Curr Opin Plant Biol. 2015; 26:141–146. pmid:26247121
13. Weiberg A, Bellinger M, Jin H. Conversations between

kingdoms: small RNAs. Curr Opin Biotechnol. 2015; 32:207–15. pmid:25622136

14. Knip M, Constantin ME, Thordal-Christensen H. Trans-kingdom cross-talk: Small RNAs on the move. PLoS Genet. 2014; 10(9):e1004602. pmid:25188222
15. Weiberg A, Wang M, Lin FM, Zhao H, Zhang Z, Kaloshian I, Huang HD, Jin H. Fungal small RNAs suppress plant immunity by hijacking host RNA interference pathways. Science. 2013; 342(6154):118–23. pmid:24092744
16. Nowara D, Gay A, Lacomme C, Shaw J, Ridout C, Douchkov D, Hensel G, Kumlehn J, Schweizer P. HIGS: Host-Induced Gene Silencing in the obligate biotrophic fungal pathogen *Blumeria graminis*. Plant Cell. 2010; 22(9):3130–41. pmid:20884801
17. Price DRG, Gatehouse JA. RNAi-mediated crop protection against insects. *Trends Biotechnol.* 2008; 26(7):393–400. pmid:18501983
18. Koch A, Kogel KH. New wind in the sails: improving the agronomic value of crop plants through RNAi-mediated gene silencing. Plant Biotechnol J. 2014; 12(7):821–31. pmid:25040343
19. Koch A, Kumar N, Weber L, Keller H, Imani J, Kogel KH. Host-induced gene silencing of cytochrome P450 lanosterol C14α-demethylase-encoding genes confers strong resistance to Fusarium species. Proc Natl Acad Sci U S A. 2013; 110(48):19324–9. pmid:24218613
20. Jansen C, von Wettstein D, Schäfer W, Kogel KH, Felk A, Maier FJ. Infection patterns in barley and wheat spikes inoculated with wild-type and trichodiene synthase gene disrupted *Fusarium graminearum*. Proc Natl Acad Sci U S A. 2005; 102(46):16892–7. pmid:16263921
21. Melnyk CW, Molnar A, Baulcombe DC. Intercellular and systemic movement of RNA silencing signals. EMBO J. 2011; 30(17):3553–3563. PMCID: PMC3181474 pmid:21878996
22. Dunoyer P, Melnyk C, Molnar A, Slotkin RK. Plant Mobile Small RNAs. Cold Spring Harb Perspect Biol. 2013; 5(7):a017897. PMCID: PMC3685890 pmid:23818501
23. Pyott DE, Molnar A. Going mobile: Non-cell-autonomous small RNAs shape the genetic landscape of plants. Plant Biotechnol J.

2015; 13(3):306–18. pmid:25756494

24. Gantier MP, Williams BRG. The response of mammalian cells to double-stranded RNA. Cytokine Growth Factor Rev. 2007; 18(5–6):363–71. pmid:17698400
25. Karpala AJ, Dorant TJ, Bean AGD. Immune responses to dsRNA: Implications for gene silencing technologies. Immunol Cell Biol. 2005; 83(3):211–6. pmid:15877597
26. Jones JDG, Dangl JL. The plant immune system. Nature. 2006; 444(7117):323–9. pmid:17108957
27. Schäfer P, Pfiffi S, Voll LM, Zajic D, Chandler PM, Waller F, Scholz U, Pons-Kühnemann J, Sonnewald S, Sonnewald U, Kogel KH. Manipulation of plant innate immunity and gibberellin as factor of compatibility in the mutualistic association of barley roots with *Piriformospora indica*. Plant J. 2009; 59(3):461–74. pmid:19392709
28. Chen Y, Gao Q, Huang M, Liu Y, Liu Z, Liu X, Ma Z. Characterization of RNA silencing components in the plant pathogenic fungus *Fusarium graminearum*. Scientific Reports. 2015; 5, article number: 12500. pmid:26212591
29. El Gueddari NE, Rauchhaus U, Moerschbacher BM, Deising HB. Developmentally regulated conversion of surface-exposed chitin to chitosan in cell walls of plant pathogenic fungi. New Phytol. 2002; 156:103–112.
30. Oliveira-Garcia E, Deising HB. Infection structure-specific expression of ß-1,3-glucan synthase is essential for pathogenicity of *Colletotrichum graminicola* and evasion of ß-glucan–triggered immunity in maize. Plant Cell. 2013; 25(6):2356–78. pmid:23898035
31. Molnar A, Molnar A, Melnyk CW, Bassett A, Hardcastle TJ, Dunn R, Baulcombe DC. Small silencing RNAs in plants are mobile and direct epigenetic modification in recipient cells. Science. 2010; 328:872–875. pmid:20413459
32. Lewsey MG, Hardcastle TJ, Melnyk CW, Molnar A, Valli A, Urich MA, Nery JR, Baulcombe DC, Ecker JR. Mobile small RNAs regulate genome-wide DNA methylation. Proc Natl Acad Sci U S A. 2016; 113(6): E801–E810. pmid:26787884
33. Ding B. The Biology of Viroid-Host Interactions. Annual Review of Phytopathol. 2009; 47: 105–131. pmid:19400635

34. Buhtz A, Springer F, Chappell L, Baulcombe DC, Kehr J. Identification and characterization of small RNAs from the phloem of *Brassica napus*. Plant J. 2008; 53:739–749. pmid:18005229
35. Martienssen RA (2010) Heterochromatin, small RNA and post-fertilization dysgenesis in allopolyploid and interploid hybrids of Arabidopsis. New Phytol. 2010; 186(1):46–53. pmid:20409176
36. Abdellatef E, Will T, Koch A, Imani J, Vilcinskas A, Kogel KH. Silencing the expression of the salivary sheath protein causes transgenerational feeding suppression in the aphid *Sitobion avenae*. Plant Biotech. J. 2015; 13:849–857.
37. Eamens A, Wang MB, Smith NA, Waterhouse PM. RNA silencing in plants: yesterday, today, and tomorrow. Plant Physiol. 2008; 147(2):456–68. pmid:18524877
38. Calo S, Nicolás FE, Vila A, Torres-Martínez S, Ruiz-Vázquez RM. Two distinct RNA-dependent RNA polymerases are required for initiation and amplification of RNA silencing in the basal fungus *Mucor circinelloides*. Mol Microbiol. 2012; 83(2):379–94. pmid:22141923
39. Koch A, Khalifa W, Langen G, Vilcinskas A, Kogel KH, Imani J. The antimicrobial peptide thanatin reduces fungal infections in Arabidopsis. J. Phytopathol. 2012; 160:606–610.
40. Horwitz BA, Sharon A, Lu SW, Ritter V, Sandrock TM, Yoder OC, Turgeon BG. A G protein alpha subunit from *Cochliobolus heterostrophus* involved in mating and appressorium formation. Fungal Genet Biol. 1999; 26(1):19–32. pmid:10072317
41. Maier FJ, Malz S, Lösch AP, Lacour T, Schäfer W. Development of a highly efficient gene targeting system for *Fusarium graminearum* using the disruption of a polyketide synthase gene as a visible marker. FEMS Yeast Res. 2005; 5(6–7):653–62. pmid:15780665
42. Covert SF, Kappor P, Lee M, Briley A, Nairn CJ. *Agrobacterium tumefaciens* mediated transformation of *Fusarium circinatum*. Mycol. Res. 2001; 105:259–264
43. Imani J, Li L, Schäfer P, Kogel KH. STARTS–A stable root transformation system for rapid functional analyses of proteins of the monocot model plant barley. Plant J. 2011; 67(4):726–35. pmid:21518054

44. Chen DH, Ronald PC. A rapid DNA minipreparation method suitable for AFLP and other PCR applications. Plant Mol Biol Rep. 1999; 17:53–57.

45. Livak KJ, Schmittgen TD. Analysis of relative gene expression data using realtime quantitative PCR and the 2(-Δ Δ C(T)). Methods. 2001; 25(4):402–8. pmid:11846609

46. Bolger AM, Lohse M, Usadel B. Trimmomatic: A flexible trimmer for Illumina Sequence Data. Bioinformatics. 2014; 30(15):2114–20. pmid:24695404

47. Langmead B, Salzberg S. Fast gapped-read alignment with Bowtie 2. Nat Methods. 2012; 9(4):357–9. pmid:22388286

CHAPTER 10

RNAi: HISTORICAL DEVELOPMENTS, ADVANTAGES, LIMITATIONS AND FUTURE

Prerna Pandey

Dr Prerna Pandey, a biotechnologist with several years of wet lab research experience, worked at International Center for Genetic Engineering and Biotechnology, New Delhi. Her field of research in PhD included isolation and molecular characterization of geminiviruses, genome sequencing, gene annotation, and gene silencing using the RNA interference technology. She has also worked at Transasia Biomedicals and Advance Enzyme Technologies as a scientist. Prerna has published papers in peer-reviewed journals, and has submitted a number of annotated Geminiviral genome sequences in the GenBank, including two novel ones. She has also completed her editing and proof-reading courses from Society for the promotion of Editors and Proof-readers, UK and now works as a freelance scientific editor and writer. When Prerna is not busy with her assignments, she loves to explore historical places.

INTRODUCTION

Andrew Fire and Craig Mello won the Nobel Prize in Physiology or Medicine in 2006 (Fire, 2007) for demonstrating RNA interference in *C. elegans* for the first time in 1998 (Fire et al., 1998). Subsequently, David Baulcombe's group reported posttranscriptional gene silencing (PTGS) in plants (Hamilton & Baulcombe 1999). They showed that

- PTGS can be induced by viral infection as well as transgenic expression and it can target both viral and cellular mRNA,
- that PTGS involves the role of small interfering RNAs (siRNAs) that are between twenty-one to twenty-five nucleotides long.

R. Jorgensen and his colleagues (1996) attempted to deepen the purple color of Petunia flowers by through the use of chalcone synthase gene. However, they observed diminished expression, or nearly colorless flowers instead of getting stronger purple color flower and termed the phenomenon as co-suppression. The suppression of endogenous gene expression by transformation of exogenous homologous sequences was later termed as quelling in *Neurospora crassa* (Hammond et al., 2001; Jorgensen et al., 1996)

Subsequently it was also observed that the RNAi mechanism is highly-conserved, followed by the establishment of three major stages of the RNA interference reaction: Dicer mediated long dsRNA cleavage into shorter dsRNA segments; the single stranded RNA loading on the RNA-induced silencing complex (RISC), followed by the targeting and degradation of mRNA by this complex (Tuschl et al., 1999).

Two overlapping transcripts of approximately 22 and 61 nucleotides fragments of the *C. elegans*lin-4 gene that inhibited the expression of lin-14 through 3′ untranslated region (UTR) complementarity to the lin-14 mRNA, were reported for the first time by Lee et al., (1993). The molecule that was near sixty base pairs could adopt a conformation of a hairpin. Lee et al suggested that lin-4 inhibits translation of lin-14 through an antisense RNA-RNA interaction, and that lin- 4 and another gene let-7 acted sequentially in a specific pattern that regulates the timing of *C. elegans* development (Pasquinelli et al., 2000). The year 2001 saw the discovery of micro RNAs by Lee & Ambros in *C. elegans* and later in mice (2001).

Specific gene silencing has been achieved in the embryo extracts and cultured cells of *Drosophila* flies by utilizing the RNAi tool (Zamore et al., 2000). Zamore and colleagues used lysates of *Drosophila melanogaster* embryo to demonstrate the cleavage of long dsRNA strands into short interfering dsRNA fragments (siRNA) of ~22 nucleotides. Later Elbashir (2001) and colleagues demonstrated that chemically synthesized 21- or 22-nt-long dsRNA carrying 3′ overhangs could induce efficient RNA cleavage in embryo extracts from *Drosophila*.

In 2001 when Thomas Tuschl and colleagues demonstrated potent and specific RNAi silencing in mammalian cells, it generated a lot of interest.2 Since then, these principles used by researchers all around the world to selectively knock down the expression of genes of interest in cell lines and animal models.

The year 2005 saw the RNAi for addressing diseases lacking treatment receiving huge investments from big players in the industry. But early clinical trials could not meet the expectations: RNAi-based drugs either failed to deliver patient benefit, or induced strong innate immune reactions, or both.

The use of RNAi technology as therapeutics and for other applications has been on the rise in the past few years, and has met with encouraging results, both in humans and in plants.

ADVANTAGES OF RNAI

The merits of using such procedures can be summarized below:

- The target specificity the almost unrestricted choice of targets. The synthesis of such an antisense molecule requires only sequence information of the target mRNA.
- Researchers have been working on RNAi to target most of the human genes that have been annotated.
- RNAi is more potent compared to the other strategies such as ribozymes, and antisense DNA oligonucleotides (Bertrand et al., 2002). For a therapeutic application, a more potent RNAi indicates that the effector molecules can function even at much lower concentrations.

FUTURE PROSPECTS: HUMANS

Case Studies

RNAi naturally plays major roles providing defense to the host genome against invaders (Bender, 2014). Oh and Park, (2009) have reported that embedding the siRNAs in lipid nanoparticles could easily help in crossing the blood to reach the diseased cells. These days, many RNAi drugs targeting liver-linked diseases are exploiting these mechanisms, with numerous RNAi therapies in clinical development.

Inherited form of high cholesterol

The regulation of cholesterol in blood involves an enzyme: proprotein convertase subtilisin/kexin type 9 (PCSK9). A siRNA called ALN-PCS02 was tested in Phase 1 trial as it targeted the above gene. The results obtained were that close to 57% of patients with high levels of low-density lipoprotein cholesterol (LDL-C), who received the drug showed encouraging results (Bender, 2014).

Transthyretin-mediated (TTR) amyloidosis

The siRNA drug patisiran (ALN-TTR02), is an advanced LNP-formulated drug, delivered via infusion to treat an orphan liver disease, the transthyretin-mediated (TTR) amyloidosis. The Phase 2 trial of the drug demonstrated significantly encouraging results and the Phase 3 trial results are awaited in 2017 (Bender, 2014).

Drugs for diverse liver diseases

As liver cells possess receptors for N-acetylgalactosamine; this sugar can be linked covalently to RNAi molecules to facilitate the specific targeting of liver cells. This can circumvent the use of intravenous administration. An N-acetylgalactosamine containing molecule against PCSK9 has entered the foray that is expected to show promise (Bender, 2014).

Treating Viral Infections

Hepatitis B virus (HBV)

RNAi is also showing promise in the treatment of viral infections. By conservative estimates, hepatitis B virus (HBV) chronically infects hundreds of millions worldwide, causes liver scarring (cirrhosis), liver failure, liver cancer and eventually killing many. Although the current drugs can inhibit HBV viral replication, some amount of viral surface antigen is still produced, which can affect the immune system. With 'Through RNAi, if the viral antigen levels could be decreased, the immune system may be strengthened to help the liver remove the virus.' The ARC-520 RNAi drug combines two sequences of siRNA with dynamic polyconjugate (DPC) (See chapter 5 for details).

Hepatitis C virus (HCV)

Hepatitis C virus (HCV) causes infection in infects 150,000,000 people globally associated with morbidity such as failure or cancer of the liver. An RNAi drug, TT-034 delivers a plasmid carrying three shRNA-encoding genes to the host nucleus through an adeno-associated virus. The expression of these genes are aimed at suppressing the proteins of the viruses in host liver cells (Denise et al., 2014).

Since the patients just have to get infused and not worry about the routine intake of pills, the therapy becomes very handy, especially in areas of the world which experience the highest rates of chronic HCV infection.

Ebola Virus

As this pathogen targets the liver its rate of causing mortality is high. A siRNA containing drug: TKM-Ebola LNP is aimed the viral genome across several locations to interfere with its multiplication (Geisbert et al., 2010).

An Armor against Cancer

RNAi can offer treatment against cancer, especially against oncogenes that are currently difficult to regulate, for example, *MYC*. Through RNAi drugs, the mutant proteins expression can be inhibited that drive a tumor's growth, with great specificity. Although the treatments may seem promising, no drug has moved into Phase 3 trials.

RNAi was found in the drug CALAA-01when applied in human melanoma patients. The Phase I study revealed that the drug was found in tumors and inhibited cell division molecules such as ribonucleotide reductase RRM2 (Davis et al., 2010). But the drug was quietly dropped by the company producing and testing it.

Other RNAi inducing drugs to treat cancer, that are under clinical trials are

- DCR-MYC, based on LNP technology target several lymphomas, myelomas and cancers involving *MYC.*
- TKM-PLK1, inhibits *polo-like kinase 1* (*PLK1*) expression, and is used for treating hepatocellular carcinoma,is currently undergoing Phase 1/2 trial.

- siRNAs can be subjected to slow release when they are administered in a tumor if they are placed in a matrix: LODER (Local Drug EluteR), of small dimensions. Scientists have been trying to target mutant k-RAS with small-molecule drugs, but with little success because the drugs have interfered with other proteins (for details, please see chapter 5).

More Powerful RNAi Agents

Recent advances have been made in architecture of nanocomplexes in which the peptides are designed to envelop the siRNAs to penetrate tumors and pass through cell membranes. The initial experiments have shown that the nanocomplex validated the inhibitor of *DNA binding 4* (*ID4*) as a novel oncogene, and also helped improve the mouse models of ovarian cancer survive (Ren et al., 2012).

In another example, there are reports of generating intricately assembled LNPs. These assembled LNPs incorporate siRNAs, phospholipids, PEG-lipids, lipopeptides, cholesterol, and are nearly 100 times more potent than earlier designs (Xue, 2015).

CASE STUDIES AND FUTURE PROSPECTS: PLANTS

miRNAs that Silence Themselves: Target Mimicry

In an attempt to provide a new plant miRNA functional analysis tools, a “target mimicry” strategy used an alternative amiRNA-like technique, to inhibit the ability of an endogenous miRNA’s to cleave its specific target(s) (Pasquinelli, 2012).

The strategy employs a principle of an mRNA sequence that does not encode a protein but possesses a site complementary to a miRNA. Scientists used this approach to study the effects of miRNAs miR156 and miR319 expression knock-outs in Arabidopsis. Based on the observations, the day is not far off when target mimicry can be applied to study plant miRNA.

Promoter-Induced Silencing

The complete silencing of target genes is often difficult to achieve through silencing of coding sequences, and it could be achieved by the transcriptional inactivation of promoters. Recombinant viruses or long hpRNA constructs containing promoter sequences can trigger TGS accompanied by methylation of a target promoter in plants. The silencing of genes in plants can be achieved with efficiency using silencing of promoters using dsRNA.

The use of dsRNA was associated with activation of methylation of DNA; an ongoing process involving enzymes that facilitates silencing of genes without inducers (Zhao et al., 2014).

- hpRNA was utilized to target promoter of endogenous gene in petunia. This was a first case of silencing transcription by such an approach.
- The transgene promoters using hpRNA constructs are more amenable to silencing then the endogenous promoters.

Systemic and Transitive Silencing for Endogenous Gene Silencing

The use of gene silencing in plants especially across the entire plant or across considerable distances can offer hazards or benefits.

- Systemic silencing can be applied in plant species from which transgenic technology is challenging. For instance, the use of dsRNA can be applied on crops like grapevine (*Vitis vinifera*) and fruit trees.
- The spread of the silencing across cells has the potential to increase the complexity tissue- or organ-specific gene silencing.
- The application of a broad systemic silencing mechanism for efficient silencing of endogenous genes may be problematic (Dunoyer et al., 2005; Liu et al., 2002).
- The silencing phenomenon has the potential to nearby areas that were not intended to be silenced; a term called 'Transitivity' (Garcia-Perez et al., 2004).

The silencing spread from the 5′ UTR targeted by the IR hpRNA to downstream coding sequences of the transgene and could then lead to cross-silencing due to its sequence homology to the coding region of the specific gene.

Endogenous genes in *Arabidopsis* and tomato have been silenced by the use of sense constructs with an inverted repeat at the 3′ UTR (Brummell et al., 2003). It can be explained by the fact that the no sequence homology is shared between the 3′ IR with the target gene sequences, so the silencing must be induced by siRNAs corresponding to transgene upstream sense sequences.

Plant Defense against Non-viral Pathogens

The development of resistance in plants against viruses using hpRNA-based strategies to target RNA of viruses has shown more success than transgenic technologies. The RNA silencing technique directly targets viruses.

Apart from viruses, the use of RNA silencing has shown promise against other pathogens or insects. Thus, the use of such technologies shows promise as another approach to target such biotic stress.

When *Pseudomonas syringae* pv *tomato* was inoculated with *Arabidopsis* nat-siRNAATGB2: natsiRNA is produced that causes down regulation of a regulator of RS2 pathway: a PPRL gene. Thus, RPS2-mediated pathway is expressed or resistance against *Pseudomonas syringae* pv *tomato* is strongly induced (Katiyar-Agarwal et al., 2006).

- The infection by *P. syringae* induced the expression of AsRNA between thirty and forty nucleotides. Such long-siRNAs, for example AtlsiRNA-1 causes silences negative regulator of plant defense causing development of resistance was associated with infection. It was reported that AtlsiRNA-1, one of the lsiRNAs, contributes to plant bacterial resistance by silencing a AtRAP; negative regulator of plant defense (Katiyar-Agarwal et al., 2007).
- The miRNA pathway may also be involved in fungal pathogen - plant interactions. For example, *Pinus* spp produces specific miRNAs in response to fusiform rust infection (Lu et al., 2007).
- Insect resistance: RDR proteins play a key role in the biogenesis of siRNAs (See chapter 1). It was reported that when the *RDR* gene was silenced in *Nicotiana attenuate* using a transgeene or

viral vector, the resultant plants were at a higher risk of attack by insects (Pandey and Baldwin, 2007). This is suggestive of the function of siRNAs in defense of plants against insects.

The plant transformation technologies have used RNAi to study gene expression by introducing dsRNA, to bring new traits such as disease resistance, stress resistance and nutritional quality traits in various crops. The developed gene delivery techniques like transient assay systems havesped-up the study of gene loss functions in targeted plant tissues. RNAi transient delivery systems include (please see chapter 2 for details):

- virus-induced gene silencing (VIGS);
- direct gene bombardment and
- Agrobacterium -mediated transient assays.

LIMITATIONS OF RNAS

Competition with Endogenous RNAs

Like the bioactive drugs have the disadvantage of saturating cellular processing pathways and thus disturbing the equilibrium. As the introduction of siRNA/shRNA for use as RNAi require the micro RNA apparatus of cell; a similar challenge is faced by this technology.

- The use of elevated concentrations of exogenous RNA can cause an obstruction to the miRNA pathway as shRNA and siRNA are similar to miRNA precursors. Grimm et al (2006) observed morbidity in the experimental mice, associated with reduced expression of liver-specific miRNA, and that the liver toxicity might stem form saturation of the miRNA pathway.
- This warrants the use of promoters with an intermediate or moderate activity when vectors are to be employed.

Stimulation of Innate Immune Responses

- Certain siRNA sequence motifs stimulateTLR7-dependent immune reaction (Homung et al., 2005). A sequence 5'-GUCCUUCAA-3' of plasmacytoid dendritic cells is supposed to be recognized by the TLR-7 to activate immune responses.

Since then a number of 'danger motif' have been characterized. These motifs have an abundance of GU bases that elicit response from innate immune system to cause production of cytokines. These results raise safety concerns for the therapeutic use of RNAi.

- The siRNA-induced immune response seems to rely on TLR receptors that are located in endosome (Marques and Williams, 2005),
- With the risk of increased immune stimulation, the use of longer (27-29 bp) and more potent Dicer substrates siRNAs (Kim et al., 2005), should be analyzed carefully.

Use of siRNA Therapeutics: Limiting Factors

Several factors limit the utility of siRNA.

- siRNA may compete with endogenous RNA, and may saturate the miRNA processing pathways, which can lead to toxicity.
- There are reports of production of inflammatory cytokines produced by GU regions.
- The off-targets may be suppressed by the siRNA due to partial nucleotide sequence match between the siRNA and off-target Mrna and/or global degradation and inhibition of mRNA translation by siRNA and its transfection agent.
- The longer dsRNA (>30 nucleotides) can rapidly induce interferons by activating the defense mechanisms that combat the invading viral pathogens, while the dsRNA molecules shorter than 30 nucleotides do not show this problem (Sioud and Sørensen,2003).

The instability of siRNA with plasma half-life of nearly 10 min and the degradation following its transportation to the lysosomes are the critical factors limiting the utility of siRNA as therapeutics is (Patzel et al., 2005; Bagga et al., 2005).

Suppression of Off-targets

The use of microarray technology for genome-wide monitoring revealed that there were instances of off target silencing in siRNA treated cells (Jackson et al., 2003). Off targets silencing is unwanted because it results in largely unpredictable altered gene activity.

- The genes that were off targets were found to have a match of six to seven nucleotides with the siRNA seed sequence (Jackson et al., 2006; Birmingham et al., 2006; Lin et al., 2005).
- The recognition for microRNA relies on the position between bases two and eight.
- The transfection agent may also independently influence expression profiles of the siRNA (Fedorov, 2005).
- Off targets may be influenced by the expression level of mRNA and microRNA.
- Jackson et al., (2006b) have shown that 2' O-Me modified siRNAs at the second base significantly reduces off-targeting with little or no compromise in the degree of specific target silencing.

stringent tests for siRNAs are required with tolerable levels of off targeting, because off targeting remains a critical issue for RNAi therapeutic applications. Since the siRNAs work at subnanomolar concentration, off target effects can be reduced by

- the identification of a more potent siRNA.
- The design and synthesis of siRNA requires careful analysis to minimize such off targets.

A faster process can be facilitated by the design and synthesis of sense and antisense targets. In the scenario involving fixed target sites such as SNP alleles; the loading of strand cannot be altered much. This requires further steps to alter siRNA to distinguish wild type from the altered.

REFERENCES

1. 1. Bagga S, Bracht J, Hunter S, Massirer K, Holtz J, Eachus R, Pasquinelli AE. (2005) Regulation by let-7 and lin-4 miRNAs results in target mRNA degradation. Cell. 122:553-63.
2. 2. Bender, E. The Second Coming of RNAi. The Scientist. 2014 September.
3. 3. Bertrand JR, Pottier M, Vekris A, Opolon P, Maksimenko A, Malvy C. (2002 Aug) Comparison of antisense oligonucleotides and siRNAs in cell culture and in vivo. Biochem Biophys Res Commun.

296(4):1000-4.

4. 4. Birmingham A, Anderson EM, Reynolds A, Ilsley-Tyree D, Leake D, Fedorov Y, et al. (2006) 3′ UTR seed matches, but not overall identity, are associated with RNAi off-targets. Nat Methods. 3:199– 204.
5. 5. Brummell DA, Balint-Kurti PJ, Harpster MH, Palys JM, Oeller PW, Gutterson N. (2003) Inverted repeat of a heterologous 3′-untranslated region for high-efficiency, high-throughput gene silencing. Plant J. 33:793–800
6. 6. Davis ME, et al. (2010) Evidence of RNAi in humans from systemically administered siRNA via targeted nanoparticle. Nature. 464:1067-70.
7. 7. Denise H, et al. (2014) Deep sequencing insights in therapeutic shRNA processing and siRNA target cleavage precision. Mol Ther Nucleic Acids. 3:e145.
8. 8. Dunoyer P, Himber C, Voinnet O. (2005) DICER-LIKE4 is required for RNA interference and produces the 21-nucleotide small interfering RNA component of the plant cell-to-cell silencing signal. Nat Genet. 37:1356–60
9. 9. Elbashir SM, Lendeckel W, Tuschl T. (2011) RNA interference is mediated by 21- and 22-nucleotide RNAs. Genes Dev. 15:188–200.
10. 10. Fedorov Y, King A, Anderson E, Karpilow J, Ilsley D, Marshall W, Khvorova A. (2005) Different delivery methods-different expression profiles. Nat Methods. 2:241.
11. 11. Fire A, Xu S, Montgomery MK, Kostas SA, Driver SE, Mello CC. (1998) Potent and specific genetic interference by double stranded RNA in Caenorhabditis elegans. Nature. 391(6669): 806-11.
12. 12. Fire AZ. (2007) Gene silencing by double-stranded RNA (Nobel Lecture). Angew ChemInt Ed Engl. 46(37): 6966-84.
13. 13. Geisbert, T.W., et al. (2010) Postexposure protection of nonhuman primates against a lethal Ebola virus challenge with RNA interference: a proof-of-concept study. The Lancet. 375:1896-905.
14. 14. Grimm D, Streetz KL, Jopling CL, Storm TA, Pandey K, Davis CR, et al. (2006) Fatality in mice due to oversaturation of cellular

microRNA/short hairpin RNA pathways. Nature. 441(7092):537-41.

15. 15. Hamilton AJ, Baulcombe DC. (1999) A species of small antisense RNA in posttranscriptional gene silencing in plants. Science. 286(5441): 950-2.
16. 16. Hammond SM, Caudy AA, Hannon GJ. (2001) Post-transcriptional gene silencing by double-stranded RNA. Nat Rev Genet. 2:110– 19. DOI: 10.1038/35052556
17. 17. Hornung V, Guenthner-Biller M, Bourquin C, Ablasser A, Schlee M, Uematsu S, et al. (2005 Mar) Sequence-specific potent induction of IFN-alpha by short interfering RNA in plasmacytoid dendritic cells through TLR7. Nat Med. 11(3):263-70.
18. 18. Jackson AL, Bartz SR, Schelter J, Kobayashi SV, Burchard J, Mao M, et al. (2003) Expression profiling reveals off-target gene regulation by RNAi. Nat Biotechnol. 21:635-37.
19. 19. Jackson AL, Burchard J, Schelter J, Chau BN, Cleary M, Lim L, Linsley PS. (2006a) Widespread siRNA "off-target" transcript silencing mediated by seed region sequence complementarity. RNA. 12:1179-87.
20. 20. Jackson AL, Burchard J, Leake D, Reynolds A, Schelter J, Guo J, et al. (2006b) Position-specific chemical modification of siRNAs reduces "off-target" transcript silencing. RNA,12:1197–1205.
21. 21. Jorgensen RA, Cluster PD, English J, Que Q, Napoli CA. (1996) Chalcone synthase cosuppression phenotypes in petunia flowers: Comparison of sense vs. antisense constructs and single-copy vs. complex T-DNA sequences. Plant Mol Biol. 31:957–73.
22. 22. Katiyar-Agarwal S, Gao S, Vivian-Smith A, Jin H. (2007) A novel class of bacteria-induced small RNAs in Arabidopsis. Genes Dev. 21:3123–34
23. 23. Katiyar-Agarwal S, Morgan R, Dahlbeck D, Borsani O, Villegas A Jr, Zhu J-K, Staskawicz BJ, Jin H. (2006) A pathogen-inducible endogenous siRNA in plant immunity. Proc Natl Acad Sci USA. 103:18002–07
24. 24. Kim DH, Longo M, Han Y, Lundberg P, Cantin E, Rossi JJ. (2004) Interferon induction by siRNAs and ssRNAs synthesized by

phage polymerase. Nat Biotechnol. 22:321–25.

25. 25. Lee, R.C., Ambros, V. (2001) An extensive class of small RNAs in Caenorhabditis elegans. Science. 294(5543): 862-4.
26. 26. Lee RC, Feinbaum RL, Ambros V. (1993) The C. elegans heterochronic gene lin-4 encodes small RNAs with antisense complementarity to lin-14. Cell. 75(5):843-54.
27. 27. Lin X, Ruan X, Anderson MG, McDowell JA, Kroeger PE, Fesik SW, Shen Y. (2005) siRNA-mediated off-target gene silencing triggered by a 7 nt complementation. Nucleic Acids Res. 33:4527-35.
28. 28. Liu Q, Singh SP, Green AG. (2002) High-stearic and higholeic cottonseed oils produced by hairpin RNA-mediated posttranscriptional gene silencing. Plant Physiol. 129 1732-43
29. 29. Lu SF, Sun YH, Amerson H, Chiang VL. (2007) MicroRNAs in loblolly pine (Pinus taeda L.) and their association with fusiform rust gall development. Plant J. 51 1077-98.
30. 30. Marques JT, Williams BR. (2005) Activation of the mammalian immune system by siRNAs. Nat Biotechnol. 23:1399–1405.
31. 31. Oh YK, Park TG. (2009) siRNA delivery systems for cancer treatment. Adv Drug Deliv Rev. 61(10):850–62. doi: 10.1016/j.addr.2009.04.018.
32. 32. Pandey SP, Baldwin IT. (2007) RNA-directed RNA polymerase 1 (RdR1) mediates the resistance of Nicotiana attenuata to herbivore attack in nature. Plant J. 50:40–53.
33. 33. Pasquinelli AE. (2012 Apr) MicroRNAs and their targets: recognition, regulation and an emerging reciprocal relationship. Nature Reviews Genetics. 13:271-82.
34. 34. Pasquinelli AE, Reinhart BJ, Slack F, Martindale MQ, Kuroda MI, Maller B, et al. (2000) Conservation of the sequence and temporal expression of let-7 heterochronic regulatory RNA. Nature. 408(6808): 86-9.
35. 35. Patzel V, Rutz S, Dietrich I, Koberle C, Scheffold A, Kaufmann SH. (2005) Design of siRNAs producing unstructured guideRNAs results in improved RNA interference efficiency. Nat Biotechnol.

23:1440–1444.

36. 36. Ren Y, et al. (2012) Targeted tumor-penetrating siRNA nanocomplexes for credentialing the ovarian cancer oncogene ID4. Sci Transl Med. 4:147ra112.
37. 37. Sioud M, Sørensen DR. (2003 Dec) Cationic liposome-mediated delivery of siRNAs in adult mice. Biochem Biophys Res Commun. 312(4):1220-5.
38. 38. Tuschl T, Zamore PD, Lehmann R, Bartel DP, Sharp PA. (1999) Targeted mRNA degradation by double-stranded RNA in vitro. Genes Dev. 13(24):3191-7.
39. 39. Xue HY, Guo P, Wen W-C, Wong HL. (2015) Lipid-Based Nanocarriers for RNA Delivery. Current Pharmaceutical Design. 21(22):3140-3147.
40. 40. Zamore PD, Tuschl T, Sharp PA, Bartel DP. (2000) RNAi: Doublestranded RNA directs the ATP-dependent cleavage of mRNA at 21 to 23 nucleotide intervals. Cell. 101:25–33.
41. 41. Zhao M, San León D, Delgadillo MO, García JA, Simón-Mateo C. (2014 Aug) Virus-induced gene silencing in transgenic plants: transgene silencing and reactivation associate with two patterns of transgene body methylation. Plant J. 79(3):440-52.

CHAPTER 11

RNA INTERFERENCE IN THE AGE OF CRISPR: WILL CRISPR INTERFERE WITH RNAI?

Unnikrishnan Unniyampurath[1], Rajendra Pilankatta[2] and Manoj N. Krishnan[1]

[1]Program on Emerging Infectious Diseases, Duke-NUS Medical School, 8 College Road, Singapore 169857, Singapore

[2]Department of Biochemistry and Molecular Biology, School of Biological Sciences, Central University of Kerala, Nileshwar 671328, India

ABSTRACT

The recent emergence of multiple technologies for modifying gene structure has revolutionized mammalian biomedical research and enhanced the promises of gene therapy. Over the past decade, RNA interference (RNAi) based technologies widely dominated various research applications involving experimental modulation of gene expression at the post-transcriptional level. Recently, a new gene editing technology, Clustered Regularly Interspaced Short Palindromic Repeats (CRISPR) and the CRISPR-associated protein 9 (Cas9) (CRISPR/Cas9) system, has received unprecedented acceptance in the scientific community for a variety of genetic applications. Unlike RNAi, the CRISPR/Cas9 system is bestowed with the ability to introduce heritable precision insertions and deletions in the eukaryotic genome. The combination of popularity and superior capabilities of CRISPR/Cas9 system raises the possibility that this technology may occupy the roles currently served by RNAi and may even make RNAi obsolete. We performed a comparative analysis of the technical aspects and applications of the CRISPR/Cas9 system and RNAi in mammalian systems, with the purpose of charting out a predictive picture on whether the CRISPR/Cas9 system will eclipse the existence and future of RNAi. The conclusion drawn from this analysis is that RNAi will still occupy specific domains of biomedical research and clinical applications, under the current state of development of these technologies. However, further improvements in CRISPR/Cas9 based technology may ultimately enable it to dominate RNAi in the long term.

Keywords: RNAi interference, RNAi, CRISPR, Cas9, genome engineering, reverse genetic screens

INTRODUCTION

The RNA interference (RNAi), a powerful transient gene-expression repression approach discovered over a decade ago, has brought unprecedented applications in the research on gene-phenotype relationship in mammalian systems [1,2]. Parallel to the development of RNAi, several other techniques for stable genetic modifications have also evolved, collectively called gene editing techniques [3]. One such gene editing approach is based on the Clustered Regularly Interspaced Short Palindromic Repeats (CRISPR)/CRISPR-associated protein 9 (Cas9) systems present in prokaryotes for nucleic acid editing. The application of CRISPR/Cas9 for general DNA editing as well as for mammalian gene editing was established in the 2012–2013 period, and, in just three years, this technique has revolutionized the entire gene editing field [4,5,6].

While RNAi dominated the mammalian gene expression manipulation field for the past 15 years, the recent rapid growth of CRISPR/Cas9 raises the question whether RNAi will become a tool of the past or not. This review provides a concise account on the origins and mechanisms of RNAi and CRISPR/Cas9, along with a detailed analysis on whether either one of these two techniques will outweigh the other in terms of their applications. The description below pertains only to the applications of CRISPR/Cas9 and RNAi in the mammalian systems. A schematic showing currently available options of gene editing using CRISPR/Cas9 systems and RNAi are given in Figure 1.

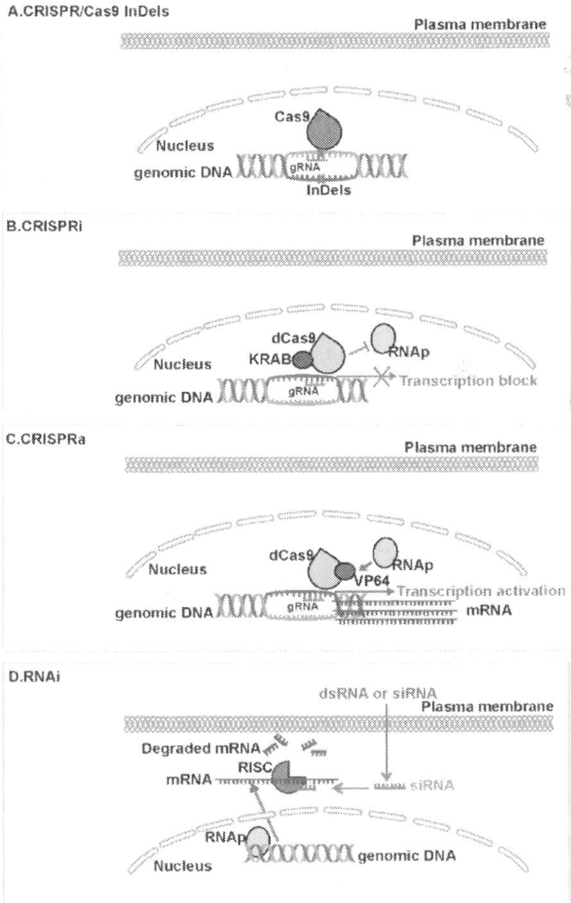

Figure 1: Schematic representation of CRISPR/Cas9 systems and RNAi. The general functioning of CRISPR/Cas9 systems and RNAi are displayed. (A) CRISPR/Cas9 induces insertions and deletions (InDel) when

targeted to a specific genomic site with the help of appropriate guide RNA (gRNA); (B) CRISPR interference (CRISPRi) down-regulates gene transcription; (C) CRISPR activation (CRISPRa) up-regulates gene transcription; (D) RNA interference (RNAi) down regulates gene transcription by acting post-transcriptionally. dCas, catalytic mutant Cas9; siRNA, small interfering RNA; dsRNA, double stranded RNA; RNAp, RNA polymerase; RISC, RNA induced silencing complex.

CRISPR: PROKARYOTIC ADAPTIVE IMMUNE RESPONSE SYSTEM TO POWERFUL EUKARYOTIC GENE EDITING TOOL

The CRISPR/Cas systems is a naturally occurring prokaryotic immune defense strategy that has been recently adapted for eukaryotic genome engineering. The potential of the CRISPR/Cas system to edit eukaryotic genomes offers a wide variety of therapeutic, biomedical, agricultural and research applications.

CRISPR: A Prokaryotic Adaptive Immune Response System

The CRISPR/Cas system confers adaptive immune protection to prokaryotes such as bacteria and archaea against non-self DNA based invasions (e.g., viruses, plasmids) in a sequence-specific manner [7,8]. The CRISPR-Cas loci are present in both bacteria and archaea (48% and 84% respectively) [9]. Although the initial observation on the presence of clustered repeats of genetic elements from the genome of the bacterium *Escherichia coli* was made in 1987, the term CRISPR was coined more than a decade later [10,11]. Subsequently, the CRISPR system was identified to target invading DNA elements [12,13,14,15].

CRISPR: Biochemical Architecture

The CRISPR/Cas systems that are present in diverse organisms essentially have a comparable core genetic organization [16,17,18]. They have several DNA repeat elements interspersed with short "spacer" sequences derived from foreign DNA, and multiple *Cas* genes, some of which are nucleases [19]. The spacer sequence constitutes a code for the respective foreign genetic element, which will be used by the host prokaryote to quickly identify any homologous sequence subsequently entering the host cell.

The CRISPR/Cas system functions in three steps [20,21]. It generates immunity by first genetically tethering invading phage and plasmid DNA segments (spacers) into CRISPR loci (called the spacer acquisition step) [22,23,24]. Subsequently, the host prokaryotic organism will transcribe and process CRISPR loci to generate mature CRISPR RNA (crRNA) containing both CRISPR repeat elements and the integrated spacer genetic segment of the foreign DNA corresponding to the previous non-self genetic invasion (called crRNA maturation step) [12,25,26,27]. Finally, the crRNA will detect homologous DNA sequences by complementary base pairing when newly introduced, such as during infection with another virus having the same sequence. Once crRNA detects foreign DNA, the invading non-self genetic material will be clipped through a complex nucleic acid processing biochemical machinery, yielding protection against foreign DNA (called the target interference step) [7,22,28,29].

The Cas proteins are involved in all three major steps of CRISPR/Cas system functioning. Among the ten superfamilies of Cas proteins (Cas1-Cas10) [30], Cas1 and 2 are primarily involved in spacer acquisition [31,32], Cas6 and 5d mediate crRNA maturation [16,25,33], and several other remaining Cas proteins are part of the interference complex.

The CRISPR/Cas systems have been classified into three major types based on the conservation and composition of the Cas genes [30,34]. The type II is present exclusively in bacteria, while types I and III systems are present in both bacteria and archaea. All three types of CRISPR/Cas systems are capable of editing DNA. The two major differences between the three types of CRISPR/Cas systems are: (i) target: Type II and III systems have DNA and RNA targeting subclasses. The type I system targets only DNA; and (ii) subunit composition. The type II system has two different RNA subunits in complex with a single Cas9 protein. Type I and III systems have multiple Cas proteins in complex with a single RNA.

CRISPR/Cas9, CRISPRi and CRISPRa: Tools for Eukaryotic Gene Editing

Recently, CRISPR/Cas system has been adapted for RNA-directed DNA editing of eukaryotic genomes including mammalian systems. The original identification that the CRISPR/Cas system can edit DNA subsequently catalyzed the thinking that Cas proteins with nuclease function could be potentially adapted for genome engineering of higher eukaryotes [5]. Validating this line of thinking, it was eventually established in 2013 that the CRISPR/Cas type II system, specifically the CRISPR/Cas9 system, could be engineered to edit mammalian genomes [4,6].

Among the three types of CRISPR/Cas systems, only type II has been widely adapted for eukaryotic gene editing [4,6,20]. A major reason for this is the simplicity of the type II system, in which only the Cas9 was needed for both recognizing target DNA regions and creating the gene modification using an appropriate crRNA. The type II CRISPR/Cas system required only the tracrRNA cofactor, the crRNA and Cas9 to induce gene editing. All other CRISPR/Cas types needed multiple components, with still poorly understood regulatory mechanisms [8,20,34].

The basal CRISPR/Cas9 machinery was originally adapted for introducing stable insertions and deletions (InDels) in the mammalian genome. However, there could be several applications that will only require transient repression or activation of gene expression. Subsequently, Cas9 was engineered to also perform transient repression and activation of mammalian genes. This was based on the principle that gRNA can target Cas9 to any destination segment of the genome. The ability of gRNA to target Cas9 to any destination segment of the genome could be exploited to alter gene expression levels if repressor or activator functionalities are conjoined to this machinery. On this line, it was later determined that fusion of transcription repression (e.g., Krüppel associated box KRAB) or activation (e.g., VP64) domains to catalytically inactive Cas9 molecule (called dCas9) could result in respectively transcription repression (called CRISPR interference, CRISPRi) or activation (called CRISPR activation, CRISPRa) of a gene targeted using appropriate gRNA [35,36,37]. Essentially, the gRNA could guide the dCas9 to the promoter regions of the intended target gene, and the repression or activation domains could modify the transcription. In many ways, CRISPRi works analogous to RNAi, although the underlying mechanisms are dissimilar. Both CRISPRi and RNAi are transient gene expression repression approaches. While CRISPRi transiently represses gene expression at the level of DNA, RNAi employs post-transcriptional mechanisms. The duration of transcript repression exerted by both of these approaches directly corresponds to the time the targeting components (dCas9-fusion domain + gRNA for CRISPRi, and siRNA or shRNA for RNAi) are present in the cells at the adequate doses.

The mammalian cells can respond to CRISPR/Cas9 mediated DNA breaks by adopting any one of the two repair mechanisms. The host cell will fix the DNA damage caused by CRISPR/Cas9 through either non-homologous end joining method (NHEJ) or homology directed repair (HDR) [38,39]. In NHEJ, the host cellular DNA repair systems will tether the DNA double strand break by random insertion or deletion of short stretches of oligonucleotide bases. This could potentially lead to the disruption of the codon-reading frame, resulting in erroneous transcripts and ablation of gene expression. In HDR, introduction of a segment of DNA with regions having homology to the sequences flanking both sides of the DNA double strand break will lead to the repair by host machinery through the incorporation of the extra segment of the DNA fragment. Both these types of host DNA repair response mechanisms to CRISPR/Cas9 mediated DNA double strand break offers various types of applications for CRISPR/Cas9 in mammalian gene editing. NHEJ can lead to ablation of gene mutations, and hence can be used to create loss of function effects. HDR can be used for introducing specific point mutations, or introducing DNA segments of varying length.

RNAI: EVOLUTIONARILY CONSERVED EUKARYOTIC POST-TRANSCRIPTIONAL GENE EXPRESSION REGULATION MECHANISM

RNAi: Eukaryotic Post-Transcriptional Gene Expression Regulation Mechanism

As a mature field, RNAi has been well reviewed elsewhere [40,41,42]. Here, we will provide a very brief overview of the RNAi principles. RNA interference is a double stranded RNA mediated homology based mechanism evolved to post-transcriptionally regulate eukaryotic gene expression and to serve in host immunity. The discovery of the phenomenon of RNAi happened over a long stretch of time. Phenotypic effects of RNAi, and RNA induced homologous endogenous mRNA suppression were initially observed in plants while experimenting for flower color alterations [43]. Plant scientists refereed to this phenomenon as cosuppression. The RNAi, as it stands today, was first observed in *Caenorhabditis elegans*, through the discovery that gene expression could be modified using double stranded RNA [1]. The core component of RNAi is the formation of small interfering RNAs (siRNAs) with regions homologous to the intended cellular RNA targets [2]. The siRNAs are generated endogenously by processing long double stranded RNAs in the cytoplasm, leading to the formation of 21-nt double stranded RNA with a 3' overhang of two nucleotides, and a 5' phosphate group. This ~21-mer will finally pair with the intended homologous sequence containing target mRNAs, and induce their degradation through a biochemical complex called RNA induced silencing complex (RISC). RNAi could also be induced by direct exogenous supply of short double stranded RNAs that are ~21 nucleotides in length, called small interfering RNAs (siRNA). Alternatively, RNAi can also be introduced into cells using plasmid delivery of short hairpin RNAs (shRNAs), which will be processed similarly to generate siRNAs inside the cytoplasm [44]. In mammalian systems, RNAi is best achieved by the delivery of small double stranded RNAs, because long double stranded RNAs have been shown to induce potent toxic interferon response. As most of the applications of RNAi are achieved through exogenous supply of chemically synthesized siRNAs, the rest of this review will discuss on the mechanism involved in siRNA based induction of RNAi to repress eukaryotic genes [45].

RNAi: Biochemical Mechinery

The core operation of the RNAi involves two steps. In Step 1, the double stranded RNA will be processed in the cytoplasm by an RNAse III type endoribonuclease called Dicer to generate ~21 nucleotide long siRNAs [46]. Subsequently, siRNA bound Dicer will transfer small dsRNA to Argonaute, with the help of double stranded RNA binding proteins (dsRNABP) [47]. Subsequently, in Step 2, Argonaute will bind one strand of the duplex (guide strand), while the other strand will be displaced

[48,49,50]. This whole protein-RNA complex is called RISC. The fully charged RISC will survey for long RNAs that can pair with the single stranded guide RNA bound to the Argonaute. Upon engagement with a long homologous RNA target that pairs with the guide RNA, Argonaute will be activated and degrade the target RNA though its RNAse-H like activity.

COMPARISON OF THE APPLICATIONS OF CRISPR/CAS9 AND RNAI IN THE MAMMALIAN BIOMEDICAL FIELD: WILL CRISPR DELETE RNAI OR ONLY MODERATELY REPRESS?

Recently the CRISPR/Cas9 system has received wide acceptance as a robust and versatile tool for gene editing, primarily because of both its simplicity and wider capabilities. Until now, RNAi has been the major dominating genetic tool for several applications in mammalian biomedical field requiring gene expression modification. The rapid growth and use of CRISPR/Cas9 in mammalian systems warrants an analysis on how it will impact on the future of the uses and applications of RNAi in biomedical research. Will CRISPR/Cas9 replace RNAi, or, will these two applications have unique space in the biomedical research? On the mechanistic side, these two technologies are fundamentally different. While the primary use of the CRISPR/Cas9 system is to induce gene editing, RNAi is a post-transcriptional gene expression modifier process. This fundamental difference seems to indicate that these two technologies may occupy unique non-overlapping spaces in biomedical research. However, a closer analysis of the modifications and applications of these two technologies reveals that CRISPR/Cas9 and RNAi also occupy several overlapping domains of biomedical research, raising the possibility that one may selectively dominate over the other.

In general, the applications of both CRISPR/cas9 and RNAi can be classified into two major domains: (i) research tool applications; and (ii) clinical applications. We have provided below a comparative analysis of the scope of CRISPR/Cas9 and RNAi in the research tool field and clinical applications. In addition, we have also discussed some of the "technical considerations" associated with each of these technologies that may also impact on their applications. This analysis is presented with an emphasis on gauging how each technology may affect the scope of the other.

Comparative Analysis of the Applications of CRISPR/Cas9 and RNAi as Research Tools

Both RNAi and CRISPR/Cas9 are widely used as gene expression modifying research tools for various scientific discovery purposes. Some of the major applications are discussed below, by highlighting the advantages and disadvantages of each one of these two technologies.

Mammalian in vitro reverse genetic screens: Reverse genetic screens were used heavily over the past decade to understand several biological processes. RNAi has revolutionized reverse genetic screens to an unprecedented level, emerging as the most valuable tool for conducting gene-function studies in diverse aspects such as cell signaling, host-pathogen interactions, immune response, and cancer mechanisms, among several others. Recently, CRISPR/Cas9 based human genome wide genetic screens have also been reported, indicating that CRISPR/Cas9 could also be used for this purpose [51,52,53,54]. However, there are several advantages for RNAi over CRISPR/Cas9 as a tool of choice for reverse genetic screens.

Cell based genome wide mammalian genetic screens involving gene expression modification tools can be performed theoretically in two different ways: either by transient introduction of gene expression modification tools (e.g., siRNA, shRNA, CRISPR/Cas9 systems) into cells followed by rapid completion of various phenotypic assays, or by using cell clones already having each one of the genes in the genome individually suppressed stably. In reality, all genetic screens conducted till now in cell culture models were mostly performed by transient suppression of gene expression. Although shRNA offers the ability to create cells with stable suppression of individual genes, genome wide collection of individual gene-silenced cells is currently unavailable for any mammal. Similarly, currently no genome wide collection of CRISPR/Cas9 based gene-deleted cells is available for any mammal. Transient use of gene expression modification tools for genetic screens requires that majority of the cells in the treated population of an experimental set up should show suppression of the targeted gene. Tools of both CRISPR (gRNA/Cas9 carrying plasmids or gRNA as synthetic oligonucleotide along with Cas9 separately through plasmids) and RNAi (siRNA, or shRNA as plasmid or virus) can be introduced into nearly all mammalian cells *in vitro*, by optimizing the delivery of the respective reagents. Therefore both technologies have comparable delivery efficiency in *in vitro* cell culture systems. However, when it comes to their efficiency to alter gene expression, both these technologies may perform differently as they stand today. RNAi has a relatively very high efficiency in suppressing gene expression. Optimally designed RNAi tools offers >70% knockdown of the intended target mRNAs upon delivery into the target cells. The ability of CRISPR/Cas9 to edit mammalian genes varies widely, with some of the studies reporting varying efficiency in the range of 1% to 79% [55,56,57]. While NHEJ mediated repair has a high frequency of occurrence, HDR based repair happens with low efficiency. Recently, efforts to enhance the rate of homology directed repair response have been reported [58,59]. Considering various confounding factors such as differences in delivery efficiency, and the probability of all alleles of a gene in a given cell being deleted simultaneously, the total efficiency of CRISPR/Cas9 mediated gene editing becomes a constrained phenomenon. Such a varying editing efficiency mandates that CRISPR/Cas9 tool treated cells should be selected for gene modification before it could be used for genetic screens. However, the relatively high efficiency of RNAi to attenuate transcript levels in each one of the cells that received the appropriate RNAi tool (siRNA or shRNA) makes its effect rather uniform across all cells in an experimental set up. In this regard, at present, RNAi is has some advantage over CRISPR for conducting genetic screens. Another advantage of RNAi is that it can be adopted for genetic screening in both immortalized cell lines, and short lived primary cells that often do not live longer than a week *ex vivo*. The use of varying efficiency CRISPR/Cas9 systems for conducting genetic screens in short lived primary cells will be less feasible because there may not be enough time for selecting gene edited population and performing phenotypic assays. Yet another major limitation of CRISPR/Cas9 system is its inability to simultaneously target all alleles of a given gene in the targeted cells. Because CRISPR/Cas9 targets the DNA element, in order for it to completely modify the expression of a gene, all alleles of the genes may have to be modified. Most of the widely used immortalized human cell lines are aneuploid with triploid or greater chromosome complements, making gene deletion process complex. Contrary to this, because RNAi targets transcripts, it can cause efficient and uniform gene repression. However, generation of validated collection of CRISPRi tools targeting the entire genome of mammals of interest may change this technological imbalance between RNAi and CRISPR for genetic screen applications [35,36,37,60]. Improved algorithms for CRISPRi design have been developed recently, further enhancing the potential of this application [35]. It remains to be established how successful and universally applicable will be the CRISPRi based approach for knocking down the expression of large sets of genes. Whether all genes can be easily suppressed through CRISPRi, or whether different genes might need tedious scanning with large numbers of gRNA to find the one that effectively suppresses gene expression needs to be determined. In this background, it is reasonable to predict that RNAi will most likely remain as the preferred tool for mammalian cell based reverse genetic screens. It is anticipated that newer discoveries will enable enrichment of the on-target specificity of RNAi, enhancing its reliability.

Transient in vitro gene expression suppression: There could be instances in which only transient perturbation of gene expression is possible or preferred. For example, essential genes could not be knocked out using CRISPR/Cas9 as this will lead to cell death. In such scenarios, RNAi based transient knockdown of gene transcripts will be the technique of choice for gene expression manipulation. In another set of contexts, one may only need transient alteration of gene expression to deal with or investigate a transient phenomenon. Examples include cellular response to external factors such as infection by pathogens, and drug treatment. Most known pharmacological drugs exert their effects by transiently interacting with the intended cellular target, often proteins. If one has to genetically mimic the effect of transient exposure of cells to a drug, RNAi will be the ideal choice because conventional CRISPR/Cas9 imparts stable genetic change. However, early studies have demonstrated that CRISPRi could also be adapted for transient gene knockdown, without permanently modifying DNA.

Introduction or correction of mutations: Studies on the role of genes in cell physiology and cell-environment interactions often needs engineering precise mutations or insertion of DNA segments (e.g., reporter gene) in specific parts of the genes in the endogenous context, as opposed to their complete deletion or repression. Examples include introduction of inactivating point mutations of catalytic activity, deletion of protein domains and so forth. Conversely, there could be situations in which one may want to reverse a mutation in experimental systems. Only the CRISPR/Cas9 system, which targets the DNA, has the ability to introduce or correct mutations into genes. RNAi cannot be used for such precision genome engineering applications because it functions primarily as transcript degradation machinery.

Enhancement of gene expression: While both CRISPR/Cas9 and RNAi can down-regulate the expression of genes (respectively irreversibly and transiently), there may arise conditions in which one may want to selectively upregulate the expression of a gene permanently or transiently. RNAi is incapable of directly upregulating the expression of genes because it is a post-transcriptional suppressor of transcript availability. However, CRISPR/Cas9 could be adapted to upregulate the expression of genes. In one approach, CRISPR/Cas9 could be used to stably introduce constitutively active promoter elements to a gene to enhance its expression levels. Alternatively, dCas9 fused to a transcription activator protein by various means (e.g., VP64 protein) could be targeted to any gene of interest using appropriate gRNA, to drive gene expression at an enhanced level transiently (called CRISPR activation, CRISPRa) [35,60,61]. However, the ease and success rate of CRISPRa is yet to be evaluated by testing against large sets of genes. Thus, CRISPR but not RNAi offers the ability to enhance gene expression.

Comparative Analysis of the Clinical Applications of CRISPR/Cas9 and RNAi

Gene modifying technologies hold the promise of therapeutic applications to treat both heritable genetic as well as non-genetic diseases such as infectious diseases [62,63,64]. Accordingly, both RNAi and CRISPR are predicted to have several human therapeutic applications. We selectively present here some of the major core concepts by which RNAi and CRISPR could potentially be used for clinical applications, along with a discussion on how each technology will affect the acceptance of the other.

In vivo gene therapy has always been viewed as a very promising approach for curing many monogenic, complex genetic and non-genetic medical conditions. Gene therapy tools could be either directly used *in vivo*, or used *ex vivo* to generate gene corrected cells for adoptive therapeutic transfer [65]. The curative effects of gene therapy are often achieved by introducing either a desired gene, correction of a mutation, ablation of the expression of a gene containing gain of function mutation, or by activating a compensatory gene, into the tissues of interest, depending on the context. Introduction of genes and correction of point mutations in target host cell genomes can be best achieved though HDR. Ablation of gene expression can be achieved by NHEJ. Examples for diseases that would benefit from therapeutic correction of mutation include gene therapy for severe combined immunodeficiency and cystic fibrosis [66,67]. An example for curative activation of a gene other than the disease causing mutation-harboring gene is the selective upregulation of γ-globin for treating sickle cell anemia arising from mutation in β-thalassemia gene [68].

Many of the genetic diseases will benefit from gene therapy through stable gene editing. As RNAi is unable to either stably introduce gene segments/mutations or induce activation of genes, CRISPR/Cas9 system will be the method of choice for such gene therapy applications. Because CRISPR based gene editing is heritable, technically, one will have to introduce the genomic changes into host cells only a single time, and the physiological effect might last as long as the targeted cells are viable, or will be inherited to daughter cells in the case of dividing cells. In contrast to CRISPR/Cas9, *in vivo* application of RNAi is limited only to the instances in which the expression of genes has to be suppressed post-transcriptionally. RNAi can be made stable and heritable only by using viral delivery platforms such as a lentiviral platform which can integrate into host genome. Because lentiviral integration into mammalian cells *in vivo* can give rise to undesirable medical outcomes in the long term, this approach may be less desirable [69].

Successful *in vivo* use of CRISPR/Cas9 requires an adequate delivery platform, coupled with efficient DNA repair response. Due to their potential tumorigenic properties, retroviruses and lentiviruses may not be the ideal choice for direct *in vivo* delivery of CRISPR/Cas9 systems. Alternative delivery platforms include those based on adenovirus and adeno-associated virus. However, the large size of Cas9 protein is often a limitation for the *in vivo* delivery of CRISPR/Cas9 systems using some of the well-established vectors such as those based on adenovirus. Recently, adeno-associated viruses have been reported as efficient vector systems for *in vivo* gene editing using a smaller variant of Cas9 [3]. Similarly, *in vivo* delivery of RNAi has also been achieved using adeno-associated virus vectors [70]. However, unlike as in CRISPR/Cas9 systems, the size of the targeting insert that needed to be carried in the delivery vector is very small for RNAi.

Another important aspect impacting on the applications of various gene editing technologies in gene therapy *in vivo*is the host immune response to both the delivery vehicle and editing associated proteins. Many times, a host mounts immune response to delivery vehicles such as adenoviruses and adeno-associated viruses. Similarly, Cas9 itself may be immunogenic. For example, a recent study that used adenovirus mediated delivery of CRISPR/Cas9 to target the *Pten*gene in mouse liver to treat nonalcoholic steatohepatitis observed that the murine host developed adaptive immune response to both the vector and the Cas9 protein [71]. On the other hand, in the case of RNAi, only the immune response to vector is prevalent during *in vivo* delivery using viral delivery platforms.

The efficiency of DNA repair mechanisms activated by Cas9 mediated DNA break will also impact its *in vivo*applications. Due to its inherently slow rate, HDR mediated gene editing might yield only fewer desired editing events in the target tissues when applied *in vivo*. However, if the gene edited cells have any selective advantage over the unedited disease cells, even a small starting population of gene edited cells might be enough to eventually display a curative effect.

There are many human diseases that could be treated by taking cells, including stem cells, from the patient and correcting the mutations *ex vivo* before re-introducing cells back into the patients [65,66,67,72]. Examples for this include therapeutic use of CRISPR/Cas9 mediated *ex vivo* gene edited cells to treat Duchenne muscular dystrophy and cystic fibrosis [73,74]. Such applications will require the generation of cells with stable genetic changes. Similarly, stem cells or induced pluripotent stem cells could be engineered to have heritable curative genetic modifications for therapeutic purposes [66]. CRISPR would be the ideal choice to create such stable genetic changes. Another similar growing area of medical research that will benefit from genome engineering involves the generation of chimeric antigen receptor T cells (CART), to train T cells to detect specific epitopes [75,76]. The CRISPR/Cas system will be ideal for generating CART cells. None of these therapeutic applications can be achieved through RNAi. CRISPR could in fact be used to edit genes without having the need to leave the delivery vehicle (e.g., potentially dangerous lentivirus vector) in the target cells indefinitely, because the genetic change introduced by Cas9 through NHEJ/HDR is stable and irreversible. For example, a gRNA can be delivered as synthesized oligonucleotides and Cas9 could be delivered as purified protein [77,78]. This evidence points to CRISPR as a supreme technology over RNAi for gene therapy using *ex vivo* gene edited cells.

On the other hand, there may also arise specific instances in which transient depletion of a mammalian gene *in vivo* is needed for therapeutic purpose. For example, transient attenuation of expression of an essential pro-pathogen host factor may help to clear an acutely infecting pathogen, or brief attenuation of the expression of a host gene may alleviate an acute clinical condition. Because these or similar applications only need transient repression of the expression of host gene to treat the disease, both RNAi and CRISPRi could be used alike, both for direct *in vivo* modification and adoptive transfer of *ex vivo* gene edited cells. However, while RNAi is proven to work efficiently in most mammalian cell types if efficient delivery can be achieved, whether CRISPRi would work in specific cell subtypes *in vivo* is yet to be determined. Whether chromosomal topology or prior formation of regulatory protein complexes at a particular genomic location *in vivo* affects the ability of CRISPRi to repress transcription is currently not well defined. Moreover, RNAi could be delivered *in vivo* through the delivery of siRNAs. In this regard, RNAi may offer advantages over CRISPRi for such transient *in vivo* gene expression suppression purposes.

Targeting viral pathogens: There are several viruses that infect humans with severe consequences. In principle, gene editing technologies could be used to directly edit the genomic DNA of viruses such as hepatitis B virus, resulting in the elimination of the virus [79]. Similarly, HIV could also be targeted by mutating its genetic elements integrated into the host genome. Such applications could only be achieved through CRISPR, and not through RNAi because RNAi will only target the viral transcripts. Besides DNA viruses, there are also several RNA viral pathogens infecting humans. The ability of CRISPR to edit RNA is still not fully understood and widely exploited. However, there were recent reports attempting to edit RNAs, including that of an RNA virus, using CRISPR, hinting that eventually RNA targeting applications of CRISPR may become more efficient [80]. These studies used multicomponent CRISPR systems, unlike the popularly used simple Cas9 based systems, making their clinical use a major challenge. As the technologies stand today, adaptation of CRISPR systems as a versatile tool to edit RNA is yet to be established. Contrary to this, RNAi has been widely used for directly targeting the RNA of RNA viruses [81,82]. In conclusion, while CRISPR offers enormous potential to directly inactivate DNA viral pathogens, RNAi currently stands as the most promising tool to target RNA viruses.

Technical Considerations Impacting the Applications of CRISPR/Cas9 and RNAi

There are several technical and methodological aspects of CRISPR/Cas9 and RNAi that may also contribute to the choice of applications that each one of them could be associated with. Some of these are listed below.

Time: The time needed for each one of these techniques to manifest the intended effect on gene structure or expression varies greatly, and this may profoundly affect the choice for specific applications. RNAi exhibits rapid effects on gene expression. In around 24 h, RNAi treatment can result in significant attenuation of gene expression repression uniformly in the majority of the treated cells. However, the low to moderate efficiency of CRISPR requires selection of cells having desired InDels in all alleles, and this process may take up to a month or more depending on the specific needs. However, CRISPRi and CRISPRa may alter gene expression relatively faster, by 2–5 days [36].

Nature of the phenotype: CRISPR induces a absolute null phenotype that is irreversible [20]. However, RNAi, CRISPRi and CRISPRa impart reversible and incomplete modification of gene expression [35,37,40,60].

Genetic information needed for designing targeting systems: Design of RNAi tools need information on the sequence of the corresponding gene transcript, and may not require substantial information of the sequence of the corresponding gene. On the contrary, CRISPR requires the gene sequence information [4,20]. Furthermore, CRISPRi and CRISPRa design requires proper information on the transcription start site of the target gene [35,37,40,60].

Target specificity: It is widely established that RNAi has off-target effects [83]. Although newer design principles have greatly contributed to reduce off-target effects, there still exists no methodology to completely eliminate RNAi induced off targets [33]. Currently, the off-target effect generation potential of CRISPR/Cas9 system in mammalian cells is not yet fully appreciated. However, there are several recent reports that identified noticeable off-target potential of CRISPR technology [84,85,86,87]. Scientists have also modified CRISPR/Cas9 methodology to minimize its potential off-target effects [88]. Delivery of Cas9 as mRNA, and gRNA-Cas9 as ribonucleoprotein complex, have been proposed to reduce off-target due to the short cellular life of these formulation, as opposed to plasmid based delivery [77,78,89]. In addition, a recent study reported the generation of a mutant Cas9 with minimal off-target potential [90]. All these approaches indicate that newer discoveries will eventually reduce the off-target potential of CRISPR/Cas9 systems.

Targetable genes: RNAi can be used to target literally any protein-coding gene. The requirement of a short protospacer-adjacent motif at the target gene site makes the application of CRISPR/Cas9 tied to the presence of this sequence [5]. This raises the question of how many human genes could be edited by CRISPR/Cas9. Use of other CRISPR/Cas systems may help to increase the targetable mammalian gene coverage. Recently, Cas9 has been adapted to tolerate variations in protospacer-adjacent motif requirement through *in vitro* evolution [91]. RNAi cannot be used to edit genetic elements such as non-transcribed noncoding regulatory regions. Unlike RNAi, CRISPR/Cas9 can be used to target non-transcribed genomic segments [92]. Another factor that affects the efficacy of the gene modulation by RNAi and CRISPR/Cas9 is the state of the cell. While in theory, RNAi can target transcripts in any phase of the cell cycle, host response to CRISPR mediated DNA damage has dependence on the cell cycle phase. While NHEJ can operate in all phases of cell cycle, HDR based DNA repair system happens predominantly during S/G2 phase, limiting their applications in genome editing [93]. In such a scenario, NHEJ can dominate over HDR dependent repair, and may yield potentially unintended results.

CONCLUSIONS

As extensively described in this analysis, both CRISPR/Cas9 and RNAi are powerful tools for gene manipulations. CRISPR/Cas9 is certainly more versatile and superior to RNAi as it can be used to induce InDels, both repress or activate gene expression, and cause both heritable and non-heritable genomic changes. In addition, the emergence of newer gene editing tools such as the Cpf1 enzyme may eventually strengthen the portfolio of applications that can be achieved by CRISPR mediated genome engineering [94]. These qualitative advantages of CRISPR/Cas9 poise it to effectively dominate RNAi in diverse genetic applications in both clinical and research tool applications fields. However, based on the analysis of the currently available facts, it is reasonable to conclude that RNAi will also have a certain unique space in both biomedical research and clinical applications, at least in the near future. The biggest threat to RNAi is from CRISPRi. However, wider application of CRISPRi in the mammalian system is yet to be established through extensive experimental validations at a genome scale. Therefore, it is reasonable to predict that the fate of RNAi will greatly depend on the success of CRISPRi. One of the biggest technical applications of RNAi that might be unaffected by CRISPR in the short term future will be in the field of arrayed reverse genetic screens. Another area where RNAi may find its own niche will be for targeting RNA-viral pathogens. In conclusion, while CRISPR/Cas9 will predictably dominate applications of gene modifying technologies in the long term, RNAi will likely continue to survive with restricted domains of applications.

ACKNOWLEDGMENTS

Manoj N. Krishnan was funded by Ministry of Education, Singapore. The authors thank Sajith Kumar Nair for critical reading of the manuscript.

AUTHOR CONTRIBUTIONS

Unnikrishnan Unniyampurath, Rajendra Pilankatta and Manoj N. Krishnan wrote the manuscript.

REFERENCES

1. Fire, A.; Xu, S.; Montgomery, M.K.; Kostas, S.A.; Driver, S.E.; Mello, C.C. Potent and specific genetic interference by double-stranded RNA in *Caenorhabditis elegans*. *Nature* 1998, *391*, 806–811.
2. Wilson, R.C.; Doudna, J.A. Molecular mechanisms of RNA interference. *Annu. Rev. Biophys.* 2013, *42*, 217–239.
3. Boettcher, M.; McManus, M.T. Choosing the right tool for the job: RNAi, TALEN, or CRISPR. *Mol. Cell* 2015, *58*, 575–585.
4. Cong, L.; Ran, F.A.; Cox, D.; Lin, S.; Barretto, R.; Habib, N.; Hsu, P.D.; Wu, X.; Jiang, W.; Marraffini, L.A.; *et al.* Multiplex genome engineering using CRISPR/Cas systems. *Science* 2013, *339*, 819–823.
5. Jinek, M.; Chylinski, K.; Fonfara, I.; Hauer, M.; Doudna, J.A.; Charpentier, E. A programmable dual-RNA-guided DNA endonuclease in adaptive bacterial immunity. *Science* 2012, *337*, 816–821.
6. Mali, P.; Yang, L.; Esvelt, K.M.; Aach, J.; Guell, M.; DiCarlo, J.E.; Norville, J.E.; Church, G.M. RNA-guided human genome engineering via Cas9. *Science* 2013, *339*, 823–826.
7. Barrangou, R.; Fremaux, C.; Deveau, H.; Richards, M.; Boyaval, P.; Moineau, S.; Romero, D.A.; Horvath, P. CRISPR provides acquired resistance against viruses in prokaryotes. *Science* 2007, *315*, 1709–1712.
8. Sorek, R.; Kunin, V.; Hugenholtz, P. CRISPR—A widespread system that provides acquired resistance against phages in bacteria and archaea. *Nat. Rev. Microbiol.* 2008, *6*, 181–186.
9. Grissa, I.; Vergnaud, G.; Pourcel, C. The CRISPRdb database and tools to display CRISPRs and to generate dictionaries of spacers and repeats. *BMC Bioinform.* 2007, *8*, 172.
10. Ishino, Y.; Shinagawa, H.; Makino, K.; Amemura, M.; Nakata, A. Nucleotide sequence of the iap gene, responsible for alkaline phosphatase isozyme conversion in *Escherichia coli*, and identification of the gene product. *J. Bacteriol.*1987, *169*, 5429–5433.

11. Jansen, R.; Embden, J.D.; Gaastra, W.; Schouls, L.M. Identification of genes that are associated with DNA repeats in prokaryotes. *Mol. Microbiol.* 2002, *43*, 1565–1575.
12. Brouns, S.J.; Jore, M.M.; Lundgren, M.; Westra, E.R.; Slijkhuis, R.J.; Snijders, A.P.; Dickman, M.J.; Makarova, K.S.; Koonin, E.V.; van der Oost, J. Small CRISPR RNAs guide antiviral defense in prokaryotes. *Science* 2008, *321*, 960–964.
13. Han, D.; Krauss, G. Characterization of the endonuclease SSO2001 from Sulfolobus solfataricus P2. *FEBS Lett.*2009, *583*, 771–776.
14. Makarova, K.S.; Grishin, N.V.; Shabalina, S.A.; Wolf, Y.I.; Koonin, E.V. A putative RNA-interference-based immune system in prokaryotes: Computational analysis of the predicted enzymatic machinery, functional analogies with eukaryotic RNAi, and hypothetical mechanisms of action. *Biol. Direct* 2006, *1*, 7.
15. Marraffini, L.A.; Sontheimer, E.J. CRISPR interference limits horizontal gene transfer in staphylococci by targeting DNA. *Science* 2008, *322*, 1843–1845.
16. Haurwitz, R.E.; Jinek, M.; Wiedenheft, B.; Zhou, K.; Doudna, J.A. Sequence- and structure-specific RNA processing by a CRISPR endonuclease. *Science* 2010, *329*, 1355–1358.
17. Kunin, V.; Sorek, R.; Hugenholtz, P. Evolutionary conservation of sequence and secondary structures in CRISPR repeats. *Genome Biol.* 2007, *8*, R61.
18. Wiedenheft, B.; Sternberg, S.H.; Doudna, J.A. RNA-guided genetic silencing systems in bacteria and archaea. *Nature* 2012, *482*, 331–338.
19. Haft, D.H.; Selengut, J.; Mongodin, E.F.; Nelson, K.E. A guild of 45 CRISPR-associated (Cas) protein families and multiple CRISPR/Cas subtypes exist in prokaryotic genomes. *PLoS Comput. Biol.* 2005, *1*, e60.
20. Jiang, W.; Marraffini, L.A. CRISPR-Cas: New tools for genetic manipulations from bacterial immunity systems. *Annu. Rev. Microbiol.* 2015, *69*, 209–228.
21. Van der Oost, J.; Jore, M.M.; Westra, E.R.; Lundgren, M.; Brouns, S.J. CRISPR-based adaptive and heritable immunity in prokaryotes. *Trends Biochem. Sci.* 2009, *34*, 401–407.

22. Garneau, J.E.; Dupuis, M.E.; Villion, M.; Romero, D.A.; Barrangou, R.; Boyaval, P.; Fremaux, C.; Horvath, P.; Magadán, A.H.; Moineau, S. The CRISPR/Cas bacterial immune system cleaves bacteriophage and plasmid DNA. *Nature* 2010, *468*, 67–71.

23. Mojica, F.J.; Diez-Villasenor, C.; Garcia-Martinez, J.; Almendros, C. Short motif sequences determine the targets of the prokaryotic CRISPR defence system. *Microbiology* 2009, *155 Pt 3*, 733–740.

24. Mojica, F.J.; Diez-Villasenor, C.; Garcia-Martinez, J.; Soria, E. Intervening sequences of regularly spaced prokaryotic repeats derive from foreign genetic elements. *J. Mol. Evol.* 2005, *60*, 174–182.

25. Carte, J.; Wang, R.; Li, H.; Terns, R.M.; Terns, M.P. Cas6 is an endoribonuclease that generates guide RNAs for invader defense in prokaryotes. *Genes Dev.* 2008, *22*, 3489–3496.

26. Deltcheva, E.; Chylinski, K.; Sharma, C.M.; Gonzales, K.; Chao, Y.; Pirzada, Z.A.; Eckert, M.R.; Vogel, J.; Charpentier, E. CRISPR RNA maturation by trans-encoded small RNA and host factor RNase III. *Nature* 2011, *471*, 602–607.

27. Gesner, E.M.; Schellenberg, M.J.; Garside, E.L.; George, M.M.; Macmillan, A.M. Recognition and maturation of effector RNAs in a CRISPR interference pathway. *Nat. Struct. Mol. Biol.* 2011, *18*, 688–692.

28. Semenova, E.; Jore, M.M.; Datsenko, K.A.; Semenova, A.; Westra, E.R.; Wanner, B.; van der Oost, J.; Brouns, S.J.; Severinov, K. Interference by clustered regularly interspaced short palindromic repeat (CRISPR) RNA is governed by a seed sequence. *Proc. Natl. Acad. Sci. USA* 2011, *108*, 10098–10103.

29. Wiedenheft, B.; van Duijn, E.; Bultema, J.B.; Waghmare, S.P.; Zhou, K.; Barendregt, A.; Westphal, W.; Heck, A.J.; Boekema, E.J.; Dickman, M.J.; *et al.* RNA-guided complex from a bacterial immune system enhances target recognition through seed sequence interactions. *Proc. Natl. Acad. Sci. USA* 2011, *108*, 10092–10097.

30. Makarova, K.S.; Haft, D.H.; Barrangou, R.; Brouns, S.J.; Charpentier, E.; Horvath, P.; Moineau, S.; Mojica, F.J.; Wolf, Y.I.; Yakunin, A.F.; *et al.* Evolution and classification of the CRISPR-Cas systems. *Nat. Rev. Microbiol.* 2011, *9*, 467–477.

31. Datsenko, K.A.; Pougach, K.; Tikhonov, A.; Wanner, B.L.;

Severinov, K.; Semenova, E. Molecular memory of prior infections activates the CRISPR/Cas adaptive bacterial immunity system. *Nat. Commun.* 2012, *3*, 945.

32. Yosef, I.; Goren, M.G.; Qimron, U. Proteins and DNA elements essential for the CRISPR adaptation process in *Escherichia coli*. *Nucleic Acids Res.* 2012, *40*, 5569–5576.
33. Naito, Y.; Ui-Tei, K. Designing functional siRNA with reduced off-target effects. *Methods Mol. Biol.* 2013, *942*, 57–68.
34. Tsui, T.K.; Li, H. Structure principles of CRISPR-Cas surveillance and effector complexes. *Annu. Rev. Biophys.*2015, *44*, 229–255.
35. Gilbert, L.A.; Horlbeck, M.A.; Adamson, B.; Villalta, J.E.; Chen, Y.; Whitehead, E.H.; Guimaraes, C.; Panning, B.; Ploegh, H.L.; Bassik, M.C.; *et al.* Genome-scale CRISPR-mediated control of gene repression and activation. *Cell*2014, *159*, 647–661.
36. Gilbert, L.A.; Larson, M.H.; Morsut, L.; Liu, Z.; Brar, G.A.; Torres, S.E.; Stern-Ginossar, N.; Brandman, O.; Whitehead, E.H.; Doudna, J.A.; *et al.* CRISPR-mediated modular RNA-guided regulation of transcription in eukaryotes. *Cell* 2013, *154*, 442–451.
37. Qi, L.S.; Larson, M.H.; Gilbert, L.A.; Doudna, J.A.; Weissman, J.S.; Arkin, A.P.; Lim, W.A. Repurposing CRISPR as an RNA-guided platform for sequence-specific control of gene expression. *Cell* 2013, *152*, 1173–1183.
38. Choulika, A.; Perrin, A.; Dujon, B.; Nicolas, J.F. Induction of homologous recombination in mammalian chromosomes by using the I-SceI system of Saccharomyces cerevisiae. *Mol. Cell. Biol.* 1995, *15*, 1968–1973.
39. Rouet, P.; Smih, F.; Jasin, M. Introduction of double-strand breaks into the genome of mouse cells by expression of a rare-cutting endonuclease. *Mol. Cell. Biol.* 1994, *14*, 8096–8106.
40. Bosher, J.M.; Labouesse, M. RNA interference: Genetic wand and genetic watchdog. *Nat. Cell Biol.* 2000, *2*, E31–E36.
41. Dillon, C.P.; Sandy, P.; Nencioni, A.; Kissler, S.; Rubinson, D.A.; van Parijs, L. RNAi as an experimental and therapeutic tool to study and regulate physiological and disease processes. *Annu. Rev. Physiol.* 2005, *67*, 147–173.

42. Mohr, S.; Bakal, C.; Perrimon, N. Genomic screening with RNAi: Results and challenges. *Annu. Rev. Biochem.* 2010, *79*, 37–64.

43. Napoli, C.; Lemieux, C.; Jorgensen, R. Introduction of a chimeric chalcone synthase gene into petunia results in reversible co-suppression of homologous genes in trans. *Plant Cell* 1990, *2*, 279–289.

44. Paddison, P.J.; Hannon, G.J. RNA interference: The new somatic cell genetics? *Cancer Cell* 2002, *2*, 17–23. [Google Scholar] [CrossRef]

45. Elbashir, S.M.; Harborth, J.; Lendeckel, W.; Yalcin, A.; Weber, K.; Tuschl, T. Duplexes of 21-nucleotide RNAs mediate RNA interference in cultured mammalian cells. *Nature* 2001, *411*, 494–498.

46. Bernstein, E.; Caudy, A.A.; Hammond, S.M.; Hannon, G.J. Role for a bidentate ribonuclease in the initiation step of RNA interference. *Nature* 2001, *409*, 363–366.

47. Lee, Y.; Hur, I.; Park, S.Y.; Kim, Y.K.; Suh, M.R.; Kim, V.N. The role of PACT in the RNA silencing pathway. *EMBO J.* 2006, *25*, 522–532.

48. Hammond, S.M.; Boettcher, S.; Caudy, A.A.; Kobayashi, R.; Hannon, G.J. Argonaute2, a link between genetic and biochemical analyses of RNAi. *Science* 2001, *293*, 1146–1150.

49. Janowski, B.A.; Huffman, K.E.; Schwartz, J.C.; Ram, R.; Nordsell, R.; Shames, D.S.; Minna, J.D.; Corey, D.R. Involvement of AGO1 and AGO2 in mammalian transcriptional silencing. *Nat. Struct. Mol. Biol.* 2006, *13*, 787–792.

50. Kim, D.H.; Villeneuve, L.M.; Morris, K.V.; Rossi, J.J. Argonaute-1 directs siRNA-mediated transcriptional gene silencing in human cells. *Nat. Struct. Mol. Biol.* 2006, *13*, 793–797.

51. Parnas, O.; Jovanovic, M.; Eisenhaure, T.M.; Herbst, R.H.; Dixit, A.; Ye, C.J.; Przybylski, D.; Platt, R.J.; Tirosh, I.; Sanjana, N.E.; *et al.* A Genome-wide CRISPR screen in primary immune cells to dissect regulatory networks. *Cell*2015, *162*, 675–686.

52. Shalem, O.; Sanjana, N.E.; Zhang, F. High-throughput functional genomics using CRISPR-Cas9. *Nat. Rev. Genet.*2015, *16*, 299–311.

53. Wang, T.; Birsoy, K.; Hughes, N.W.; Krupczak, K.M.; Post, Y.; Wei,

J.J.; Lander, E.S.; Sabatini, D.M. Identification and characterization of essential genes in the human genome. *Science* 2015, *350*, 1096–1101.

54. Wang, T.; Wei, J.J.; Sabatini, D.M.; Lander, E.S. Genetic screens in human cells using the CRISPR-Cas9 system. *Science* 2014, *343*, 80–84.

55. Duda, K.; Lonowski, L.A.; Kofoed-Nielsen, M.; Ibarra, A.; Delay, C.M.; Kang, Q.; Yang, Z.; Pruett-Miller, S.M.; Bennett, E.P.; Wandall, H.H.; *et al.* High-efficiency genome editing via 2A-coupled co-expression of fluorescent proteins and zinc finger nucleases or CRISPR/Cas9 nickase pairs. *Nucleic Acids Res.* 2014, *42*, e84.

56. Liang, X.; Potter, J.; Kumar, S.; Zou, Y.; Quintanilla, R.; Sridharan, M.; Carte, J.; Chen, W.; Roark, N.; Ranganathan, S.; *et al.* Rapid and highly efficient mammalian cell engineering via Cas9 protein transfection. *J. Biotechnol.* 2015, *208*, 44–53.

57. Swiech, L.; Heidenreich, M.; Banerjee, A.; Habib, N.; Li, Y.; Trombetta, J.; Sur, M.; Zhang, F. *In vivo* interrogation of gene function in the mammalian brain using CRISPR-Cas9. *Nat. Biotechnol.* 2015, *33*, 102–106.

58. Chu, V.T.; Weber, T.; Wefers, B.; Wurst, W.; Sander, S.; Rajewsky, K.; Kühn, R. Increasing the efficiency of homology-directed repair for CRISPR-Cas9-induced precise gene editing in mammalian cells. *Nat. Biotechnol.*2015, *33*, 543–548.

59. Lin, S.; Staahl, B.T.; Alla, R.K.; Doudna, J.A. Enhanced homology-directed human genome engineering by controlled timing of CRISPR/Cas9 delivery. *eLife* 2014, *3*, e04766.

60. Konermann, S.; Brigham, M.D.; Trevino, A.E.; Joung, J.; Abudayyeh, O.O.; Barcena, C.; Hsu, P.D.; Habib, N.; Gootenberg, J.S.; Nishimasu, H.; *et al.* Genome-scale transcriptional activation by an engineered CRISPR-Cas9 complex. *Nature* 2015, *517*, 583–588.

61. Mali, P.; Aach, J.; Stranges, P.B.; Esvelt, K.M.; Moosburner, M.; Kosuri, S.; Yang, L.; Church, G.M. CAS9 transcriptional activators for target specificity screening and paired nickases for cooperative genome engineering. *Nat. Biotechnol.* 2013, *31*, 833–838.

62. Borna, H.; Imani, S.; Iman, M.; Azimzadeh Jamalkandi, S.

Therapeutic face of RNAi: *In vivo* challenges. *Expert Opin. Biol. Ther.* 2015, *15*, 269–285.

63. Okamoto, K.; Murawaki, Y. The therapeutic potential of RNA interference: Novel approaches for cancer treatment. *Curr. Pharm. Biotechnol.* 2012, *13*, 2235–2247.
64. Yao, S.; He, Z.; Chen, C. CRISPR/Cas9-mediated genome editing of epigenetic factors for cancer therapy. *Hum. Gene Ther.* 2015, *26*, 463–471.
65. O'Connor, T.P.; Crystal, R.G. Genetic medicines: Treatment strategies for hereditary disorders. *Nat. Rev. Genet.*2006, *7*, 261–276.
66. Chang, C.W.; Lai, Y.S.; Westin, E.; Khodadadi-Jamayran, A.; Pawlik, K.M.; Lamb, L.S., Jr.; Goldman, F.D.; Townes, T.M. Modeling human severe combined immunodeficiency and correction by CRISPR/Cas9-enhanced gene targeting. *Cell Rep.* 2015, *12*, 1668–1677.
67. Lee, C.M.; Flynn, R.; Hollywood, J.A.; Scallan, M.F.; Harrison, P.T. Correction of the DeltaF508 mutation in the cystic fibrosis transmembrane conductance regulator gene by zinc-finger nuclease homology-directed repair. *Biores. Open Access* 2012, *1*, 99–108.
68. Graslund, T.; Li, X.; Magnenat, L.; Popkov, M.; Barbas, C.F., 3rd. Exploring strategies for the design of artificial transcription factors: Targeting sites proximal to known regulatory regions for the induction of gamma-globin expression and the treatment of sickle cell disease. *J. Biol. Chem.* 2005, *280*, 3707–3714.
69. Thrasher, A.J.; Gaspar, H.B.; Baum, C.; Modlich, U.; Schambach, A.; Candotti, F.; Otsu, M.; Sorrentino, B.; Scobie, L.; Cameron, E.; *et al.* Gene therapy: X-SCID transgene leukaemogenicity. *Nature* 2006, *443*, E5–E6.
70. Borel, F.; Kay, M.A.; Mueller, C. Recombinant AAV as a platform for translating the therapeutic potential of RNA interference. *Mol. Ther.* 2014, *22*, 692–701.
71. Wang, D.; Mou, H.; Li, S.; Li, Y.; Hough, S.; Tran, K.; Li, J.; Yin, H.; Anderson, D.G.; Sontheimer, E.J.; *et al.* Adenovirus-mediated somatic genome editing of pten by CRISPR/Cas9 in mouse liver in spite of Cas9-Specific immune responses. *Hum. Gene Ther.* 2015, *26*,

432–442.

72. Oakland, M.; Sinn, P.L.; McCray, P.B., Jr. Advances in cell and gene-based therapies for cystic fibrosis lung disease. *Mol. Ther.* 2012, *20*, 1108–1115.
73. Ousterout, D.G.; Kabadi, A.M.; Thakore, P.I.; Majoros, W.H.; Reddy, T.E.; Gersbach, C.A. Multiplex CRISPR/Cas9-based genome editing for correction of dystrophin mutations that cause Duchenne muscular dystrophy. *Nat. Commun.* 2015, *6*, 6244.
74. Schwank, G.; Koo, B.K.; Sasselli, V.; Dekkers, J.F.; Heo, I.; Demircan, T.; Sasaki, N.; Boymans, S.; Cuppen, E.; van der Ent, C.K.; *et al.* Functional repair of CFTR by CRISPR/Cas9 in intestinal stem cell organoids of cystic fibrosis patients. *Cell Stem Cell* 2013, *13*, 653–658.
75. Liechtenstein, T.; Perez-Janices, N.; Escors, D. Lentiviral vectors for cancer immunotherapy and clinical applications. *Cancers* 2013, *5*, 815–837.
76. Lloyd, A.; Vickery, O.N.; Laugel, B. Beyond the antigen receptor: Editing the genome of T-cells for cancer adoptive cellular therapies. *Front. Immunol.* 2013, *4*, 221.
77. Kim, S.; Kim, D.; Cho, S.W.; Kim, J.; Kim, J.S. Highly efficient RNA-guided genome editing in human cells via delivery of purified Cas9 ribonucleoproteins. *Genome Res.* 2014, *24*, 1012–1019.
78. Schumann, K.; Lin, S.; Boyer, E.; Simeonov, D.R.; Subramaniam, M.; Gate, R.E.; Haliburton, G.E.; Ye, C.J.; Bluestone, J.A.; Doudna, J.A.; *et al.* Generation of knock-in primary human T cells using Cas9 ribonucleoproteins. *Proc. Natl. Acad. Sci. USA* 2015, *112*, 10437–10442.
79. Kennedy, E.M.; Kornepati, A.V.; Cullen, B.R. Targeting hepatitis B virus cccDNA using CRISPR/Cas9. *Antivir. Res.*2015, *123*, 188–192.
80. Price, A.A.; Sampson, T.R.; Ratner, H.K.; Grakoui, A.; Weiss, D.S. Cas9-mediated targeting of viral RNA in eukaryotic cells. *Proc. Natl. Acad. Sci. USA* 2015, *112*, 6164–6169.
81. Jahan, S.; Khaliq, S.; Samreen, B.; Ijaz, B.; Khan, M.; Ahmad, W.; Ashfaq, U.A.; Hassan, S. Effect of combined siRNA of HCV E2 gene and HCV receptors against HCV. *Virol. J.* 2011, *8*, 295.

82. Kumar, P.; Lee, S.K.; Shankar, P.; Manjunath, N. A single siRNA suppresses fatal encephalitis induced by two different flaviviruses. *PLoS Med.* 2006, *3*, e96.

83. Ui-Tei, K. Optimal choice of functional and off-target effect-reduced siRNAs for RNAi therapeutics. *Front. Genet.*2013, *4*, 107.

84. Cho, S.W.; Kim, S.; Kim, Y.; Kweon, J.; Kim, H.S.; Bae, S.; Kim, J.S. Analysis of off-target effects of CRISPR/Cas-derived RNA-guided endonucleases and nickases. *Genome Res.* 2014, *24*, 132–141.

85. Frock, R.L.; Hu, J.; Meyers, R.M.; Ho, Y.J.; Kii, E.; Alt, F.W. Genome-wide detection of DNA double-stranded breaks induced by engineered nucleases. *Nat. Biotechnol.* 2015, *33*, 179–186.

86. Fu, Y.; Foden, J.A.; Khayter, C.; Maeder, M.L.; Reyon, D.; Joung, J.K.; Sander, J.D. High-frequency off-target mutagenesis induced by CRISPR-Cas nucleases in human cells. *Nat. Biotechnol.* 2013, *31*, 822–826.

87. Lin, Y.; Cradick, T.J.; Brown, M.T.; Deshmukh, H.; Ranjan, P.; Sarode, N.; Wile, B.M.; Vertino, P.M.; Stewart, F.J.; Bao, G. CRISPR/Cas9 systems have off-target activity with insertions or deletions between target DNA and guide RNA sequences. *Nucleic Acids Res.* 2014, *42*, 7473–7485.

88. Ran, F.A.; Hsu, P.D.; Lin, C.Y.; Gootenberg, J.S.; Konermann, S.; Trevino, A.E.; Scott, D.A.; Inoue, A.; Matoba, S.; Zhang, Y.; *et al.* Double nicking by RNA-guided CRISPR Cas9 for enhanced genome editing specificity. *Cell* 2013, *154*, 1380–1389.

89. Chang, N.; Sun, C.; Gao, L.; Zhu, D.; Xu, X.; Zhu, X.; Xiong, J.W.; Xi, J.J. Genome editing with RNA-guided Cas9 nuclease in zebrafish embryos. *Cell Res.* 2013, *23*, 465–472.

90. Kleinstiver, B.P.; Pattanayak, V.; Prew, M.S.; Tsai, S.Q.; Nguyen, N.T.; Zheng, Z.; Joung, J.K. High-fidelity CRISPR-Cas9 nucleases with no detectable genome-wide off-target effects. *Nature* 2016, *529*, 490–495.

91. Kleinstiver, B.P.; Prew, M.S.; Tsai, S.Q.; Nguyen, N.T.; Topkar, V.V.; Zheng, Z.; Joung, J.K. Broadening the targeting range of Staphylococcus aureus CRISPR-Cas9 by modifying PAM recognition. *Nat. Biotechnol.* 2015, *33*, 1293–1298.

92. Han, J.; Zhang, J.; Chen, L.; Shen, B.; Zhou, J.; Hu, B.; Du, Y.; Tate, P.H.; Huang, X.; Zhang, W. Efficient *in vivo*deletion of a large imprinted lncRNA by CRISPR/Cas9. *RNA Biol.* 2014, *11*, 829–835.

93. Shrivastav, M.; De Haro, L.P.; Nickoloff, J.A. Regulation of DNA double-strand break repair pathway choice. *Cell Res.* 2008, *18*, 134–147.

94. Zetsche, B.; Gootenberg, J.S.; Abudayyeh, O.O.; Slaymaker, I.M.; Makarova, K.S.; Essletzbichler, P.; Essletzbichler, P.; Volz, S.E.; Joung, J.; van der Oost, J.; *et al.* Cpf1 Is a Single RNA-Guided Endonuclease of a Class 2 CRISPR-Cas System. *Cell* 2015, *163*, 759–771.

CHAPTER 12

APPLICATION OF RNAI TO GENOMIC DRUG TARGET VALIDATION IN SCHISTOSOMES

Alessandra Guidi[1], Nuha R. Mansour[1] , Ross A. Paveley[1] , Ian M. Carruthers[2] , Jérémy Besnard[2] , Andrew L. Hopkins[2] , Ian H. Gilbert[2] , Quentin D. Bickle[1]

[1] Department of Infection and Immunity, London School of Hygiene and Tropical Medicine, London, United Kingdom

[2] Division of Biological Chemistry and Drug Discovery, College of Life Sciences, University of Dundee, Dundee, United Kingdom

ABSTRACT

Concerns over the possibility of resistance developing to praziquantel (PZQ), has stimulated efforts to develop new drugs for schistosomiasis. In addition to the development of improved whole organism screens, the success of RNA interference (RNAi) in schistosomes offers great promise for the identification of potential drug targets to initiate drug discovery. In this study we set out to contribute to RNAi based validation of putative drug targets. Initially a list of 24 target candidates was compiled based on the identification of putative essential genes in schistosomes orthologous of *C. elegans* essential genes. Knockdown of Calmodulin (Smp_026560.2) (*Sm*-Calm), that topped this list, produced a phenotype characterised by waves of contraction in adult worms but no phenotype in schistosomula. Knockdown of the atypical Protein Kinase C (Smp_096310) (*Sm*-aPKC) resulted in loss of viability in both schistosomula and adults and led us to focus our attention on other kinase genes that were identified in the above list and through whole organism screening of known kinase inhibitor sets followed by chemogenomic evaluation. RNAi knockdown of these kinase genes failed to affect adult worm viability but, like *Sm*-aPKC, knockdown of Polo-like kinase 1, *Sm*-PLK1 (Smp_009600) and p38-MAPK, *Sm*-MAPK p38 (Smp_133020) resulted in an increased mortality of schistosomula after 2-3 weeks, an effect more marked in the presence of human red blood cells (hRBC). For *Sm*-PLK-1 the same effects were seen with the specific inhibitor, BI2536, which also affected viable egg production in adult worms. For *Sm*-PLK-1 and *Sm*-aPKC the *in vitro* effects were reflected in lower recoveries *in vivo*. We conclude that the use of RNAi combined with culture with hRBC is a reliable method for evaluating genes important for larval development. However, in view of the slow manifestation of the effects of *Sm*-aPKC knockdown in adults and the lack of effects of *Sm*-PLK-1 and *Sm*-MAPK p38 on adult viability, these kinases may not represent suitable drug targets.

AUTHOR SUMMARY

Schistosomiasis is a neglected tropical disease affecting 200 million people in the poorest areas of the world. To date no vaccine exists against this parasitosis, and the only possible cure relies on the use of the one available drug, praziquantel. Praziquantel is a safe and effective drug active against adult worms in a single dose administration. However the increasing threat of emerging resistance makes the search for novel drugs a compelling priority. *S. mansoni* possesses a quite large and complex genome so new *in silico* strategies to select genes which might represent potential novel drug target candidates are an important approach to drug development. Here, we report application of RNA interference, which knocks down gene function, to validate a number of putative drug targets for schistosomiasis. Some of these genes were selected because they had been shown to be essential for life in another worm species, *C. elegans*, and others because they were potential targets of compounds shown to be active against schistosomes *in vitro*. Although none of the identified genes looked promising as drug targets, our results showed important roles for some of the genes in schistosome development.

INTRODUCTION

Schistosomiasis is a parasitic disease caused by a trematode of the genus *Schistosoma*, affecting around 200 million people in the poorest areas of the world and despite progress in development of a vaccine against schistosomiasis, none is yet available [1–3]. Therefore, currently, both treatment and most disease control initiatives [4] rely on chemotherapy using a single drug, praziquantel (PZQ), which is active against adult worms of all the medically important *Schistosoma* species [5]. Moreover, it has proved to be generally safe and effective using a single oral dose [6, 7]. However, there have been a number of reports of poor *in vivo*efficacy of PZQ [8–11] and strains isolated from such cases have shown lower susceptibility to PZQ experimentally [12] although as yet there is no convincing evidence of development and selection of heritable resistance even following repeated rounds of treatment [9, 13–15]. However, its increasingly extensive use, especially in mass drug administration programs, raises concerns about drug resistance emerging and this has led to renewed interest in research into drug discovery including: development and application of whole organism screens for compound testing [16]; rational drug discovery [17]; and identification of putative molecular targets by analysis of the annotated schistosome sequences [18].

Following the availability of transcriptome data and the complete genome sequence of *Schistosoma mansoni* [19–23] a number of groups have set out to identify likely drug targets from amongst the >11,000 genes predicted for the *S. mansoni* genome and to prioritise them for validation by molecular/biochemical techniques that can offer a quicker and more selective tool to identify a subset of possible essential genes. As recently reviewed [24] this *in silico*analysis has highlighted various "druggable" targets [25] in schistosomes [19, 26–28] and the application of comparative genomics has identified orthologues in schistosomes of druggable genes shown to be essential in other organisms. Such approaches have led to the compilation of a number of partially overlapping lists of putative targets [19, 20, 29]. The demonstration that effective gene knockdown can be achieved in a variety of life cycle stages of schistosomes using RNA interference (RNAi) [24, 30–36] has led to this method being applied to validate putative drug targets based on changes to phenotype and/or viability in culture or *in vivo* [32, 36–44].

In parallel with our ongoing whole organism high throughput screen (HTS) approach [45] we have also undertaken studies aimed at target identification and validation using RNAi. At the outset we were interested to compare the use of adult and larval schistosomes with a view to possible use of our automated image-based HTS [45] for drug-induced damage to larval schistosomes to develop RNAi HTS. Studies were initiate by selecting putative essential genes in *S. mansoni* which were orthologues of genes previously shown, through RNAi methodology, to be essential in *C. elegans* [46]. Moreiver the RNAi-induced phenotype of *Sm*-Protein Kinase C together with the highlighted potential of kinases as key targets for drug design in schistosomes [28, 47, 48] led us to focus on kinases in our bioinformatics list. We also identified some other potential targets known to interact with known small molecule, drug-like compounds (i.e. "druggable" targets). Early detection was based on compounds found to be active in whole organism screening of a focused kinase inhibitor library (from GlaxoSmithKline) and a diverse collection of lead-like compounds from the Division of Biological Chemistry and Drug Discovery, University of Dundee.

Thus, a number of additional potential kinases were selected and targeted for silencing in adult and larval stages to establish their essentiality. Overall, phenotypic effects were seen following the dsRNA-mediated knockdown of 4 out of 16 genes (Smp_026560.2, Smp_096310, Smp_009600 and Smp_133020 in SchistoDB) coding for: Calmodulin, atypical Protein kinase C, Polo-like kinase 1 and p38-MAPK family member, respectively.

MATERIALS AND METHODS

Essentiality Predictions by Orthology

Orthologues were detected using OrthoMCL [49] version 1.4. The program was installed locally and modified to use the NCBI BLAST+ program [50]. The default OrthoMCL parameters, for e-value and MCL inflation index were used, as it had been shown previously [51] that tightening these parameters only increased specificity at a cost on sensitivity. OrthoMCL produced clusters of genes within the two given genomes representing co-orthologous groups. Where the cluster contained a single gene from each genome, these genes were considered true orthologs.

Where a cluster contained one gene from a species and multiple genes from the other species, the single gene was considered the ortholog, and the multiple genes considered paralogs. Where a cluster contained multiple genes from both genomes, all the genes were considered paralogs to each other.

Identification of Putative Targets Kinases from Phenotypic Screening of Kinase Inhibitor Libraries

Datasets

We also used an additional chemocentric approach to identify potential drug targets in the schistosome kinome. We tested the GlaxoSmithKline Published Kinase Inhibitor Set (PKIS) against adult schistosomes [16]. The PKIS comprises 367 published small molecule kinase inhibitors that inhibit a range of kinases and represent a number of different chemical scaffolds. A number of compounds found to be active against larval and adult schistosome worms during screening of a collection of compounds from the Drug Discovery Unit (DDU), University of Dundee [45] were also used to explore potential drug targets. However, the DDU set does not contain experimental information on target activity.

Predicted Target Assignment for the DDU Set

In order to obtain some ideas about the potential targets that the DDU set may be targeted a multi categorical Laplacian-modified naïve Bayesian model, as described by Nidhi et al. [52] with ChEMBL version 6 as training set [53], using Pipeline Pilot v8.0 (Biovia) was built. Compounds from ChEMBL were considered active if they displayed an activity (Ki, EC50, Kd, Kb) below 10 μM and targets were kept if they had at least 10 active compounds, the dataset contains 254,253 compounds and 1,028 targets. The descriptors used in the model were molecular weight, ALogP, numbers of hydrogen bond donors, acceptors and rotatable bonds, fractional Topological Polar Surface Area (TPSA/total surface area) [54] and Extended-connectivity fingerprints of depth 6 (ECFP_6) [55]. For every compound the top prediction from the model was calculated and assigned as potential target of interest for that particular compound.

Analyzing and Visualizing Experimental Results

To visualize the results for the 2 sets of compounds, similarity networks were drawn where nodes are compounds and edges linked two similar compounds (one network per set). The similarity cut-off used is a Tanimoto score [56] higher than 0.7 using ECFP_4 as descriptors [55]. The networks were displayed in Cytoscape [57]. Individual sub networks were then analyzed to identify those which were enriched in in-vivo active molecules and to identify known or predicted targets of interest.

Ethics Statement & Animals

Experimentation was carried out using the NC3Rs and ARRIVE guidelines. It was approved following local ethical review by the LSHTM Animal Welfare and Ethical Review Board and was performed in strict accordance with the U. K Home Office Animals (Scientific Procedures) Act 1986 (approved H.O. Project License 600456).

Female CD1 mice (aged 6–8 weeks) supplied by Charles River, UK were maintained at St Mary's Hospital, Imperial College London in SPF conditions with access to food and water *ad libitum*.

Parasite Preparation & Maintenance

Experiments were performed using the Puerto Rican strain of *S. mansoni* maintained in *Biomphalaria glabrata* and CD1 mice. Schistosomula were mechanically prepared as previously described [45] using medium 169 (M169) [58] with 5% FCS. For some experiments the M169 was supplemented with 0.25% packed A+ human red blood cells (hRBC) (National blood transfusion service, Collindale, UK).

Mice were infected subcutaneously under mild isoflurane (Merial Animal Health Ltd (UK) anaesthesia with 400 cercariae in 100μl water. Adult worms were recovered from infected mice using sterile techniques by portal perfusion 6–8 weeks post-infection using warm perfusion medium (Dulbecco's Modified Eagle's Medium [DMEM], 2mM L-glutamine, 100u/ml penicillin, 100μg/ml streptomycin, 20mM Hepes, 10units/ml heparin [Sigma, UK] [16].

We used DMEM since this was the medium used in our previous studies [16] and has also been used by others for long term culture [59, 60]. As in these earlier studies we found good viability using cDMEM but would also expect M169 which we used for the schistosomula to work equally well [24, 61].

In some experiments the mice were infected with male only worms. These were produced by infection of snails with single miracidia and the cercariae produced screened by PCR using the female-specific W probe [62] and only male cercariae were used for mice infections. Adult parasites were washed free of red blood cells using the perfusion medium and finally placed in culture in complete medium (cDMEM: DMEM, 2mM L-glutamine, 100u/ml penicillin, 100μg/ml streptomycin, 10% foetal calf serum (FCS) at 37°C, in an atmosphere of 5% CO_2 [59].

dsRNA Synthesis

For all the targets above, specific T7 promoter-tagged primers were designed in order to amplify a PCR product of ~500 base pair (bp) for each gene. We also produced an exogenous (irrelevant [IRR]) non-schistosome dsRNA fragment from the yeast expression vector pPIC9K as previously described [42]. The sequences of the primers were tested against the *S. mansoni* genome using the BLAST program at National Center for Biotechnology Information (NCBI) to avoid the risk of possible off-targeting. PCR products were used to synthesize *in vitro* the long dsRNAs using the T7 Megascript RNAi Kit (Ambion) following the instructions reported by the manufacturer. Briefly, all synthesis reactions were carried out for 4 hours at 37°C incubation followed by DNAse and RNAse treatments. At the end dsRNA products were run in a 1% agarose gel to check their integrity and concentrations assessed at OD_{260} using the ND 1000-Spectrophotometer.

In addition, for each target short interfering RNAs (siRNAs) were synthesized commercially from both Applied Biosystem (AB) and Integrated DNA Technology (IDT) which use a different algorithm to design them. From both companies two siRNAs were designed based on different regions of the sequence in order to avoid any possible annealing problems due to mRNA secondary structure. A siRNA irrelevant control was obtained from IDT based on pPIC9K sequence, as mentioned above.

dsRNA Delivery in Schistosomes

Adult worms

Fifteen adult worms (males or pairs) were electroporated in 100 μl of electroporation buffer (Applied Biosystem) in 4 mm cuvettes and at specific conditions (125 V, 20 ms and room temperature) as previously described [34]. Worms were treated either with long dsRNA (100μg/ml) or using the synthetic siRNA (2.5μg/ml). After electroporation parasites were cultured overnight in cDMEM with 5% FCS and the following day the medium was replaced with cDMEM with 10% FCS which was subsequently changed once a week.

Schistosomula

In preliminary experiments simple incubation was compared with electroporation for delivery of dsRNA [63] but the incubation method was chosen for the experiments described herein. For this 10,000 schistosomula were incubated in 50 μl of M169 or RPMI medium without FCS for 30 minutes at 37°C in presence of long dsRNA (100 μg/ml). After the treatment parasites were transferred to 1 ml of complete M169 alone or supplemented with hRBC provided by addition of 2.5μl/ml packed, washed human whole blood cells. Medium±hRBC was changed once a week. The level of mRNA suppression was measured 7 days after the silencing by qRT-PCR. Adult worms were kept in culture for observation up to 8 weeks and schistosomula up to 6 weeks. During this time parasites were checked weekly for any phenotypic change (motility, viability and tegumental damage). All of the silencing/phenotype monitoring experiments were repeated at least twice and several three times.

Gene Expression Analysis

To assess the level of gene knock down RNA was extracted from parasites using the Trizol method (Invitrogen) following the instructions reported by the manufacturer. Parasites were homogenized on ice using a sterile electronic pestle and the RNA precipitated overnight in ethanol, sodium acetate (3M, pH 5.2) and glycogen (1 μg/ml) and subsequently treated with DNase (Invitrogen) to remove any possible genomic DNA residues. cDNA was synthesized using 500 ng of total extracted RNA from each sample in the Super Script III kit (Invitrogen) and an oligo $(dT)_{20}$ primer. Quantitative real-time PCR was performed in triplicate using custom TaqMan Gene Expression Assays including a specific set of primers and a probe labeled with 6-carboxyfluorescin (FAM), obtained from Applied Biosystem. All TaqMan probes have been designed using a gene sequence outside the region in which the long dsRNA was synthesized. PCR reactions were carried out in triplicate on a 7500 ABI PRISM Sequence Detection System Instrument using an equivalent of 10 ng of parasite RNA according to the manufacturer's instructions. For relative quantification, the $\Delta\Delta C_t$ method was employed, using alpha tubulin as the endogenous standard for each sample [64]. Results obtained from parasites treated with irrelevant dsRNA were used as calibrators [65]. For graphical representation, the $\Delta\Delta C_t$ values were normalized to controls and expressed as percent difference.

Infection of Mice with dsRNA-treated Schistosomula

Newly transformed schistosomula were soaked in presence of specific long dsRNA or control as described above. After the treatment parasites were maintained in culture overnight in 1 ml of complete M169 without blood. The following morning parasites were washed three times with M169 without serum, counted and resuspended in an appropriate volume to contain 1000 schistosomula/100 µl of medium. This was injected into CD1 mice (5 age and weight matched mice per group) maintained under mild isoflurane (Merial Animal Health Ltd (UK) anaesthesia. In order to increase the chance of administration of comparable numbers of larvae to each mouse, two separate intramuscular injections each of 50 µl suspension were administered into different thigh sites of each mouse using a 30 g needle (BD Micro-Fine™+). Samples of the treated parasites were kept in culture in order to assess gene knock down 7 days and 28 days after the dsRNA treatment. Mature worms were recovered from mice 4 weeks later by mesenteric vein perfusion [66] and checked for number and morphological changes. Parasites were then used to assess the gene suppression levels by qRT-PCR as described above.

Inhibitor Treatment of Parasites

500 newly transformed schistosomula were cultured in wells of 48 well plates (Nunc,UK) in 1 ml of M169 medium supplemented with 100 U/ml Penicillin, 100 µg/ml Streptomycin and 5% foetal calf serum (Sigma, UK) in the presence of 100 nM of BI2536 inhibitor (Axon MedChem, Netherlands) dissolved in DMSO (final DMSO concentration was 0.1%). Control schistosomula were cultured in medium with 0.1% DMSO. For adult worms, cultures were set up in 24 well plates (Nunc, UK) and 5 worm pairs recovered from mice 6 weeks post infection added to each well in 2 ml cDMEM. As above, 100nM of Bl12536 was added to test wells and DMSO to controls. Cultures were set up with or without 2.5µl/ml packed washed human hRBC. Parasites were maintained in 5% CO_2 at 37°C and the medium ± inhibitor ± hRBC renewed twice a week.

Statistical Analysis

Differences between groups were assesses for statistical significance using unpaired Student's t-test using GraphPad Prism 4.0 Software. Results are considered significant if p value < 0.05.

RESULTS

Selection and Target Validation for Putative Drug Targets

Genes predicted to be both essential and druggable (*i.e.* interact with small molecule ligands) in *S. mansoni*, were selected for study using RNAi. Due to the lack to whole genome essentiality studies in *S. mansoni* essentiality was inferred from *C. elegans* data, as the genome sequence [67] and a systematic functional analysis of the *C. elegans* genome, using RNAi, are available [46]. The latter study assigned each gene to a functional class based on the phenotype, and the functional classes used to define "essential" genes here were: (i) Embryonic lethality (Emb), defined as >10% dead embryos; (ii) Sterile (Ste), required a brood size of <10 (wild-type worms under similar conditions typically have >100 progeny); (iii) Sterile progeny (Stp), progeny brood size of <10. Gene essentiality in *S. mansoni* was inferred from the *C. elegans* data [46] using a logical model: the predicted gene has an ortholog or out-paralogs, but no in-paralogs and any one of the predictor co-orthologs has been experimentally verified as essential. The assumptions of the model were that if a gene was retained after a speciation event, despite a large evolutionary time and adaption to new conditions, then its function was potentially essential. However, if the gene had subsequently been duplicated there was a potential redundancy of function and the gene was not individually essential.

Of the 11,809 genes in the *S. mansoni* genome, 323 were predicted to be essential using this method. The 323 putative essential *S. mansoni* genes were searched against the protein targets in ChEMBL [53] (http://www.ebi.ac.uk/chembl/) using BLAST+, (E-value cut-off of 1×10^{-03} and target coverage of >50%). Only ChEMBL targets that had at least one potent compound (<10nM) were considered. This filter reduced the *S. mansoni* set to just 24 genes. Interestingly these genes cover a variety of functional categories including kinases, phosphatase and proteasome subunits. One of those selected genes, glycogen synthase kinase 3, GSK-3, (Smp_008260.1) was previously proposed as a potential drug target in *S. mansoni* [68].

Selection of Putative Targets Kinases from Phenotypic Screening of Kinase Inhibitor Libraries

The 2 libraries of compounds were tested and the DDU set similarity network with active and inactive compound, while the activity of the PKIS dataset in-vivo and the network similarity network. Some sub networks were identified as being enriched in active compounds and the associated known or predicted targets for those compounds were investigated. Some of the corresponding genes in the *S. mansoni* genome were selected by searching for the homologous gene directly in the schisto DB database (http://schistodb.net/schisto/) (i.e. EGFR). Other genes have been identified via information relative to the genes function and/or mechanism of action available in literature (i.e. Insulin receptor and PLK1). For all the other genes, the protein sequences of the human genes were used to query the schistosome genome via a BLAST search and the genes having the highest sequence similarity were selected. In this way a total of 10 different genes were selected for RNAi. Despite effective mRNA knockdown no differences between dsRNA-treated and controls were observed for any of the genes tested in adult worms. However the silencing of 2 of the genes (Smp_133020 and Smp_009600) led to phenotypic changes and loss of viability when performed in larvae as described below.

In the first part of this work we present the results obtained from the silencing of several genes selected. In order to establish and confirm the optimal conditions to perform the RNAi in adult and larval parasites we started with the gene at the top of the predicted essential list, a putative calmodulin gene (Smp_026560.2).

Then, in the second part of this study we focused on kinases selected from phenotypic screening of kinase inhibitor libraries.

Suppression of Calmodulin (Smp_026560.2) (*Sm*-Calm) affects Adult Worm Morphology

Initial experiments involved electroporation of adult male worms in the presence of dsRNA since we and others had previously found this to be successful [34, 69]. We tested the ability of specific long *Sm*-Calm dsRNA (dsRNA) versus synthetic small interfering RNA (siRNA) (designed and supplied by Applied Biosystems [AB] and Integrated DNA Technology [IDT]) to efficiently suppress *Sm*-Calm expression. A week after the RNAi treatment relative gene expression was determined using qRT-PCR. We found that all dsRNA constructs exerted strong transcript level suppression, ranging from 70 to 95% compared to controls. Comparable knockdown was achieved with both ds and siRNA from two different companies used separately or in combination. This and subsequent comparisons of ds and siRNAs with seven other genes showed both to be effective but in general the dsRNA gave somewhat higher levels of inhibition and, therefore, in later experiments we only used long dsRNA (normally ~500bp). Fig 1A shows data representative of 3 repeat experiments.

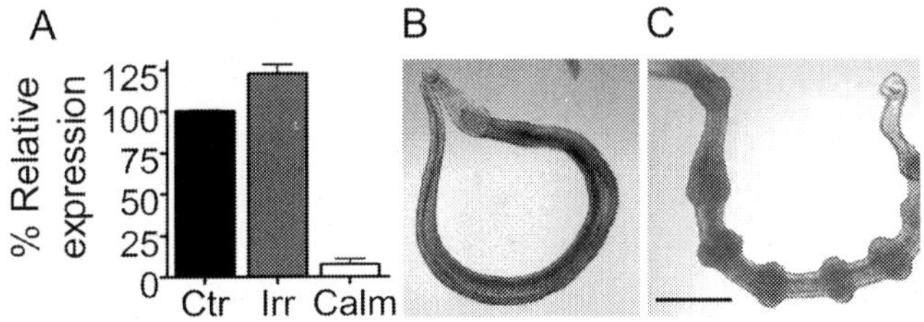

Figure 1: Effect of *Sm*-Calm suppression on adult schistosomes.

(A) *Sm*-Calm expression in adult schistosomes 7 days after electroporation with *Sm*-Calm long dsRNA (CALM) relative to that in untreated control worms (CTR) or worms similarly electorporated with irrelevant dsRNA (IRR). (B-C) Light microscope images of adult worms electroporated 3 weeks earlier with IRR dsRNA (B) or *Sm*-Calm long dsRNA (C). Note the waves of contraction along the bodies of worms following *Sm*-Calm knockdown (scale bar = 100μm). Data are representative of 3 independent experiments.

https://doi.org/10.1371/journal.pntd.0003801.g001

Relative expression of *Sm*-Calm a week after the electroporation with dsRNA was reduced by 90%. Between 1–2 weeks after treatment the worms started to manifest the characteristic waves of contraction/dilation (Fig 1C). In contrast control worms treated with irrelevant dsRNA (IRR) had a smooth and regular shape (Fig 1B). This phenotype did not seem to affect worm viability initially but by 5 weeks, when the waving phenotype had reduced, the treated worms appeared darkened, had damage to the tegument and showed lower motility compared with controls. Two months after the silencing *Sm*-Calm was still 50% lower compared to controls. Electroporation of a mixture of male and female worms, some in copula, resulted in mRNA suppression of 89% in male worms but only 26% in females. The contraction phenotype was not seen in the females but did occur in the males even when *in copula* although this did not seem to affect pairing. The same phenotype was seen with the siRNA suppression of *Sm*-Calm suppression.

Optimisation of the Conditions to Perform RNAi in Schistosomula

Whilst evaluating the effects of *Sm*-Calm RNAi in schistosomula we carried out numerous experiments to confirm the optimal protocols based on various methods reported in the literature: use of dsRNA delivery (by soaking or electroporation), age of parasites (newly transformed or one week old cultured schistosomula), the effect of different culture media on efficiency of knockdown and parasite survival) and culture with and without human red blood cells (hRBC). These findings, replicated with several other genes, are reported in brief since similar conclusions have been reached by others [24, 34, 39, 40] i.e. markedly better survival in M169 than in RPMI, better survival following delivery of dsRNA by soaking than electroporation, comparable mRNA knockdown in newly transformed and 7 days old schistosomula whether by electroporation or soaking. Based on these results the soaking method with newly transformed schistosomula using M169 was used for all further experiments. Although marked *Sm*-Calm suppression (up to >95%) was consistently observed, no phenotypic changes were observed in terms of larval survival or development when silenced parasites were cultured in M169 with or without hRBC.

Suppression of a typical Protein Kinase C (Smp_096310) (*Sm*-aPKC) affects Viability in both Adult and Larval Schistosomes

Adult Worm Silencing

The second gene was an atypical protein kinase C (*Sm*-aPKC). Fig 2A shows robust suppression (95%) of *Sm*-aPKC mRNA levels assessed by qRT-PCR a week after electroporation with dsRNA.

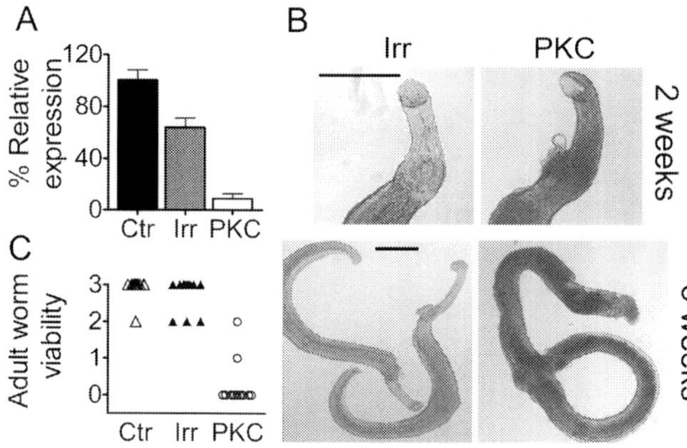

Figure 2: Effect of *Sm*-aPKC suppression on adult schistosomes.

(A) *Sm*-aPKC expression in adult schistosomes 7 days after electroporation with *Sm*-aPKC long dsRNA (PKC) relative to that in untreated control worms (CTR) or worms similarly electroporated with irrelevant ds RNA (IRR). (B) Light microscope images of adult worms 2 or 8 weeks following electroporation with IRR or *Sm*-aPKC (PKC) long dsRNA (scale bars = 100µm). (C) Visual score of worm viability 3 weeks after silencing:- 3-healthy, attached; 2-healthy, not attached; 1- darkened/low motility; 0-severely damaged. Data are representative of 3 independent experiments.

https://doi.org/10.1371/journal.pntd.0003801.g002

Suppression (range of 67–75%) was also achieved with siRNA from AB and IDT. *Sm*-aPKC silenced and control worms treated with IRR dsRNA or siRNA were visually similar during the first two weeks of culture but by day 14 the *Sm*-aPCK treated worms were noticeably darker than the controls, did not attach to the plate with their suckers and showed blebbing notably at the anterior end (Fig 2B). By the third week the presence of vesicles was more generalised and some worms showed severe damage. At this stage parasites were visually scored, assigning to each worm in culture a score from 0 to 3 (Fig 2C). By the 8th week from the silencing the majority of the *Sm*-aPKC dsRNA-treated parasites appeared dead or severely damaged (Fig 2C). The same phenotype and kinetics were observed with the siRNAs. The data shown is representative of three independent repeats.

Schistosomula Silencing

High levels of mRNA knockdown were also achieved with schistosomula as seen in Fig 3A.

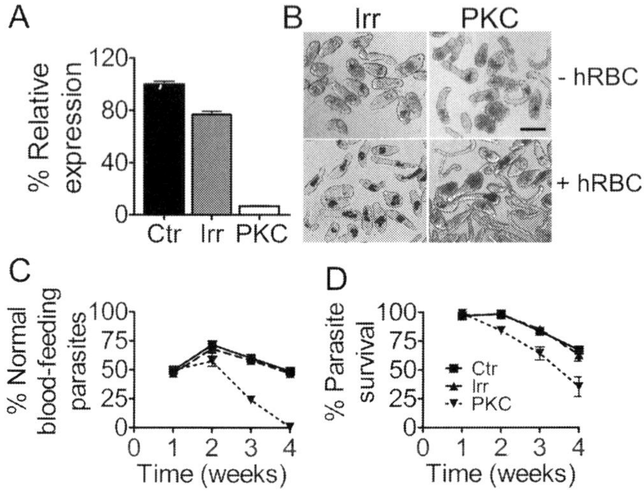

Figure 3: Effect of *Sm*-aPKC suppression in schistosomula.

(A) *Sm*-aPKC expression in schistosomula 7 days after soaking with *Sm*-aPKC long dsRNA, irrelevant ds RNA (IRR) or untreated controls (CTR). (B) Light microscope images of schistosomula 4 weeks after the RNAi treatment. Top images:—wells without hRBC; bottom images:—wells with hRBC (scale bar = 100μm). Percent normal development (C) and percent survival (D) in cultures with hRBC (mean±SE triplicate cultures). Data are representative of 3 independent experiments.

https://doi.org/10.1371/journal.pntd.0003801.g003

Initial experiments showed modest effects on schistosomula viability and so further experiments tested cultures with or without hRBC, since this triggers larval development and transcriptional changes in gene expression [70]. In the presence of hRBC the phenotypic effects were much more marked (Fig 3B). The control and *Sm*-aPKC treated schistosomula looked similar up to one week of culture with many ingesting hRBC and showing haematin pigment in their developing guts. After 2 weeks there was increased death of the *Sm*-aPKC treated schistosomula (Fig 3D) and a marked drop in the proportion of those with developing guts so that virtually all remaining by 4 weeks showed the elongate shape of the lung stage schistosomula (Fig 3C). This indicated high mortality once development had been initiated by onset of blood feeding *in vitro*. The data shown is representative of three independent experiments.

Due to the effects induced by silencing *Sm*-aPKC and the fact that kinases are considered to be an important class of potential drug targets, owing to the essential roles many of them play [28, 47] we focused our RNAi experiments on some other kinases in the list of putative essential genes i.e. caseine kinase (Smp_180400), serine/threonine protein kinase (Smp_080730), putative serine/threonine-protein kinase vrk (Smp_141380) and glycogen synthase kinase 3 (Smp_008260.1). Table high levels of mRNA reductions were observed in adults for each of the genes although levels for the larvae were more variable. However, knockdown of none of these genes caused any phenotypic changes in either adult or larval stages.

RNAi Suppression of Polo-like kinase 1, *Sm*-PLK1 (Smp_009600), has no effect on Adult Viability but affects Viability and Development of Schistosomula

Sm-PLK1 suppression experiments were performed three times using adult male and female worms cultured with or without RBC. Despite high levels of silencing (>90% mRNA reduction) no phenotypic changes were observed during culture for 5 weeks.

RNAi treatment of schistosomula induced >95% suppression of mRNA for *Sm*-PLK1 as assessed one week later. As shown in Fig 4 this silencing led to severe morphological abnormalities and reduced survival between two and three weeks of culture.

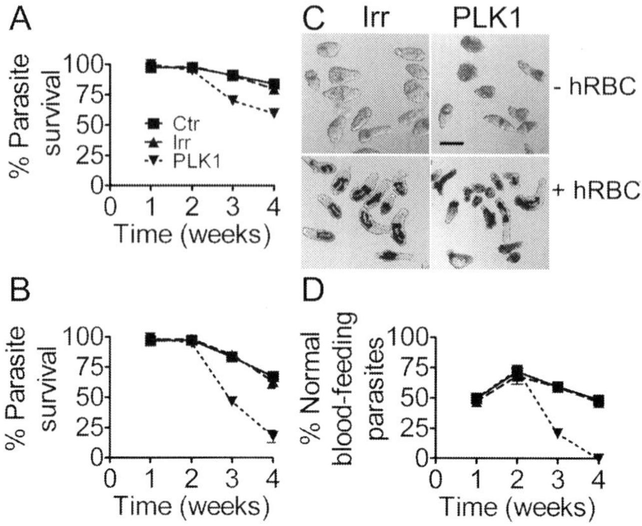

Figure 4: Effect of *Sm*-PLK1 silencing on schistosomula viability.

(A-B) Survival of parasites treated with *Sm*-PLK1 long dsRNA compared with untreated controls (CTR) or controls treated with irrelevant dsRNA (IRR) in the absence (A) or presence (B) of hRBC. (C-D) Reduced viability and a lower percentage of blood feeding parasites in the presence of hRBC following *Sm*-PLK1 knockdown. The photomicrographs (C) are of schistosomula cultured for 3 weeks after RNAi treatment (scale bar = 100μm).

https://doi.org/10.1371/journal.pntd.0003801.g004

As with *Sm*-PKCa, this effect was markedly more pronounced with schistosomula cultured with hRBC, with only 15% of those in hRBC surviving at 4 weeks (Fig 4B) compared with 60% cultured without hRBC (Fig 4A).

Inhibition of *Sm*-PLK1 with the Specific Inhibitor BI2356 affects Schistosomula Development/Viability and also Egg Development

The above evidence of involvement of *Sm*-PLK1 in early worm development was corroborated by using the specific inhibitor, BI2536, as previously used with adult worms [71]. Triplicate cultures of 500 schistosomula with or without hRBC were set up as above in 48-well plates with or without 100nM Bl2536. During the first 2 weeks the inhibitor caused no obvious effect but by 3 weeks a proportion of larvae in the cultures with inhibitor showed morphological damage and death. This was markedly more pronounced in the cultures with hRBC (Fig 5) such that by 4 weeks very few remained alive and those that did were larvae which retained the lung form and had not started to feed (e.g. Fig 5E).

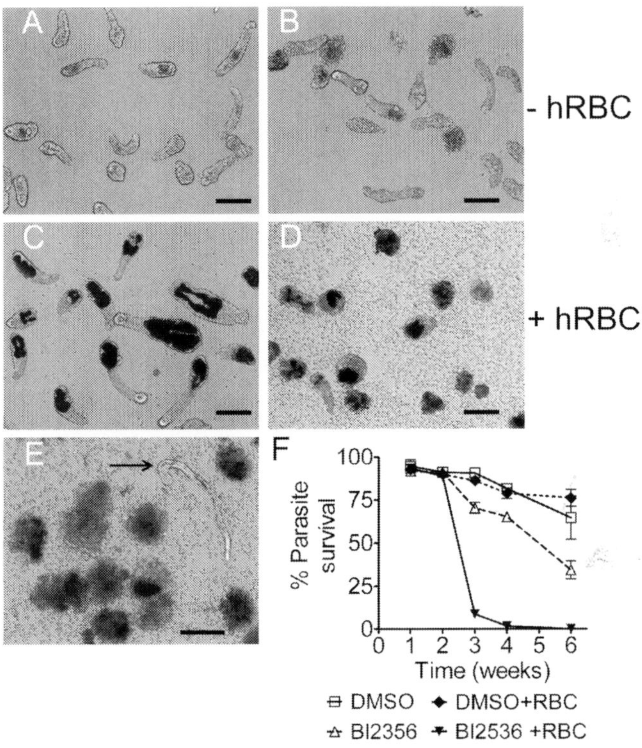

Figure 5: Effect of the PLK-1 inhibitor BI2536 on larval development.

Schistosomula were treated with 100 nM BI2356 (B, D and E) or carrier DMSO alone (A and C). A and B: cultures without hRBC; C, D and E: cultures with hRBC. Images A-D: after culture for 20 days; image E: after culture for 30 days (arrow shows a viable lung-form schistosomulum) (scale bar = 100μm). (F) Mean ± SE larval survival based on triplicate cultures.

https://doi.org/10.1371/journal.pntd.0003801.g005

BI2536 at 100nM was also used in adult worm cultures. During culture for 7 weeks no differences were seen in the morphology, behaviour or viability of the worms with or without inhibitor. However, in contrast to the controls the eggs produced in cultures with the inhibitor failed to embryonate. To investigate if this could be due to effects of the inhibitor on egg development, the unembryonated eggs produced by *ex-vivo* adults maintained for 3 days in control medium were recovered from the cultures and incubated with or without inhibitor in quadriplicate. After 10 days 41±9.3% of eggs in control cultures contained viable miracidia or had hatched whereas none of the eggs in the inhibitor had developed to contain miracidia (Fig 6). This clearly indicates that PLK-1 is necessary for embryonation of viable eggs.

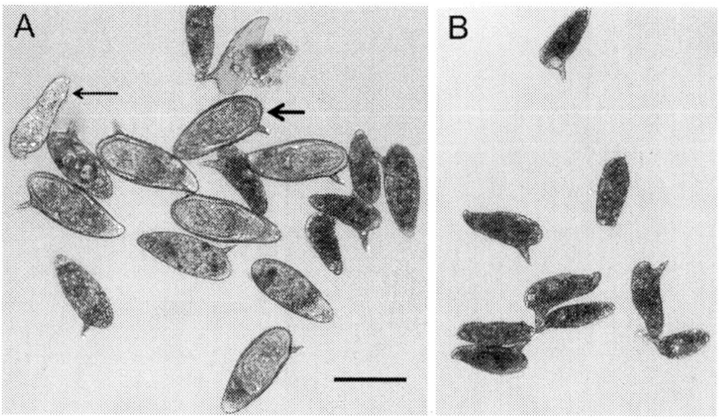

Figure 6: Effect of the PLK-1 inhibitor BI2536 on egg development.

Unembryonated eggs recovered from *ex-vivo* adult worms and cultured in vitro for 3 days failed to embryonate when subsequently cultured in 100 nM BI2536 (B) in contrast most eggs cultured with DMSO carrier alone developed normally (A). Thin arrow: free miracidium; thick arrow: embryonated egg containing viable miracidium. Scale bar = 100μm.

https://doi.org/10.1371/journal.pntd.0003801.g006

Schistosomula Viability and Development is Inhibited by Suppression of the p38-MAPK Family Member (Smp_133020)

Marked suppression (>85%) of p38-MAPK mRNA was observed a week after the long dsRNA treatment in both adult worms and schistosomula. Again no phenotype was seen in the adult worms but the RNAi silencing consistently reduced viability of schistosomula starting after 2 weeks and, as previously observed for *Sm*-aPKC and *Sm*-PLK1, such effects on parasite viability were more evident when p38-MAPK silenced schistosomula where cultured in the presence of hRBC (Fig 7).

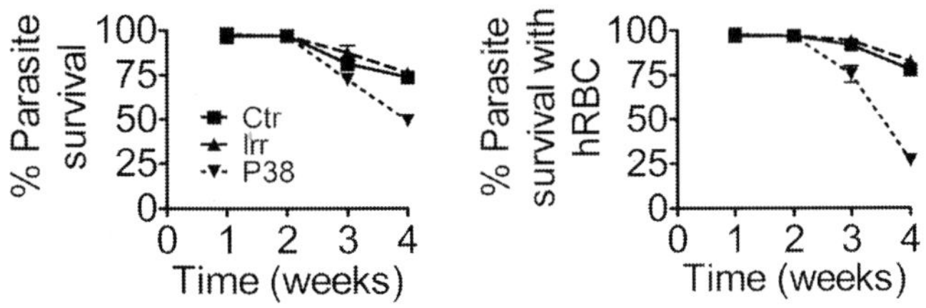

Figure 7: Effect of RNAi silencing of *Sm*-MAPK p38.

Graphs showing percentage survival of schistosomula following treatment with long dsRNA corresponding to the *Sm*-MAPK p38 family member (Smp_133020) (P38) compared with untreated controls (CTR) or controls treated with irrelevant dsRNA (IRR). Schistosomula were cultured in the absence (A) or presence (B) of hRBC.

https://doi.org/10.1371/journal.pntd.0003801.g007

Suppression of a typical Protein Kinase C (Smp_096310) (*Sm*-aPKC) and Polo-like kinase 1 (Smp_009600) (*Sm*-PLK1) affects Viability of Schistosomula *in vivo*

Groups of 5 mice were injected intramuscularly with 1,000 control, *Sm*-aPKC or *Sm*-PLK1 treated schistosomula/mouse and perfused 4 weeks later. Aliquots of the same batches of schistosomula were cultured *in vitro*. Data representative of two independent experiments is shown in Fig 8.

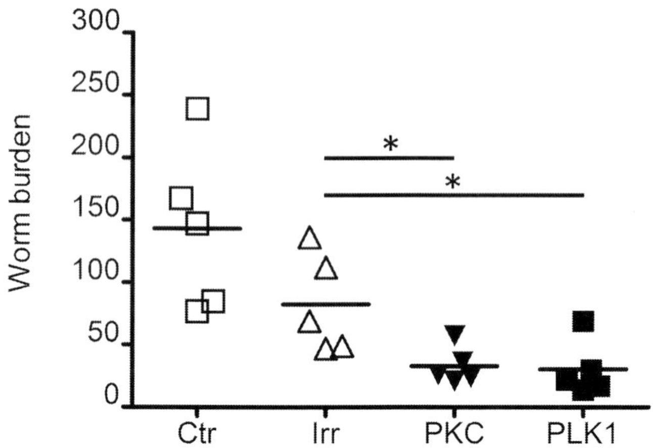

Figure 8: Worm recovery following *in vivo* **injection of** schistosomula silenced for *Sm*-aPKC or *Sm*-PLK1.

Worms recovered by portal perfusion 4 weeks after i.m. injection of 1000 schistosomula treated with *Sm*-aPKC (aPKC), *Sm*-PLK1 (PLK1), irrelevant long dsRNA (IRR) or untreated (CTR). Significant differences were assessed by unpaired Student's t-test; * = $P<0.05$.

https://doi.org/10.1371/journal.pntd.0003801.g008

Mice injected with both *Sm*-aPKC and *Sm*-PLK1 silenced schistosomula yielded significantly fewer parasites compared with untreated controls or mice injected with IRR dsRNA. However, none of the surviving suppressed parasites showed any differences in size or shape compared to the controls. The recovered worms also showed no persisting gene suppression in contrast to the *Sm*-aPKC or *Sm*-PLK1 treated larvae maintained in culture for 4 weeks which showed significant knockdown (~70% and 80% respectively) comparable to that assessed one week after treatment.

DISCUSSION

In this study we set out to contribute to RNAi based validation of putative drug targets for schistosomiasis. *S. mansoni* possesses a quite large (270 Mb) and complex genome containing between 15,000 and 25,000 genes [22, 72]. It is thus crucial to find new *in silico* strategies to select genes which have a potential to be target candidates for new drugs or vaccines.

In looking for essential genes in pathogens a variety of characteristics can be considered. Crowther et al., [20] used the targets database (TDR) (http://tdrtargets.org/) as a tool to illustrate how to prioritize proteins which have a potential to be considered drug targets in neglected-disease causing pathogens. Such target selection has been applied to several parasitic diseases such as Chagas disease, parasitic worm infections and malaria. The strategy is based on a multi-criteria search through the use of the open-access resource in TDR in which it is possible to manually select proteins that fit specific desired traits (druggability, assayability, essentiality). In *S. mansoni* proteins are not yet scored for druggability in the TDR database, so their druggability was assessed by comparing the proteins' amino acid sequences to those of known targets in the protChEMBL database. In addition to this, proteins' functional importance have been compared to gene knockout data of their orthologs in *C. elegans* and *D. melanogaster*. This method yielded a top list of 170 *Schistosoma* targets.

Caffrey et al. [68], focused their search on a comparative chemogenomics approach based on the proteome of two model organisms, the nematode *Caenorhabditis elegans* and the fruit fly, *Drosophila melanogaster* in which gene knockout studies had identified proteins which appear to be essential or producing a deleterious phenotype in both organisms. The search aimed to highlight proteins in these organisms having clear sequence similarities to orthologues in schistosomes. This type of manual screening led to a selection of 35 *S. mansoni* druggable targets 8 of which were also present in the earlier list [20].

Our strategy was similar to the above and aimed at identifying potential essential and druggable targets for *S. mansoni* by inference from the ChEMBL database of drug targets and orthology essential data from *C. elegans*, as a one of the few model systems where whole genome functional genomics studies using RNAi have been performed. This analysis produced a list of 24 target candidate genes to be analysed by RNAi of which only the glycogen synthase kinase 3-related (gsk3) gene (Smp_008260) was also present in the earlier lists [68].

In the case of *Sm*-Calm, the first gene on our list of 24 genes, robust suppression was induced in adult worms and resulted in regular waves of contraction and dilation in male worms but not females. Interestingly, siRNA knockdown of a 24kDa calcium-regulated heat-stable protein of *S. japonicum* (CRHSP-24) induces what appears to be a similar contraction wave phenotype in juvenile *S. japonicum* worms [73]. Calmodulin as a transporter of Ca^{2+} plays a key role in calcium signalling, affecting numerous functions including muscle contraction and metabolism. Selective inhibitors of calmodulin in schistosomes block egg hatching [74] and both inhibitors [75] and siRNA knockdown [76] inhibit miracidial transformation to sporocysts. Exposure of worms to PZQ results in disruption of Ca^{2+} homeostasis mediating its action by inducing Ca^{2+}influx [5] and it has been suggested that molecules in the calcium signalling pathway may be potential targets for drug discovery. After 5 weeks the contraction phenotype waned but by this time the worms were slow, darkened and had tegumental damage. Despite the important role that calmodulin plays during different stages of *S. mansoni* development the very slow induction of damage to adults and the fact that the gene exhibits more than 90% identity with mammalian calmodulin [76] reduce its suitability as a target for drug development.

Silencing of the atypical protein kinase C (*Sm*-aPKC) similarly had no immediate effect on adult worms but after 2 weeks in culture the worms became darkened, non-adherent and developed blebs. By the third week they were severely damaged. *Sm*-aPKC silencing in schistosomula also induced decreased viability and development after 2 weeks of *in vitro* culture especially when cultured in the presence of hRBC. Reduced viability was also seen when *Sm*-aPKC silenced schistosomula were injected *in vivo* supporting a role for this gene in schistosomula development.

Interestingly we found that parasites recovered from mice *in vivo* showed no more silencing resulting in mRNA levels comparable to controls. Such phenomenon is not surprising as it has been previously observed in other studies [32, 42, 63].

As suggested by Krautz-Peterson et al., one hypothesis is that different schistosomula can receive variable amounts of dsRNA or, alternatively, this disparity could be due to a more active metabolism of parasites *in vivo* that can determine a shorter half life of the dsRNA [77].

Members of the family of atypical PKCs are considered to be key components of downstream signalling pathways activated by the EGF receptors which can activate diverse signal transduction cascades [78]. Earlier studies on another *S. mansoni* protein kinase C (*Sm*PKC1), with high homology to mammalian PKCβ, was found to be differentially expressed during the schistosome life cycle, with the highest mRNA and protein levels being found in miracidia and sporocysts, respectively. The protein was also immunolocalized to the acetabular glands of newly transformed schistosomula suggesting a possible role in larval transformation [48]. A further study demonstrated activated schistosome PKCβ expression during postembryonic development of the miracidium to mother sporocyst suggesting its role in regulating this transformation [79]. The effect that *Sm*-aPKC silencing had on both adult and larvae schistosomes in our studies indicates an important role in the parasite although the slow onset of effects does not support it as a likely drug target.

Given the importance of kinases as potential drug targets and having shown some phenotypic effects following knockdown of the atypical protein kinase C (*Sm*-aPKC) we focused our attention on additional kinases. Andrade et al., [28] identified and classified 252 *S. mansoni*eukaryotic protein kinases (ePKs), describing some of them as good drug candidates as they may perform an essential function in the parasite. Although they often had high sequence similarity to their human homologues, new and effective drugs can overcome such similarity by binding protein kinases close but not in the ATP site occluding ATP access to the kinase to retard enzyme activity [28]. Initially we carried out RNAi suppression on four other kinases in our list of putative essential genes alongside three other non-kinase genes but failed to demonstrate any phenotype with these knockdowns. So attention was switched to selecting genes through computational prediction of putative targets based on either searching the ChEMBL database for compounds found to be active against adult *S. mansoni* in previous whole organism screens [16, 45, 80] or from the focused PKIS kinase inhibitor collection and the kinase inhibitors represented in the DDU compound collection we had tested previously [45]. A similar study was recently conducted in zebrafish by Laggner *et al.* that sought to identify putative targets of 681 neuroactive molecules from a 14,000-compound phenotypic screen in zebrafish embryos [81]. Putatively homologous genes for the identified kinases were mapped in the *S. mansoni* genome and thereafter 10 kinases highlighted to be validated by RNAi phenotypic screening. From this screening we obtained detectable phenotype in schistosomula but not in adults following the knockdown of 2 of the 10 genes tested, *Sm*-PLK1 and *Sm*-MAPK p38. As for *Sm*-aPKC the loss in schistosomula viability with *Sm*-PLK-1 and *Sm*-MAPK p38 started after 2–3 weeks and were more marked in cultures with hRBC in which the larvae were developing rapidly indicating a key role for these kinases during development. In the case of PLK-1 the RNAi effects were exactly comparable when the PLK-1 specific inhibitor BI2536 was used.

The failure to demonstrate activity in the adults with the RNAi knockdowns despite the adult stage having been used initially to screen the kinase inhibitors may be explained by the relatively modest effects seen against adults in the initial *in vitro* testing i.e. the highest percentage inhibition of adult motility achieved with the five GSK kinase inhibitors identified to target PLK-1 was 65% and the next highest only 25%. P38 MAPK was implicated by two inhibitors inducing 82% and 76% inhibition of motility. So none of these inhibitors actually killed the adults which combined with the only partial, albeit high, RNAi knockdown may have explained why no effect on phenotype was seen with the RNAi in adults. By comparison the phenotypic effects seen in schistosomula may reflect greater sensitivity of this developing stage to knockdown of these two genes. It also cannot be ruled out that the compounds inhibit additional targets (possibly protein kinases) within the schistosomes, complicating the analysis.

PLKs constitute a family of serine threonine kinases which have a highly conserved function in all eukaryotes. Along with other mitotic kinases, they play a crucial role in coordinating cell cycle progression and are potent regulators of M phase [82]. In vertebrates, there are four well-known PLKs (PLK 1–4) whilst schistosomes only possess two, *Sm*-PLK1 and *Sm*-Sak, both well characterized [71, 83, 84]. *Sm*-PLK1 has been shown to be abundantly expressed in germinal and vitelline cells and more abundant in sporocysts than in other parasite stages, consistent with its potential role in mitosis and/or meiosis in schistosomes. Our data suggest that *Sm*-PLK1 might play an important role also at the schistosomula stage which is characterized by rapid growth and body remodeling particularly when hRBC are added, processes which would require active mitotic machinery. Such an important role in schistosomula can also be observed *in vivo* where a reduced viability was also seen in silenced parasites injected in mice thus supporting the involvement for this gene in schistosomula development. In addition, previous studies have shown that adult worms treated with BI2536, a specific inhibitor of *Sm*-PLK1, which showed altered morphology of reproductive organs with significant effects on oogenesis and spermatogenesis [83]. In our studies neither RNAi knockdown or BI2536 affected adult worm viability but BI2536 inhibited embryonation of eggs produced *in vitro* as would be expected from inhibition of mitosis.

Similar effects in terms of morphology and kinetics were seen with silencing of the *Sm*-MAPK p38 family member (Fig 7). It has been previously shown how p38 MAPK inhibition with pyridinylimidazole can have crucial effects on survival and replication of some pathogens such as *P. falciparum* and *T. gondii* [85, 86]. The active form of this gene in *S. mansoni* has been associated with the cilia of miracidia, playing a role in the ciliary motion [87] and also during the early stages of post-embryonic development [88]. Treatment performed with a specific inhibitor (SB203580) is able to retard the development of the miracidium to post-miracidium stage and also from post-miracidium to mother sporocyst stage, without affecting larval viability [88]. We demonstrated here that *Sm*-MAPK p38 might also play an important role in schistosomula development and that its suppression affects growth and survival of parasites *in vitro*.

Despite the coincidence of the nature and timing of the effects seen *in vitro* with these different RNAi gene knockdowns in schistosomula we believe the effects to be specific for the following reasons: (i) IRR dsRNA prepared in exactly same way as that for the specific genes was used in all experiments and cause no significant phenotypes; (ii) the effect was reproducible for the genes in question whereas no effects were seen for a number of other genes in which the RNAi was prepared in exactly the same way; (iii) in a recent study of RNAi knock-down of 3-hydroxy-3-methyl-glutaryl-CoA reductase (HMGR), the target for statins, there was, similarly little effect on schistosomula viability until 2–3 weeks of culture [36]; (iv) in the case of the *Sm*-PLK-1 knockdown the phenotype and time of its onset were exactly replicated with the PLK-1 inhibitor, BI2356; (v) in the cultures with hRBC the loss of viability is seen once larval development characterized by gut formation is initiated whereas schistosomula which retain a "lung-form" morphology remain unaffected.

Although we were able to demonstrate phenotypic effects on both larval and adult schistosomes, the earliest effects seen were at 2 weeks (calmodulin in adults) whilst most of the effects on larvae were not seen until the second/third week. Furthermore, with some of the gene knockdowns effects were seen with larvae but not adults and vice versa. In addition the deleterious effects on schistosomula were markedly enhanced and visualized when hRBC were added and the larvae induced to feed. Indeed it was commonly seen that larvae which did not feed and retained the lung form phenotype seemed unaffected (e.g. Fig 5E). It is not surprising that the induction of development following initiation of red blood cells is associated with expression of genes vital for survival. Microarray studies on schistosomula cultured for 5 days have characterized the transcriptional changes in genes encoding surface and secreted proteins during the first 5 days in *in vitro* culture and show upregulation of a number of genes in the presence of hRBC [70].

We conclude that RNAi of schistosomula combined with culture in the presence of hRBC is a valuable approach to understanding key roles played by individual genes. However, as mentioned at the outset one of our intentions was to assess whether the automated image analysis of larval motility and morphology which we have developed for drug screening [45] could be developed for HTS RNAi. Effects which take weeks to manifest, vary markedly between individual worms, and rely on addition of hRBC are not conducive to this sort of analysis not least because once hRBC are added to schisostomula, the individual worms develop at different rates some developing guts quickly and others remaining like lung forms. So there is even more variation in shape, size, motility and appearance than with newly transformed parasites making development of segmentation and image analysis of morphological changes more challenging. Nevertheless, inhibition of motility would be easier to develop although the presence of hRBC would interfere with segmentation of the parasites for image analysis. This proof of concept study illustrates the potential of combining bioinformatics target prioritization with RNAi viability studies as a method to identify putative drug targets in *S. mansoni*.

AUTHOR CONTRIBUTIONS

Conceived and designed the experiments: AG QDB ALH. Performed the experiments: AG QDB RAP NRM. Analyzed the data: AG JB QDB. Contributed reagents/materials/analysis tools: IMC JB IHG. Wrote the paper: AG IHG JB ALH QDB.

REFERENCES

1. Bethony J. M. , et al. , Can schistosomiasis really be consigned to history without a vaccine? Vaccine, 2008. 26(27–28): p. 3373–6.
2. Abath F. G. , Development of vaccines against human parasitic diseases: tools, current status and perspectives. Expert Opin Investig Drugs, 2000. 9(2): p. 301–10. pmid:11060678
3. McManus D. P. and Loukas A. , Current status of vaccines for schistosomiasis. Clin Microbiol Rev, 2008. 21(1): p. 225–42. pmid:18202444
4. Fenwick A. , et al. , The Schistosomiasis Control Initiative (SCI): rationale, development and implementation from 2002–2008. Parasitology, 2009. 136(13): p. 1719–30. pmid:19631008
5. Cioli D. and Pica-Mattoccia L. , Praziquantel. Parasitol Res, 2003. 90 Supp 1: p. S3–9. pmid:12811543
6. Doenhoff M. J. , Cioli D. , and Utzinger J. , Praziquantel: mechanisms of action, resistance and new derivatives for schistosomiasis. Curr Opin Infect Dis, 2008. 21(6): p. 659–67. pmid:18978535
7. Doenhoff M. J. , et al. , Praziquantel: its use in control of schistosomiasis in sub-Saharan Africa and current research needs. Parasitology, 2009. 136(13): p. 1825–35. pmid:19281637
8. Alonso D. , et al. , Failure of standard treatment with praziquantel in two returned travelers with Schistosoma haematobium infection. Am J Trop Med Hyg, 2006. 74(2): p. 342–4. pmid:16474094
9. Melman S. D. , et al. , Reduced susceptibility to praziquantel among naturally occurring Kenyan isolates of Schistosoma mansoni. PLoS Negl Trop Dis, 2009. 3(8): p. e504. pmid:19688043
10. Guisse F. , et al. , Therapeutic evaluation of two different dose regimens of praziquantel in a recent Schistosoma mansoni focus in Northern Senegal. Am J Trop Med Hyg, 1997. 56(5): p. 511–4.

pmid:9180600

11. Stelma F. F. , et al. , Efficacy and side effects of praziquantel in an epidemic focus of Schistosoma mansoni. American Journal of Tropical Medicine and Hygiene, 1995. 53(2): p. 167–70. pmid:7677219
12. Cioli D. , et al. , Determination of ED50 values for praziquantel in praziquantel-resistant and—susceptible Schistosoma mansoni isolates. International Journal for Parasitology, 2004. 34(8): p. 979–87. pmid:15217737
13. Black C. L. , et al. , Impact of intense, longitudinal retreatment with praziquantel on cure rates of schistosomiasis mansoni in a cohort of occupationally exposed adults in western Kenya. Trop Med Int Health, 2009. 14(4): p. 450–7. pmid:19222824
14. Botros S. , et al. , Current status of sensitivity to praziquantel in a focus of potential drug resistance in Egypt. Int J Parasitol, 2005. 35(7): p. 787–91. pmid:15925597
15. Guidi A. , et al. , Praziquantel efficacy and long-term appraisal of schistosomiasis control in Pemba Island. Trop Med Int Health, 2010. 15(5): p. 614–8. pmid:20214757
16. Ramirez B, B. Q. , Yousif F, Fakorede FK, Mouries M-A, Nwaka S Schistosomes: challenges in compound screening. Expert Opinion on Drug Discovery, 2007. 2: p. S53–361. pmid:23489033
17. Sayed A. A. , et al. , Identification of oxadiazoles as new drug leads for the control of schistosomiasis. Nat Med, 2008. 14(4): p. 407–12. pmid:18345010
18. Caffrey C. R. and Secor W. E. , Schistosomiasis: from drug deployment to drug development. Curr Opin Infect Dis, 2011. 24(5): p. 410–7. pmid:21734570
19. Berriman M. , et al. , The genome of the blood fluke Schistosoma mansoni. Nature, 2009. 460(7253): p. 352–8. pmid:19606141
20. Crowther G. J. , et al. , Identification of attractive drug targets in neglected-disease pathogens using an in silico approach. PLoS Negl Trop Dis, 2010. 4(8): p. e804. pmid:20808766
21. Oliveira G. , The Schistosoma mansoni transcriptome: an update. Exp Parasitol, 2007. 117(3): p. 229–35. pmid:17624328

22. Zerlotini A. , et al. , SchistoDB: a Schistosoma mansoni genome resource. Nucleic Acids Res, 2009. 37(Database issue): p. D579–82. pmid:18842636
23. Verjovski-Almeida S. , et al. , Transcriptome analysis of the acoelomate human parasite Schistosoma mansoni. Nat Genet, 2003. 35(2): p. 148–57. pmid:12973350
24. Stefanic S. , et al. , RNA interference in Schistosoma mansoni schistosomula: selectivity, sensitivity and operation for larger-scale screening. PLoS Negl Trop Dis, 2010. 4(10): p. e850. pmid:20976050
25. Hopkins A. L. and Groom C. R. , The druggable genome. Nat Rev Drug Discov, 2002. 1(9): p. 727–30. pmid:12209152
26. Dissous C. and Grevelding C. G. , Piggy-backing the concept of cancer drugs for schistosomiasis treatment: a tangible perspective? Trends Parasitol, 2011. 27(2): p. 59–66. pmid:20920890
27. Fitzpatrick J. M. , et al. , Anti-schistosomal intervention targets identified by lifecycle transcriptomic analyses. PLoS Negl Trop Dis, 2009. 3(11): p. e543. pmid:19885392
28. Andrade L. F. , et al. , Eukaryotic protein kinases (ePKs) of the helminth parasite Schistosoma mansoni. BMC Genomics, 2011. 12: p. 215. pmid:21548963
29. Aguero F. , et al. , Genomic-scale prioritization of drug targets: the TDR Targets database. Nat Rev Drug Discov, 2008. 7(11): p. 900–7. pmid:18927591
30. Boyle J. P. , et al. , Using RNA interference to manipulate endogenous gene expression in Schistosoma mansoni sporocysts. Mol Biochem Parasitol, 2003. 128(2): p. 205–15. pmid:12742587
31. Geldhof P. , et al. , RNA interference in parasitic helminths: current situation, potential pitfalls and future prospects. Parasitology, 2007. 134(Pt 5): p. 609–19. pmid:17201997
32. Krautz-Peterson G. , et al. , RNA interference in schistosomes: machinery and methodology. Parasitology, 2010. 137(3): p. 485–95. pmid:19765345
33. Krautz-Peterson G. and Skelly P. J. , Schistosoma mansoni: the dicer gene and its expression. Exp Parasitol, 2008. 118(1): p. 122–8. pmid:17707814

34. Ndegwa D. , Krautz-Peterson G. , and Skelly P. J. , Protocols for gene silencing in schistosomes. Exp Parasitol, 2007. 117(3): p. 284–91. pmid:17870072
35. Magarinos M. P. , et al. , TDR Targets: a chemogenomics resource for neglected diseases. Nucleic Acids Res, 2012. 40(Database issue): p. D1118–27. pmid:22116064
36. Rojo-Arreola L. , et al. , Chemical and genetic validation of the statin drug target to treat the helminth disease, schistosomiasis. PLoS One, 2014. 9(1): p. e87594. pmid:24489942
37. Faghiri Z. and Skelly P. J. , The role of tegumental aquaporin from the human parasitic worm, Schistosoma mansoni, in osmoregulation and drug uptake. FASEB J, 2009. 23(8): p. 2780–9. pmid:19364765
38. Kasinathan R. S. , Morgan W. M. , and Greenberg R. M. , Genetic knockdown and pharmacological inhibition of parasite multidrug resistance transporters disrupts egg production in Schistosoma mansoni. PLoS Negl Trop Dis, 2011. 5(12): p. e1425. pmid:22163059
39. Kuntz A. N. , et al. , Thioredoxin glutathione reductase from Schistosoma mansoni: an essential parasite enzyme and a key drug target. PLoS Med, 2007. 4(6): p. e206. pmid:17579510
40. Mourao M. M. , et al. , Phenotypic screen of early-developing larvae of the blood fluke, schistosoma mansoni, using RNA interference. PLoS Negl Trop Dis, 2009. 3(8): p. e502. pmid:19668375
41. Rinaldi G. , et al. , RNA interference targeting leucine aminopeptidase blocks hatching of Schistosoma mansoni eggs. Mol Biochem Parasitol, 2009. 167(2): p. 118–26. pmid:19463860
42. Skelly P. J. , Da'dara A. , and Harn D. A. , Suppression of cathepsin B expression in Schistosoma mansoni by RNA interference. Int J Parasitol, 2003. 33(4): p. 363–9. pmid:12705930
43. Swierczewski B. E. and Davies S. J. , A schistosome cAMP-dependent protein kinase catalytic subunit is essential for parasite viability. PLoS Negl Trop Dis, 2009. 3(8): p. e505. pmid:19707280
44. Tran M. H. , et al. , Suppression of mRNAs encoding tegument tetraspanins from Schistosoma mansoni results in impaired tegument turnover. PLoS Pathog, 2010. 6(4): p. e1000840. pmid:20419145
45. Paveley R. A. , et al. , Whole organism high-content screening

by label-free, image-based Bayesian classification for parasitic diseases. PLoS Negl Trop Dis, 2012. 6(7): p. e1762. pmid:22860151

46. Kamath R. S. , et al. , Systematic functional analysis of the Caenorhabditis elegans genome using RNAi. Nature, 2003. 421(6920): p. 231–7. pmid:12529635
47. Bahia D. , et al. , Protein tyrosine kinases in Schistosoma mansoni. Mem Inst Oswaldo Cruz, 2006. 101 Suppl 1: p. 137–43. pmid:17308761
48. Bahia D. , et al. , SmPKC1, a new protein kinase C identified in the platyhelminth parasite Schistosoma mansoni. Biochem Biophys Res Commun, 2006. 345(3): p. 1138–48. pmid:16713993
49. Li L. , Stoeckert C. J. Jr. , and Roos D. S. , OrthoMCL: identification of ortholog groups for eukaryotic genomes. Genome Res, 2003. 13(9): p. 2178–89. pmid:12952885
50. Camacho C. , et al. , BLAST+: architecture and applications. BMC Bioinformatics, 2009. 10: p. 421. pmid:20003500
51. Chen F. , et al. , Assessing performance of orthology detection strategies applied to eukaryotic genomes. PLoS One, 2007. 2(4): p. e383. pmid:17440619
52. Nidhi , et al. , Prediction of biological targets for compounds using multiple-category Bayesian models trained on chemogenomics databases. J Chem Inf Model, 2006. 46(3): p. 1124–33. pmid:16711732
53. Gaulton A. , et al. , ChEMBL: a large-scale bioactivity database for drug discovery. Nucleic Acids Res, 2012. 40(Database issue): p. D1100–7. pmid:21948594
54. Ertl P. , Rohde B. , and Selzer P. , Fast calculation of molecular polar surface area as a sum of fragment-based contributions and its application to the prediction of drug transport properties. J Med Chem, 2000. 43(20): p. 3714–7. pmid:11020286
55. Rogers D. and Hahn M. , Extended-connectivity fingerprints. J Chem Inf Model, 2010. 50(5): p. 742–54. pmid:20426451
56. Rogers D. J. and Tanimoto T. T. , A Computer Program for Classifying Plants. Science, 1960. 132(3434): p. 1115–8. pmid:17790723
57. Shannon P. , et al. , Cytoscape: a software environment for integrated

models of biomolecular interaction networks. Genome Res, 2003. 13(11): p. 2498–504. pmid:14597658

58. Basch P. F. and Humbert R. , Cultivation of Schistosoma mansoni in vitro. III. implantation of cultured worms into mouse mesenteric veins. J Parasitol, 1981. 67(2): p. 191–5. pmid:7241279
59. Mann V. H. , et al. , Culture for genetic manipulation of developmental stages of Schistosoma mansoni. Parasitology, 2010. 137(3): p. 451–62. pmid:19765348
60. Kines K. J. , et al. , Integration of reporter transgenes into Schistosoma mansoni chromosomes mediated by pseudotyped murine leukemia virus. FASEB J, 2008. 22(8): p. 2936–48. pmid:18403630
61. Basch P. F. , Cultivation of Schistosoma mansoni in vitro. I. Establishment of cultures from cercariae and development until pairing. J Parasitol, 1981. 67(2): p. 179–85. pmid:7241277
62. Boissier J. , Durand P. , and Mone H. , PCR effectiveness for sexing Schistosoma mansoni cercariae: application for sexing clonal cercarial populations. Mol Biochem Parasitol, 2001. 112(1): p. 139–41. pmid:11166395
63. Correnti J. M. , Brindley P. J. , and Pearce E. J. , Long-term suppression of cathepsin B levels by RNA interference retards schistosome growth. Mol Biochem Parasitol, 2005. 143(2): p. 209–15. pmid:16076506
64. Krautz-Peterson G. , et al. , Optimizing gene suppression in schistosomes using RNA interference. Mol Biochem Parasitol, 2007. 153(2): p. 194–202. pmid:17420062
65. Livak K. J. and Schmittgen T. D. , Analysis of relative gene expression data using real-time quantitative PCR and the 2(-Delta Delta C(T)) Method. Methods, 2001. 25(4): p. 402–8. pmid:11846609
66. Doenhoff M. , et al. , Factors affecting the acquisition of resistance against Schistosoma mansoni in the mouse. I. Demonstration of resistance to reinfection using a model system that involves perfusion of mice within three weeks of challenge. J Helminthol, 1978. 52(3): p. 173–86. pmid:722037
67. Hillier L. W. , et al. , Genomics in C. elegans: so many genes, such a little worm. Genome Res, 2005. 15(12): p. 1651–60. pmid:16339362

68. Caffrey C. R. , et al. , A comparative chemogenomics strategy to predict potential drug targets in the metazoan pathogen, Schistosoma mansoni. PLoS One, 2009. 4(2): p. e4413. pmid:19198654

69. Valentim C. L. , et al. , Genetic and molecular basis of drug resistance and species-specific drug action in schistosome parasites. Science, 2013. 342(6164): p. 1385–9. pmid:24263136

70. Gobert G. N. , et al. , Transcriptional changes in Schistosoma mansoni during early schistosomula development and in the presence of erythrocytes. PLoS Negl Trop Dis, 2010. 4(2): p. e600. pmid:20161728

71. Dissous C. , Grevelding C. G. , and Long T. , Schistosoma mansoni Polo-like kinases and their function in control of mitosis and parasite reproduction. An Acad Bras Cienc, 2011. 83(2): p. 627–35. pmid:21670883

72. LoVerde P. T. , et al. , Schistosoma mansoni genome project: an update. Parasitol Int, 2004. 53(2): p. 183–92. pmid:15081950

73. Zou X. , et al. , RNAi silencing of calcium-regulated heat-stable protein of 24 kDa in Schistosoma japonicum affects parasite growth. Parasitol Res, 2011. 108(3): p. 567–72. pmid:21085993

74. Katsumata T. , et al. , Hatching of Schistosoma mansoni eggs is a Ca2+/calmodulin-dependent process. Parasitol Res, 1989. 76(1): p. 90–1. pmid:2622901

75. Kawamoto F. , et al. , Possible roles of cAMP and Ca2+ in the regulation of miracidial transformation in Schistosoma mansoni. Parasitol Res, 1989. 75(5): p. 368–74. pmid:2542928

76. Taft A. S. and Yoshino T. P. , Cloning and functional characterization of two calmodulin genes during larval development in the parasitic flatworm Schistosoma mansoni. J Parasitol, 2011. 97(1): p. 72–81. pmid:21348610

77. Krautz-Peterson G. , et al. , Suppressing glucose transporter gene expression in schistosomes impairs parasite feeding and decreases survival in the mammalian host. PLoS Pathog, 2010. 6(6): p. e1000932. pmid:20532163

78. Yarden Y. and Sliwkowski M. X. , Untangling the ErbB signalling network. Nat Rev Mol Cell Biol, 2001. 2(2): p. 127–37. pmid:11252954

79. Ludtmann M. H. , et al. , Protein kinase C signalling during miracidium to mother sporocyst development in the helminth parasite, Schistosoma mansoni. Int J Parasitol, 2009. 39(11): p. 1223–33. pmid:19394337

80. Mansour N. R. and Bickle Q. D. , Comparison of microscopy and Alamar blue reduction in a larval based assay for schistosome drug screening. PLoS Negl Trop Dis, 2010. 4(8): p. e795. pmid:20706580

81. Laggner C. , et al. , Chemical informatics and target identification in a zebrafish phenotypic screen. Nat Chem Biol, 2012. 8(2): p. 144–6.

82. Archambault V. and Glover D. M. , Polo-like kinases: conservation and divergence in their functions and regulation. Nat Rev Mol Cell Biol, 2009. 10(4): p. 265–75. pmid:19305416

83. Long T. , et al. , Schistosoma mansoni Polo-like kinase 1: A mitotic kinase with key functions in parasite reproduction. Int J Parasitol, 2010. 40(9): p. 1075–86. pmid:20350550

84. Long T. , et al. , SmSak, the Second Polo-Like Kinase of the Helminth Parasite Schistosoma mansoni: Conserved and Unexpected Roles in Meiosis. PLoS One, 2012. 7(6): p. e40045. pmid:22768216

85. Brumlik M. J. , et al. , Human p38 mitogen-activated protein kinase inhibitor drugs inhibit Plasmodium falciparum replication. Exp Parasitol, 2011. 128(2): p. 170–5. pmid:21338604

86. Wei S. , et al. , Pyridinylimidazole p38 mitogen-activated protein kinase inhibitors block intracellular Toxoplasma gondii replication. Int J Parasitol, 2002. 32(8): p. 969–77. pmid:12076626

87. Ressurreicao M. , et al. , A role for p38 MAPK in the regulation of ciliary motion in a eukaryote. BMC Cell Biol, 2011. 12: p. 6. pmid:21269498

88. Ressurreicao M. , et al. , A role for p38 mitogen-activated protein kinase in early post-embryonic development of Schistosoma mansoni. Mol Biochem Parasitol, 2011. 180(1): p. 51–5. pmid:21787807

CITATIONS

CHAPTER 4

Marc S. Weinberg and Kevin V. Morris, Transcriptional gene silencing in humans, doi: 10.1093/nar/gkw139.

CHAPTER 5

Thornbrough JM, Gopinath A, Hundley T, Worley MJ (2016) Human Genome-Wide RNAi Screen for Host Factors That Facilitate Salmonella Invasion Reveals a Role for Potassium Secretion in Promoting Internalization. PLoS ONE 11(11): e0166916. https://doi.org/10.1371/journal.pone.0166916.

CHAPTER 6

Trieu TA, Navarro-Mendoza MI, Pérez-Arques C, Sanchis M, Capilla J, Navarro-Rodriguez P, et al. (2017) RNAi-Based Functional Genomics Identifies New Virulence Determinants in Mucormycosis. PLoS Pathog 13(1): e1006150. https://doi.org/10.1371/journal.ppat.1006150.

CHAPTER 8

Kamthan A, Chaudhuri A, Kamthan M and Datta A (2015) Small RNAs in plants: recent development and application for crop improvement. Front. Plant Sci. 6:208. doi: 10.3389/fpls.2015.00208

CHAPTER 9

Koch A, Biedenkopf D, Furch A, Weber L, Rossbach O, Abdellatef E, et al. (2016) An RNAi-Based Control of Fusarium graminearum Infections Through Spraying of Long dsRNAs Involves a Plant Passage and Is Controlled by the Fungal Silencing Machinery. PLoS Pathog 12(10): e1005901. https://doi.org/10.1371/journal.ppat.1005901.

CHAPTER 11

Unnikrishnan Unniyampurath, Rajendra Pilankatta and Manoj N. Krishnan, RNA Interference in the Age of CRISPR: Will CRISPR Interfere with RNAi?, doi:10.3390/ijms17030291.

CHAPTER 12

Guidi A, Mansour NR, Paveley RA, Carruthers IM, Besnard J, Hopkins AL, et al. (2015) Application of RNAi to Genomic Drug Target Validation in Schistosomes. PLoS Negl Trop Dis 9(5): e0003801. https://doi.org/10.1371/journal.pntd.0003801.

INDEX

M

N

P

Q

R

S

T

U

V

W